SOLUTIONS BOOK

TO ACCOMPANY

GENERAL CHEMISTRY
PRINCIPLES AND STRUCTURE 5/E

James E. Brady
St. John's University

Larry Peck
Texas A&M University

JOHN WILEY & SONS
New York Chichester Brisbane Toronto Singapore

Copyright ©1990 by John Wiley & Sons, Inc.

All rights reserved.

Reproduction or translation of any part of this work beyond that permitted by Sections 107 and 108 of the 1976 United States Copyright Act without the permission of the copyright owner is unlawful. Requests for permission or further information should be addressed to the Permissions Department, John Wiley & Sons.

ISBN 0-471-51420-9
Printed in the United States of America

10 9 8 7 6 5 4 3 2 1

PREFACE

This Solutions Book has been prepared as a supplement to GENERAL CHEMISTRY - PRINCIPLES AND STRUCTURE, FIFTH EDITION by James Brady. The textbook by J. Brady is an excellent introduction to modern general chemistry. A thorough understanding of general chemistry should include the ability to work problems and answer questions related to the principles and concepts presented. Many students have difficulty developing the ability to utilize knowledge gained from textual material. Through the use of the textbook and this book, this author hopes that students will improve their problem-solving skills while learning a considerable amount of chemistry. Knowledge gained and problem-solving skills acquired in a general chemistry course should prove valuable in subsequent science and technical courses as well as assisting in preparing students for a wide variety of professions.

Many problems can be answered (solved) in more than one way. In this book I have encouraged students to develop solutions that they understand. For consistency I have striven to follow the methods used by the textbook. Where practical, I have tried to include dimensions and to show only the correct number of significant figures; in some problems an extra digit is carried in intermediate steps.

I believe that students will find this book beneficial during their study of general chemistry. To gain the most benefit from this book, one should view it as a learning resource and not just as a place to go for answers. Students should always strive to solve each question first without using this book. Then, when the solution given in this book is referred to, the student should develop not only a better understanding of the chemisty associated with the question, but also increase his skills in recalling and applying scientific facts.

I wish to thank Dr. Frank Kolar for his tremendous assistance in proofreading and checking solutions in this book. Special thanks go to my family (Sandra, Molly and Marci) for their typing and support during the production of this book.

Larry Peck
Chemistry Department
Texas A&M university

CONTENTS

1 INTRODUCTION

1.1 A chemical is any of the multitude of things that surround us. "Chemicals are everywhere!" In nature chemicals are all the things one can touch, see or smell. A chemical may be a compound or an element. Mixtures contain more than a single chemical.

1.2 The science of chemistry studies what chemicals are composed of and how the characteristics or properties of chemicals are determined by their composition.

1.3 We know that a chemical reaction has occurred between sodium and chlorine when they interact to give a new substance with remarkable changes in properties. Not all chemical reactions are as dramatic as the reaction of sodium and chlorine.

1.4 When iron rusts, one observes changes in appearance as well as changes in magnetic and metallic properties.

1.5 A laws summarizes facts while a theory is a tested explanation of the behavior of nature.

1.6 The doctor first gathers as many facts as practical by observing symptoms and performing tests. That is the observation stage in the scientific method. He then makes a diagnosis; that is the hypothesis stage. Next, he prescribes treatment; that is the testing by experiment.

1.7 Qualitative observations lack the numbers associated with quantitative observations. Quantitative observations are more useful because they contain more information.

1.8 Data are observed information.

1.9 (a) mass: **Kilogram, kg** (b) time: **Second, s** (c) temperature: **Kelvin, K**
(d) length: **Meter, m** (e) amount of substance: **Mole, mol**

1.10 **Kilometers per liter**

1.11 N= mass x distance/time^2 = (kilograms x meters)/(seconds)^2
= **kg•m/s^2**

1.12 (a) pico: **10^{-12}, p** (b) mega: **10^6, M** (c) centi: **10^{-2}, c**
(d) nano: **10^{-9}, n** (e) kilo: **10^3, k**

1.13 (a) milligram: **mg** (b) decimeter: **dm** (c) kilosecond: **ks**
(d) microsecond: **μs** (e) centigram: **cg**

1.14 (a) 3.4 kg = **3.4 x 10^3 g**
(b) 5.7 ns = **5.7 x 10^{-9} s**
(c) 4.0 Mmol = **4.0 x 10^6 mol**
(d) 6.4 mg = **6.4 x 10^{-3} g**
(e) 7.2 cm = **7.2 x 10^{-2} m**

1.15 (a) 3.2 **nm** = 3.2 x 10^{-9} m
(b) 42 **mm** = 4.2 x 10^{-2} m
(c) 7.3 **mg** = 0.0073 g
(d) 12.5 **cm** = 125 mm
(e) 3.5 **μL** = 3.5 x 10^{-3} mL
(f) 0.84 **dam** = 840 cm

1.16 (a) 1.40 m = ? cm

$$1.40\text{ m} \times \left(\frac{1\text{ cm}}{10^{-2}\text{m}}\right) = \mathbf{1.40 \times 10^2\text{ cm}}$$

(b) 2855 mm = ? m

$$2855\text{ mm} \times \left(\frac{10^{-3}\text{ m}}{1\text{ mm}}\right) = \mathbf{2.855\text{ m}}$$

(continued)

1.16 (continued)

(c) 185 mL = ? L

$$185\text{ mL} \times \left(\frac{10^{-3}\text{ L}}{1\text{ mL}}\right) = \mathbf{0.185\ L}$$

(d) 0.0253 L = ? mL

$$0.0253\text{ L} \times \left(\frac{1\text{ mL}}{10^{-3}\text{ L}}\right) = \mathbf{25.3\ mL}$$

(e) 195 mm = ? cm

$$195\text{ mm} \times \left(\frac{10^{-3}\text{ m}}{1\text{ mm}}\right) \times \left(\frac{1\text{ cm}}{10^{-2}\text{ m}}\right) = \mathbf{19.5\ cm}$$

(f) 1885 nm = ? µm

$$1885\text{ nm} \times \left(\frac{10^{-9}\text{ m}}{1\text{ nm}}\right) \times \left(\frac{1\ \mu\text{m}}{10^{-6}\text{ m}}\right) = \mathbf{1.885\ \mu m}$$

(g) 0.385 kg = ? mg

$$0.385\text{ kg} \times \left(\frac{10^{3}\text{ g}}{1\text{ kg}}\right) \times \left(\frac{1\text{ mg}}{10^{-3}\text{ g}}\right) = \mathbf{3.85 \times 10^{5}\ mg}$$

(h) 48.5 cm^2 = ? mm^2

$$48.5\text{ cm}^2 \times \left(\frac{10^{-2}\text{ m}}{1\text{ cm}}\right)^2 \times \left(\frac{1\text{ mm}}{10^{-3}\text{ m}}\right)^2 = \mathbf{4.85 \times 10^{3}\ mm^2}$$

(i) 143 mL = ? mm^3

$$143\text{ mL} \times \left(\frac{1\text{ cm}^3}{1\text{ mL}}\right) \times \left(\frac{10^{-2}\text{ m}}{1\text{ cm}}\right)^3 \times \left(\frac{1\text{ mm}}{10^{-3}\text{ m}}\right)^3 = \mathbf{1.43 \times 10^{5}\ mm^3}$$

(j) 345 cm^3 = ? L

$$345\text{ cm}^3 \times \left(\frac{1\text{ L}}{1000\text{ cm}^3}\right) = \mathbf{0.345\ L}$$

1.17 In the laboratory we usually use the units of grams and milliliters because they are conveniently-sized units for most laboratory measurements of mass and volume. In some laboratories, milligrams and microliters may better reflect the size of measurements most often used but for most laboratories the units of grams and milliliters are the appropriate units.

1.18 (a) 112 lb = ? kg

$$112 \text{ lb} \times \left(\frac{1\text{kg}}{2.204 \text{ lb}}\right) = \mathbf{50.8\ kg} \text{ (3 significant figures)}$$

(b) 45 m/s = ? mi/hr

$$45 \text{ m/s} \times \left(\frac{39.37 \text{ in.}}{\text{m}}\right) \times \left(\frac{1 \text{ ft}}{12 \text{ in.}}\right) \times \left(\frac{1 \text{ mi}}{5280 \text{ ft}}\right) \times \left(\frac{60 \text{ s}}{\text{min}}\right) \times \left(\frac{60 \text{ min}}{\text{hr}}\right) =$$

$$\mathbf{1.0 \times 10^2\ mi/hr} \text{ (only 2 significant figures)}$$

(c) 1.00 mi^3 = ? m^3

$$1.00 \text{ mi}^3 \times \left(\frac{5280 \text{ ft}}{\text{mi}}\right)^3 \times \left(\frac{12 \text{ in.}}{\text{ft}}\right)^3 \times \left(\frac{1 \text{ m}}{39.37 \text{ in.}}\right)^3 = \mathbf{4.17 \times 10^9\ m^3}$$

(d) 48 mi/hr = ? cm/s

$$48 \text{ mi/hr} \times \left(\frac{5280 \text{ ft}}{1 \text{ mi}}\right) \times \left(\frac{12 \text{ in.}}{1 \text{ ft}}\right) \times \left(\frac{2.540 \text{ cm}}{1 \text{ in.}}\right) \times \left(\frac{1 \text{ hr}}{60 \text{ min}}\right) \times \left(\frac{1 \text{ min}}{60 \text{ sec}}\right)$$

$$= \mathbf{2.1 \times 10^3\ cm/s}$$

(e) 15 yd^2 = ? m^2

$$15 \text{ yd}^2 \times \left(\frac{36 \text{ in.}}{\text{yd}}\right)^2 \times \left(\frac{1 \text{ m}}{39.37 \text{ in.}}\right)^2 = \mathbf{13\ m^2}$$

1.19 **For the Pferdburper:**

$$\left(\frac{10 \text{ km}}{1\text{L}}\right) \times \left(\frac{1 \text{ L}}{1.057 \text{ qt}}\right) \times \left(\frac{4 \text{ qt}}{1 \text{ gal}}\right) \times \left(\frac{1000 \text{ m}}{\text{km}}\right) \times \left(\frac{39.37 \text{ in.}}{1 \text{ m}}\right) \times$$

$$\left(\frac{1 \text{ ft}}{12 \text{ in.}}\right) \times \left(\frac{1 \text{ mi}}{5280 \text{ ft}}\right) = \mathbf{24\ miles/gal.}$$

Therefore, the Pferdburper is better than the Smokebelcher and its 21 miles/gal.

1.20 $\frac{35\text{ mi}}{\text{hr}} \times \left(\frac{5280\text{ ft}}{1\text{ mile}}\right) \times \left(\frac{12\text{ in.}}{1\text{ ft}}\right) \times \left(\frac{1\text{ m}}{39.37\text{ in.}}\right) \times \left(\frac{1\text{ km}}{1000\text{ m}}\right) = \mathbf{56\ km/hr}$

1.21 $4.3\text{ miles} \times \left(\frac{5280\text{ ft}}{1\text{ mile}}\right) \times \left(\frac{12\text{ in.}}{1\text{ ft}}\right) \times \left(\frac{1\text{ m}}{39.37\text{ in.}}\right) \times \left(\frac{1\text{ km}}{1000\text{ m}}\right) = \mathbf{6.9\ km}$

1.22 $\left(\frac{142\text{ thrubs}}{1\text{ wk}}\right) \times \left(\frac{1}{14}\right) \times \left(\frac{1\text{ lb potatoes}}{2\text{ thrubs}}\right) \times \left(\frac{52\text{ wks}}{1\text{ yr}}\right) = \mathbf{\frac{260\ lb\ potatoes}{year}}$

1.23 The freezing point and boiling point of water are chosen as the reference temperatures for the definition of the Fahrenheit and Celsius temperature scales because they are convenient, reproducible constant temperatures.

1.24 0 K is also known as **absolute zero.**

1.25 (a) 50°F = ?°C $\quad (50°\text{F} - 32) \times \frac{5}{9} = \mathbf{10°C}$

(b) 25°C = ?°F $\quad \left(25°\text{C} \times \frac{9}{5}\right) + 32 = \mathbf{77°F}$

(c) 80 K = ?°C $\quad$ 80 K - 273 = **-193°C**

(d) -40°C = ?°F $\quad \left(-40°\text{C} \times \frac{9}{5}\right) + 32 = \mathbf{-40°F}$

(e) 0 K = ?°F $\quad$ 0 K - 273 = -273°C

$\left(-273°\text{C} \times \frac{9}{5}\right) + 32 = \mathbf{-459°F}$

1.26 Melting point $\left(30°\text{C} \times \frac{9}{5}\right) + 32 = \mathbf{86°F}$

Boiling point $\left(1983°\text{C} \times \frac{9}{5}\right) + 32 = \mathbf{3601°F}$

1.27 $\frac{5}{9} \times (6152°\text{F} - 32) = \mathbf{3400°C} \qquad 3400°\text{C} + 273 = \mathbf{3673\ K}$

1.28 $\left(\frac{9}{5} \times -78°C\right) + 32 = \mathbf{-108°F}$ $\quad$ $-78°C + 273 = \mathbf{195\ K}$

1.29 $\frac{5}{9} \times (98.6°F - 32.0) = \mathbf{37.0°C}$ $\quad$ $\left(\frac{9}{5} \times 39°C\right) + 32 = \mathbf{102°F}$

1.30 One can derive the formula °N = (°C - 80) x (100°N/138°C). Using this formula, the f.p. of water (0°C) would be equal to (0-80) x (100/138) or -58°N and the b.p. of water (100°C) would be equal to (100 - 80) x (100/138) or 14°N.

1.31 It is important for scientists to indicate the reliability of measured and calculated quantities so others will know the quality of the experimental data. The reliability is reflected by the number of justifiable significant figures that one reports and usually reflects how the values were obtained.

1.32 Precision refers to how closely grouped a set of measurements happens to be. Accuracy refers to how close an experimental measurement is to the true value.

1.33 (a) 1.0370 g has **5** significant figures.
(b) 0.000417 m has **3** significant figures.
(c) 0.00309 cm has **3** significant figures.
(d) 100.1°C has **4** significant figures.
(e) 9.0010 g has **5** significant figures.

1.34 (a) $\text{Speed} = \frac{346.2 \text{ mi}}{6.27 \text{ hr}} = \mathbf{55.215311\ mi/hr}$
(b) The minimum uncertainty in the distance traveled is **± 0.1 miles.**
(c) The minimum uncertainty in the time is **±0.01 hr.**
(d) If the distance is 346.3, the calculated speed would be 55.231. This shows a difference in the second decimal place. If the time is 6.26, the calculated speed would be 55.127. This shows a difference in the first decimal place. Either error is possible; therefore, the uncertainty of the calculated speed cannot be less than ±0.1.

1.35 (a) 1,250 g = $\mathbf{1.25 \times 10^3\ g}$
(b) 13,000,000 m = $\mathbf{1.3 \times 10^7\ m}$
(c) 60,230,000,000,000,000,000,000 atoms = $\mathbf{6.023 \times 10^{22}\ atoms}$
(d) $\mathbf{2.1457 \times 10^5\ mg}$
(e) $\mathbf{3.147 \times 10\ g}$

1.36 (a) $\mathbf{4.0 \times 10^{-4}°C}$ (b) $\mathbf{3 \times 10^{-10}\ km}$ (c) $\mathbf{2.146 \times 10^{-3}\ g}$
(d) $\mathbf{3.28 \times 10^{-5}\ g}$ (e) $\mathbf{9.1 \times 10^{-13}\ m}$

1.37 (a) 3×10^{10} m = **30,000,000,000 m**
(b) 2.54×10^{-5} m = **0.0000254 m**
(c) 122×10^{-2} g = **1.22 g**
(d) 3.4×10^{-7} g = **0.00000034 g**
(e) 0.0325×10^{6} cm = **32,500 cm**

1.38 (a) **3.0×10^{3} m**
(b) **3.00×10^{3} m**
(c) 3,000 m x 100 cm/m = 300,000 cm or **3.0×10^{5} cm** (to 2 sign. figures)

1.39 (a) **14.7 cm** (b) **18.3 cm** (c) **27.5 cm** (d) **33.4 cm** (e) **8.4 cm**

1.40 (a) **7.7 cm^2** (b) **73.3 m^2** (c) **0.781 g/cm^3** (d) **3.478 g** (e) **81.4 g**

1.41 (a) **1.638×10^{9} m** (b) **2.17×10^{9} m** (c) **4.1970×10^{6} m**
(d) **8.5 x 10 mol** (e) **1.00°C**

1.42 (a) ($341.7\ cm^2 - 22\ cm^2$) + (0.00224 cm x 814,050 cm)
$= 3.20 \times 10^{2}\ cm^2 + 1.82 \times 10^{3}\ cm^2 =$ **$2.14 \times 10^{3}\ cm^2$**

(b) ($82.7\ cm^2 \times 143$ cm) + ($274\ cm^3 - 0.00653\ cm^3$)
$= 1.18 \times 10^{4}\ cm^3 + 2.74 \times 10^{2}\ cm^3 =$ **$1.21 \times 10^{4}\ cm^3$**

c) ($3.53\ cm^3 \div 0.084$ cm) - (14.8 cm x 0.046 cm)
$= 42\ cm^2 - 0.68\ cm^2 =$ **$4.1 \times 10\ cm^2$**

(d) (324 cm x 0.0033 m) + (214.2 m x 0.0225 m)
$= 1.1\ m^2 + 4.82\ m^2 =$ **$5.9\ m^2$**

(e) (4.15 mm + 82.3 mm) x (0.024 mm + 3.000 mm)
= 86.4 mm x 3.024 mm = **$2.61 \times 10^{2}\ mm^2$**

(f) 0.2510 m x (15.50 m - 12.75 m) = 0.2510 m x 2.75 m = **0.690 m^2**

1.43 (a) ($12.45 \times 10^{6}\ cm^2$) $\div$ (2.24×10^{3} cm) = **5.56×10^{3} cm**
(b) 822 m $\div$ 0.028 hr = **2.9×10^{4} m/hr**
(c) (635.4×10^{-5} cm) $\div$ (42.7×10^{-4} s) = **1.49 cm/s**
(d) (31.3×10^{-12} m) $\div$ (8.3×10^{-6} m/s) = **3.8×10^{-6} s**
(e) (0.74×10^{-9} mol) $\div$ ($825.3 \times 10^{18}\ m^3$) = **$9.0 \times 10^{-31}\ mol/m^3$**

1.44 (a) $(8.3 \times 10^{-6}\ km) \times (4.13 \times 10^{-7}\ km) \div (5.411 \times 10^{-12}\ km)$
$= \mathbf{0.63\ km}$ **or** $\mathbf{6.3 \times 10^{-1}\ km}$
(b) $[(3.125 \times 10^{-6}\ km/s) + (5.127 \times 10^{-5}\ km/s)] \times (6.72 \times 10^{8}\ s)$
$= (5.440 \times 10^{-5}\ km/s) \times (6.72 \times 10^{8}\ s) = \mathbf{3.66 \times 10^{4}\ km}$
(c) $[14.39\ m^2 + (2.43 \times 10^{1}\ m^2)] \div 1275\ m$
$= 38.7\ m^2 \div 1275\ m = \mathbf{3.04 \times 10^{-2}\ m}$
(d) $[(1.583 \times 10^{-2}\ km) - (0.00255\ km)] \div [142.3\ s + (0.257 \times 10^{2}\ s)]$
$= (1.328 \times 10^{-2}\ km) \div (1.680 \times 10^{2}\ s) = \mathbf{7.905 \times 10^{-5}\ km/s}$
(e) $(0.00425\ g) \div [(0.0008137\ cm^3) + (2.65 \times 10^{-5}\ cm^3)]$
$= (0.00425\ g) \div (8.402 \times 10^{-4}\ cm^3) = \mathbf{5.06\ g/cm^3}$

1.45 The equality 5280 ft = 1 mile is an exact relationship and, therefore, one can think of it as having an infinite number of significant figures.

1.46 0.2263 mi = ? in. $0.2263\ mi \times \left(\frac{5280\ ft}{1\ mi}\right) \times \left(\frac{12\ in.}{1\ ft}\right) = \mathbf{1.434 \times 10^{4}\ in.}$

1.47 An **extensive property** is one that depends on the size of the sample used. An **intensive property** is one that is independent of the size of the sample used.

Extensive Properties	Intensive Properties
Force (weight), length, number of atoms, moles	Freezing point, specific gravity, specific heat

1.48 Some properties that differ for a copper penny and a piece of window glass are density, color, luster, transparency, electrical conductivity, malleability, reactivity with sulfur and reactivity with nitric acid. Both have the property of retaining a constant volume and are incompressible. Although both may seem to be solids, glass is actually a liquid which flows very slowly requiring years at common temperatures to show a change in shape.

1.49 The **mass** of an object is a measure of its resistance to a change in velocity and is a measure of the amount of matter in that object. **Weight** is the force with which an object of a certain mass is attracted by gravity to the earth (or some other body such as the moon). The mass of an object does not vary from place to place; it is the same regardless of where it is measured.

1.50 (a) A physical property is one that can be observed without changing the chemical make-up of a substance.
b) A chemical property is the tendency of a substance to undergo a particular chemical reaction.
A physical change does not involve a change in chemical make-up while a chemical change does.

1.51 (a) Physical properties of heptane include: liquid, forms a heterogeneous mixture with water and floats on water because it has a density that is less than that of water.
(b) Chemical properties of heptane include: flammable and vapors burn explosively when mixed with air (oxygen) if a spark is provided.

1.52 Density is the ratio of an object's mass to its volume. **Specific gravity** is the ratio of the object's density to the density of water at a specified temperature. Units for density are g/mL for liquids and solids; specific gravity has no units.

1.53 Density = 14.3 g/8.46 cm^3 = **1.69 g/cm^3**

1.54 $D = \left(\frac{\text{mass}}{\text{vol}}\right) = \left(\frac{50.8\text{ g}}{36.2\text{ mL} - 25.0\text{ mL}}\right) = \mathbf{4.54\ g/mL}$

1.55 Vol. of cyl. = area x length = π r^2 x length = 3.14(1.24 cm)^2 x 4.75 cm =22.9 cm^3. Therefore, density = 104.2 g/ 22.9 cm^3 = **4.55 g/cm^3**

1.56 (a) 12.0 cm^3 lead ⇔ ? g

$$12.0\text{ cm}^3 \times \left(\frac{11.35\text{ g}}{1\text{ cm}^3}\right) \Leftrightarrow \mathbf{136\ g}$$

(b) 155 g lead ⇔ ? cm^3

$$155\text{ g} \times \left(\frac{1\text{ cm}^3}{11.35\text{ g}}\right) \Leftrightarrow \mathbf{13.7\ cm^3}$$

1.57 (a) 10.00 g $CHCl_3$ ⇔ ? cm^3

$$10.00\text{ g} \times \left(\frac{1\text{ mL}}{1.492\text{ g}}\right) \Leftrightarrow \mathbf{6.702\ mL}$$

(b) 10.00 mL $CHCl_3$ ⇔ ? g

$$10.00\text{ mL} \times \left(\frac{1.492\text{ g}}{1\text{ mL}}\right) \Leftrightarrow \mathbf{14.92\ g}$$

1.58 (a) Mass of water in the pycnometer is 34.914 g - 25.296 g = 9.618 g. Therefore, the volume of the pycnometer can be calculated from the mass of water and its density. D = mass/volume 0.9970 g/mL = 9.618 g/? mL Volume = **9.647 mL**
(b) Density of unknown = (33.485 g - 25.296 g) ÷ 9.647 mL = **0.8489 g/mL**

1.59 (a) **500 numerical values** (b) **105 numerical values**

1.60 (a) Specific gravity of isopropyl alcohol = density of isopropyl alcohol ÷ density of water.
Specific gravity of isopropyl alcohol = 6.56 lb/gal ÷ 8.34 lb/gal = **0.787**
(b) Specific gravity of isopropyl alcohol x density of water = density of isopropyl alcohol = 0.787 x 1.00 g/mL = **0.787 g/mL**

1.61 Specific gravity of propylene glycol times the density of water = the density of propylene glycol. Density of propylene glycol = 1.04 x 8.34 lb/gal = 8.67 lb/gal. The weight of 10,000 gal of propylene glycol would be 8.67 lb/gal x 10,000 gal = **8.67×10^4 lb**

1.62 The density of each is: #1 = 275 g ÷ 275 mL = 1.00 g/mL, #2 = 389 g ÷ 245 mL = 1.59 g/mL and #3 = 299 g ÷ 265 mL = 1.13 g/mL. Since the density of the contents of the first beaker is the closest to that of water, the detective should have selected the **first beaker**.

1.63 **Elements** are the simplest forms of matter that can exist under ordinary chemical conditions. **A compound** consists of two or more elements which are always present in the same proportions. **Mixtures** consist of two or more compounds that do not react chemically and differ from elements and compounds in that they may be of variable composition and do not undergo phase changes at constant temperature.

1.64 A physical change does not alter the chemical composition of the substances involved while a chemical change always produces one or more new substances with a different chemical composition from the original substances.

1.65 A **solution** is a homogeneous mixture and has uniform properties throughout. It consists of one phase.

1.66 The data suggest that quartz is a compound because of the constant composition between the two samples. This can be shown by the calculation of the percentage silicon in each sample.
Sample #1 [3.44 g silicon ÷ (3.44 g + 3.91 g) total] x 100 = 46.8% silicon
Sample #2 [6.42 g silicon ÷ (6.42 g + 7.30 g) total] x 100 = 46.8% silicon

1.67

Homogeneous	Heterogeneous
sea water	smog
air	smoke
black coffee	club soda (with bubbles)
	ham sandwich

1.68 There are four phases: copper pan, iron nails, glass marbles, and water.

1.69 Since the melting point changed during the melting of the solid, one would conclude that the solid sample was a mixture. The temperature of a pure compound would have stayed constant as it melted. This did not. Therefore, it must be a mixture.

1.70 A magnet could be used to separate the iron filings from the other two components. The salt and copper powder can be added to water. The salt dissolves and the copper powder can be filtered from the solution. The salt can then be recovered by evaporating the water.

1.71 (a) **Fe** (b) **Na** (c) **K** (d) **P** (e) **Br** (f) **Ca** (g) **N** (h) **Ne** (i) **Mn** (j) **Mg**

1.72 (a) **silver** (b) **copper** (c) **sulfur**
(d) **chlorine** (e) **aluminum** (f) **gold**
(g) **chromium** (h) **tungsten** (i) **nickel**
(j) **mercury**

1.73 (a) Potassium = **2 atoms**, Sulfur = **1 atom**
(b) Sodium = **2 atoms**, Carbon = **1 atom**, Oxygen = **3 atoms**
(c) Potassium = **4 atoms**, Iron = **1 atom**, Carbon and Nitrogen = **6 atoms each**
(d) Nitrogen = **3 atoms**, Hydrogen = **12 atoms**, Phosphorus = **1 atom**, Oxygen = **4 atoms**
(e) Sodium = **3 atoms**, Silver = **1 atom**, Sulfur = **4 atoms**, Oxygen = **6 atoms**

1.74 $\mathbf{CaSO_4 \cdot 2H_2O}$

1.75. Aluminum = **1 atom**, Hydrogen = **24 atoms**, Oxygen = **20 atoms**, Potassium = **1 atom, and** Sulfur = **2 atoms**

1.76 **a** and **c** are not correctly balanced.

1.77 $(3.6 \times 10^4 \text{ m}) + (5.6 \times 10^7 \text{ cm}) = (3.6 \times 10^4 \text{ m}) + \left[(5.6 \times 10^7 \text{ cm}) \times \left(\frac{10^{-2} \text{ m}}{1 \text{ cm}}\right)\right] =$

$(3.6 \times 10^4 \text{ m}) + (5.6 \times 10^5 \text{ cm}) = \mathbf{6.0 \times 10^5 \text{ m}}$

1.78 18.5 gal ⇔ ? lb
(One will need to use the density of water in the solution of this problem. The density of water can be expressed as 1.00 g/mL or (from example 1.9) as 8.34 lb/gal. If 1.00 g/mL is used, the solution to this problem would require many more steps.)
Density of gasoline = 0.684 x 8.34 lb/gal = 5.70 lb/gal
18.5 gal x 5.70 lb/gal ⇔ **105 lb**

1.79 4255 tons ⇔ ? ft^3 sea water

Density of sea water = 1.025 x 8.34 lb/gal = 8.55 lb/gal

$$4255 \text{ tons} \times \left(\frac{2000 \text{ lb}}{\text{ton}}\right) \times \left(\frac{1 \text{ gal}}{8.55 \text{ lb}}\right) \times \left(\frac{4 \text{ qt}}{\text{gal}}\right) \times \left(\frac{1 \text{ L}}{1.057 \text{ qt}}\right) \times \left(\frac{1 \text{ ft}^3}{28.32 \text{ L}}\right) \Leftrightarrow$$

1.33×10^5 ft^3

1.80 $\dfrac{28.31 \text{ kg}}{3.932 \times 10^{-3} \text{m}^3} = ? \text{ g/cm}^3$

$$\left(\frac{28.31 \text{ kg}}{3.932 \times 10^{-3} \text{ m}^3}\right) \times \left(\frac{10^3 \text{ g}}{1 \text{ kg}}\right) \times \left(\frac{10^{-2} \text{ m}}{1 \text{ cm}}\right)^3 = 7.200 \text{ g/cm}^3$$

The density suggests that the metal is chromium.

1.81 Superconducting Material

90 K= ?°C	-183°C = ?°F
90 K + (-273) = **-183°C**	[(-183)(9/5)] + 32 = **-297°F**

Boiling point of liquid nitrogen

77 K or **-196°C** or -321°F

2 ATOMS, MOLECULES, AND MOLES

2.1 1. Matter is composed of tiny indivisible particles called atoms.
2. All atoms of a given element are identical, but differ from the atoms of other elements.
3. A chemical compound is composed of the atoms of its elements in a definite fixed numerical ratio.
4. A chemical reaction merely consists of a reshuffling of atoms from one set of combinations to another. The individual atoms remain intact and do not change.

2.2 Dalton's theory explained the law of conservation of mass and the law of definite proportions. It predicted the law of multiple proportions.

2.3 Atomic mass.

2.4 **Atoms** are the fundamental particles of all matter that cannot be further subdivided by ordinary chemical means. A **molecule** is a group of atoms bound tightly enough together that they behave as one.

2.5 The currently accepted atomic mass unit is: one atomic mass unit is equal to 1/12th of the mass of one atom of carbon-12.

2.6 1 fluorine atom = 1 u and 1 carbon atom = ? u

For carbon $\frac{1\text{ u}}{18.998} \Leftrightarrow \frac{?\text{ u}}{12.011}$ carbon mass would be **0.63222 u**

For hydrogen $\frac{1\text{ u}}{18.998} \Leftrightarrow \frac{?\text{ u}}{1.0079}$ hydrogen mass would be **0.053053 u**

2.7 **No.** If the compound formed is AB, the relative masses would be one set of values; if AB_2, a different set of values; if A_2B, still a different set of values, etc. To calculate the relative masses, one would need to know the ratio of the atoms in the compound formed.

2.8 $\frac{6.92\text{ g X}}{0.584\text{ g C}} \Leftrightarrow \frac{4\text{ x atomic mass of X}}{1\text{ x atomic mass of C}} = \frac{4\text{ x atomic mass of X}}{1\text{ x }12.0\text{ u}}$

atomic mass of X = **35.5 u**

2.9 $\frac{16.0\text{ g oxygen}}{12.0\text{ g carbon}} \Leftrightarrow \frac{?\text{ u (new assigned mass of oxygen)}}{4.00\text{ u}}$

oxygen $\Leftrightarrow$ **5.33 u**

$\frac{32.1\text{ g sulfur}}{12.0\text{ g carbon}} \Leftrightarrow \frac{?\text{ u (new assigned atomic mass of sulfur)}}{4.00\text{ u}}$

sulfur $\Leftrightarrow$ **10.7 u**

2.10 (a) During a chemical reaction, mass is neither created nor destroyed.
(b) In any pure chemical substance, the same elements are always combined in the same proportions by mass.
(c) When two elements combine to form more than one compound, if the masses of one element are the same in the compounds, then the masses of the other element in the various compounds are in a ratio of small whole numbers.

2.11 $24.0\text{ g H x }\frac{6.00\text{ g C}}{1.00\text{ g H}} \Leftrightarrow$ **1.44×10^2 g C**

2.12 The law of definite composition can be demonstrated by more than one method. One method is to show that the percentage composition does not change from one sample to the next. Another method is to show a constant ratio of elements. This problem has been solved using the percentage method. Problem 2.13 uses the ratio method. Students with more chemical knowledge than has been presented thus far in the textbook may select a third method.

Sample #1 $\%X = \frac{4.31\text{ g X}}{4.31\text{ g} + 7.69\text{ g total}} \text{ x } 100 = 35.9\%\text{ X}$

Sample #2 %X = 35.9% X

Sample #3 $\%X = \frac{0.718\text{ g}}{2.00\text{ g sample}} \text{ x } 100 = 35.9\%\text{ X}$

Each sample is 35.9% X and, therefore, 64.1% Y.

2.13 (See discussion at the beginning of the solution to problem 2.12). In this solution we will compare the ratio of weights of the elements present.

For sample #1: 1.00 g C/6.33 g F = 0.158 g C/1 g F
1.00 C/11.67 g Cl = 0.0857 g C/ 1 g Cl
6.33 g F/11.67 g Cl = 0.542 g F/1 g Cl

For sample #2: 2.00 g C/12.66 g F = 0.158 g C/1 g F
2.00 g C/23.34 g Cl = 0.0857 g C/1 g Cl
12.66 g F/23.34 g Cl = 0.542 g F/1 g Cl

For both samples the ratio of mass of carbon to mass of fluorine is 0.158. The ratio of carbon to chlorine is 0.0857 and the ratio of fluorine to chlorine is 0.542 in both samples. These data support the law of definite composition.

2.14 For sample #1: 1.26 g oxygen/10.0 g copper = 0.126 g O/1.00 g Cu
For sample #2: 2.52 g oxygen/10.0 g copper = 0.252 g O/1.00 g Cu
The ratio of the two masses of oxygen is 0.126 to 0.252 or 1 to 2. These data illustrate the law of multiple proportions.

2.15 First, for each sample, calculate the amount of one of the elements per gram of the other element. Then compare the results. Grams of phosphorus per gram of oxygen will be calculated. It would be equally correct to calculate the grams of oxygen per gram of phosphorus and to compare those results.
For sample #1: 0.845 g P/(1.50 g sample - 0.845 g P) = 1.29 g P/1 g of O
For sample #2: 1.09 g P/(2.50 g sample - 1.09 g P) = 0.773 g P/1 g of O
This gives 1.29 g P (#1) to 0.773 g P (#2) or when each is divided by the smallest, they become 1.67 to 1.00; then multiplying by 3 the ratio becomes 5-to-3.

2.16 The first statement of the question shows that in MnO there are 4.00 g of O combined with 13.7 g of Mn. In MnO_2 there will be twice as much O combined with the same amount of Mn. The ratio would be: $\frac{13.7 \text{ g Mn}}{2 \times 4.00 \text{ g O}}$ or $\frac{2 \times 4.00 \text{ g O}}{13.7 \text{ g Mn}}$
Using this ratio one can solve for the number of grams of oxygen that would be combined with 7.85 g of manganese in the compound MnO_2.

$$7.85 \text{ g Mn} \times \frac{2 \times 4.00 \text{ g O}}{13.7 \text{ g Mn}} \Leftrightarrow \mathbf{4.58\ g\ oxygen}$$

2.17 All three represent a set number of objects; the mole is 6.022×10^{23} things, the dozen is 12 things and the gross means 144 things. There are 6.022×10^{23} things in a mole.

2.18 There is no difference between "1 mol C" and "one mole of carbon atoms" other than the first is an abbreviated way of writing the information.

2.19 (a) **2-to-3** (b) **2-to-3**

(c) $2 \text{ mol Al} \times \frac{3 \text{ mol O}}{2 \text{ mol Al}} \Leftrightarrow$ **3 mol O**

(d) $0.2 \text{ mol Al} \times \frac{3 \text{ mol O}}{2 \text{ mol Al}} \Leftrightarrow$ **0.3 mol O**

2.20 (a) **1-to-2** (b) **1-to-2**

2.21 The atom ratios and the mole ratios have the same values in this question.
(a) **1-to-1** (b) **1-to-1** (c) **1-to-3** (d) **1-to-1**
(e) **1-to-3** (f) **1-to-3**

2.22 (a) $\frac{1 \text{ mol Na}}{1 \text{ mol H}}$ or $\frac{1 \text{ mol H}}{1 \text{ mol Na}}$ (b) $\frac{1 \text{ mol Na}}{1 \text{ mol C}}$ or $\frac{1 \text{ mol C}}{1 \text{ mol Na}}$

(c) $\frac{1 \text{ mol Na}}{3 \text{ mol O}}$ or $\frac{3 \text{ mol O}}{1 \text{ mol Na}}$ (d) $\frac{1 \text{ mol H}}{1 \text{ mol C}}$ or $\frac{1 \text{ mol C}}{1 \text{ mol H}}$

(e) $\frac{1 \text{ mol H}}{3 \text{ mol O}}$ or $\frac{3 \text{ mol O}}{1 \text{ mol H}}$ (f) $\frac{1 \text{ mol C}}{3 \text{ mol O}}$ or $\frac{3 \text{ mol O}}{1 \text{ mol C}}$

2.23 (a) 3 (b) 3 (c) 2 (d) 2 (e) $\frac{3 \text{ mol S}}{1 \text{ mol } Al_2(SO_4)_3}$ or $\frac{1 \text{ mol } Al_2(SO_4)_3}{3 \text{ mol S}}$

(f) $\frac{2 \text{ mol Al}}{1 \text{ mol } Al_2(SO_4)_3}$ or $\frac{1 \text{ mol } Al_2(SO_4)_3}{2 \text{ mol Al}}$

2.24 (a) **2-to-5** (b) **12.0** (c) **2.50** (d) **0.240**

(e) $0.600 \text{ mol C} \times \frac{10 \text{ mol H}}{4 \text{ mol C}} \times \frac{1 \text{ mol } H_2}{2 \text{ mol H}} =$ **0.750 mol H_2**

2.25 (a) **SiO_2** (b) **50** (c) **50** (d) **4.50 moles of Si and 9.00 moles of O**

2.26 $As_2S_3 \rightarrow 2As + 3S$; $1.00 \text{ mol } As_2S_3 \times \frac{3 \text{ mol S}}{1 \text{ mol } As_2S_3} =$ **3.00 mol S**

2.27 $1.50 \text{ mol } Cr_2O_3 \times \frac{3 \text{ mol O}}{1 \text{ mol } Cr_2O_3} =$ **4.50 mol O**

2.28 $1.00 \text{ mol } CaCO_3 \times \frac{1 \text{ mol C}}{1 \text{ mol } CaCO_3} \times \frac{1 \text{ mol } CO_2}{1 \text{ mol C}} =$ **1.00 mol CO_2**

2.29 $1 \text{ mol } Al_2(SO_4)_3 \times \frac{3 \text{ mol } SO_4}{1 \text{ mol } Al_2(SO_4)_3} =$ **3 mol sulfate**

2.30 (The following calculation is based upon the amount of sulfate present.)
$1.25 \text{ mol } Al_2(SO_4)_3 \times \frac{3 \text{ mol } SO_4}{1.00 \text{ mol } Al_2(SO_4)_3} \times \frac{1 \text{ mol } BaSO_4}{1 \text{ mol } SO_4} =$ **3.75 mol $BaSO_4$**

2.31 (a) Mg 1 mol = **24.3 g** (b) C 1 mol = **12.0 g** (c) Fe 1 mol = **55.8 g**
(d) Cl 1 mol = **35.5 g** (e) S 1 mol = **32.1 g** (f) Sr 1 mol = **87.6 g**

2.32 (a) $\frac{1 \text{ mol Mg}}{24.3 \text{ g Mg}}$ or $\frac{24.3 \text{ g Mg}}{1 \text{ mol Mg}}$ (b) $\frac{1 \text{ mol C}}{12.0 \text{ g C}}$ or $\frac{12.0 \text{ g C}}{1 \text{ mol C}}$

(c) $\frac{1 \text{ mol Fe}}{55.8 \text{ g Fe}}$ or $\frac{55.8 \text{ g Fe}}{1 \text{ mol Fe}}$ (d) $\frac{1 \text{ mol Cl}}{35.5 \text{ g Cl}}$ or $\frac{35.5 \text{ g Cl}}{1 \text{ mol Cl}}$

(e) $\frac{1 \text{ mol S}}{32.1 \text{ g S}}$ or $\frac{32.1 \text{ g S}}{1 \text{ mol S}}$ (f) $\frac{1 \text{ mol Sr}}{87.6 \text{ g Sr}}$ or $\frac{87.6 \text{ g Sr}}{1 \text{ mol Sr}}$

2.33 (a) $50.0 \text{ g Na} \times \frac{1 \text{ mol Na}}{23.0 \text{ g Na}} =$ **2.17 mol Na**

(b) **0.668 mol As** (c) **0.962 mol Cr** (d) **1.85 mol Al**

(e) **1.28 mol K** (f) **0.463 mol Ag**

2.34 The term formula mass is preferred to molecular mass for substances for which formula units are used to describe a neutral combination of ions, rather than the existence of molecules.

2.35 (a) 24.3 + 16.0 = **40.3** (b) 40.1 + (2 x 35.45) = **111.0**
(c) 30.97 + (5 x 35.45) = **208.2** (d) (2 x 32.07) + (2 x 35.45) = **135.0**
(e) (3 x 22.99) + 30.97 + (4 x 16.00) = **163.9**

2.36 (a) **40.3 g** (b) **111.0 g** (c) **208.2 g** (d) **135.0 g** (e) **163.9 g**

2.37 (a) $\frac{1 \text{ mol MgO}}{40.3 \text{ g MgO}}$ or $\frac{40.3 \text{ g MgO}}{1 \text{ mol MgO}}$ (b) $\frac{1 \text{ mol } CaCl_2}{111.0 \text{ g } CaCl_2}$ or $\frac{111.0 \text{ g } CaCl_2}{1 \text{ mol } CaCl_2}$

(c) $\frac{1 \text{ mol } PCl_5}{208.2 \text{ g } PCl_5}$ or $\frac{208.2 \text{ g } PCl_5}{1 \text{ mol } PCl_5}$ (d) $\frac{1 \text{ mol } S_2Cl_2}{135.0 \text{ g } S_2Cl_2}$ or $\frac{135.0 \text{ g } S_2Cl_2}{1 \text{ mol } S_2Cl_2}$

(e) $\frac{1 \text{ mol } Na_3PO_4}{163.9 \text{ g } Na_3PO_4}$ or $\frac{163.9 \text{ g } Na_3PO_4}{1 \text{ mol } Na_3PO_4}$

2.38 (a) 28.09 x (2 x16.00) = **60.1**
(b) 24.30 x (2 x16.00) + (2 x 1.01) = **58.3**
(c) 24.30 +32.07 + (4 x16.00) + 7[(1.01 x 2) + 16.00]) = **246.5**
(d) (2x40.08)+(5x24.30)+(8x28.09)+(22x16.00)+(2x16.00)+(2x1.01) =**812.4**
(e) (6 x 12.01) + (8 x 1.01) + (6 x 16.00) = **176.1**
(f) (12 x 12.01) + (22 x 1.01) + (11 x 16.00) = **342.3**

2.39 $\frac{194.193 \text{ g caffeine}}{1 \text{ mol caffeine}}$ x 1.35 mol caffeine = **262 g of Caffeine**

2.40 $\frac{334.40 \text{ g penicillin}}{1 \text{ mol penicillin}}$ x 2.33 mol penicillin = **779 g of Penicillin**

2.41 $\frac{303.3 \text{ g lead sulfate}}{1 \text{ mol lead sulfate}}$ x 6.30 mol = 1910 g or **1.91×10^3 g lead sulfate**

2.42 $\frac{79.88 \text{ g } TiO_2}{1 \text{ mol } TiO_2}$ x 0.144 mol = **11.5 g TiO_2**

2.43 $242 \text{ g } NaHCO_3 \times \frac{1 \text{ mol } NaHCO_3}{84.01 \text{ g } NaHCO_3} = \mathbf{2.88 \text{ mol } NaHCO_3}$

2.44 $1.40 \times 10^3 \text{ g } C_4H_{10} \times \frac{1 \text{ mol } C_4H_{10}}{58.12 \text{ g } C_4H_{10}} = \mathbf{24.1 \text{ mol } C_4H_{10}}$

2.45 $85.3 \text{ g } H_2SO_4 \times \frac{1 \text{ mol } H_2SO_4}{98.08 \text{ g } H_2SO_4} = \mathbf{0.870 \text{ mol } H_2SO_4}$

2.46 $25.0 \text{ g } PbHAsO_4 \times \frac{1 \text{ mol } PbHAsO_4}{347.1 \text{ g } PbHAsO_4} = \mathbf{0.0720 \text{ mol } PbHAsO_4}$

2.47 $125 \text{ g KCl} \times \frac{1 \text{ mol K}}{74.55 \text{ g KCl}} \Leftrightarrow \mathbf{1.68 \text{ mol K}}$

2.48 $632 \text{ g } FeS_2 \times \frac{1 \text{ mol } FeS_2}{120.0 \text{ g } FeS_2} \times \frac{2 \text{ mol S}}{1 \text{ mol } FeS_2} \Leftrightarrow \mathbf{10.5 \text{ mol S}}$

2.49 $1.00 \times 10^3 \text{ g } SO_2 \times \frac{1 \text{ mol } SO_2}{64.06 \text{ g } SO_2} \times \frac{1 \text{ mol } FeS_2}{2 \text{ mol } SO_2} \Leftrightarrow \mathbf{7.81 \text{ mol } FeS_2}$

2.50 For Fe; $\frac{55.847 \text{ g Fe}}{\text{mol Fe}} \times \frac{1 \text{ mol Fe}}{6.022 \times 10^{23} \text{ atoms Fe}} = \mathbf{9.274 \times 10^{-23} \text{ g Fe/atom Fe}}$

For SO_2;

$$\frac{64.06 \text{ g } SO_2}{\text{mol } SO_2} \times \frac{1 \text{ mol } SO_2}{6.022 \times 10^{23} \text{ molecules } SO_2} = \mathbf{1.064 \times 10^{-22} \text{ g } SO_2\text{/molecule } SO_2}$$

2.51 $3.50 \times 10^{17} \text{ atoms C} \times \frac{4 \text{ atoms H}}{2 \text{ atoms C}} \Leftrightarrow \mathbf{7.00 \times 10^{17} \text{ atoms H}}$

$$3.50 \times 10^{17} \text{ atoms C} \times \frac{1 \text{ mol C}}{6.022 \times 10^{23} \text{ atoms C}} \times \frac{12.01 \text{ g C}}{1 \text{ mol C}} = \mathbf{6.98 \times 10^{-6} \text{ g C}}$$

$$7.00 \times 10^{17} \text{ atoms H} \times \frac{1 \text{ mol H}}{6.022 \times 10^{23} \text{ atoms}} \times \frac{1.008 \text{ g H}}{\text{mol H}} = \mathbf{1.17 \times 10^{-6} \text{ g H}}$$

2.52 $\frac{342.30 \text{ g}}{\text{mol}} \times \frac{1 \text{ mol}}{6.022 \times 10^{23} \text{ molecules}} = \mathbf{5.684 \times 10^{-22} \text{ g/molecule}}$

$$\left(\frac{342.30 \text{ u sucrose}}{\text{molecule}}\right) \div \left(\frac{12.011 \text{ u carbon}}{\text{atom}}\right) = \mathbf{28.50 \text{ times heavier}}$$

(Note: This is a ratio and, like all ratios, it does not have units.)

$$25.0 \text{ g sucrose} \times \frac{1 \text{ mol}}{342.30 \text{ g}} \times \frac{6.022 \times 10^{23} \text{ molecules}}{\text{mol}} = \mathbf{4.40 \times 10^{22} \text{ molecules}}$$

$$4.40 \times 10^{22} \text{ molecules} \times \frac{45 \text{ atoms}}{\text{molecule}} = \mathbf{1.98 \times 10^{24} \text{ atoms}}$$

2.53 $4.00 \times 10^{-8} \text{ g of } C_3H_8 \times \frac{1 \text{ mol}}{44.10 \text{ g}} \times \frac{6.022 \times 10^{23} \text{ molecules}}{\text{mol}} \times$

$$\frac{3 \text{ atoms C}}{\text{molecule } C_3H_8} \Leftrightarrow \mathbf{1.64 \times 10^{15} \text{ atoms C}}$$

2.54 $3.0 \text{ cm} \times \frac{1 \text{ atom}}{1.5 \times 10^{-8} \text{ cm}} \times \frac{1 \text{ mol}}{6.022 \times 10^{23} \text{ atoms}} \times \frac{12.0 \text{ g}}{\text{mol}} = \mathbf{4.0 \times 10^{-15} \text{ g C}}$

2.55 5.00×10^{20} molecules $S_8 \times \frac{8 \text{ atoms S}}{\text{molecule } S_8} \times \frac{2 \text{ atoms Cu}}{\text{atom S}} \times$

$$\frac{1 \text{ mol Cu}}{6.022 \times 10^{23} \text{ atoms}} \times \frac{63.5 \text{ g Cu}}{\text{mol Cu}} \Leftrightarrow \mathbf{0.844 \text{ g Cu}}$$

2.56 (a) % Fe in $FeCl_3$ = {55.85/[55.85 + (3 x 35.45)]} x 100 = **34.43% Fe**
65.57% Cl

(b) Na_3PO_4 **42.07% Na, 18.89% P, 39.04% O**
(c) $KHSO_4$ **28.71 % K, 0.74% H, 23.55% S, 47.00 % O**
(d) $(NH_4)_2HPO_4$ **21.21% N, 6.87% H, 23.46% P, 48.46% O**
(e) Hg_2Cl_2 **84.98% Hg, 15.02% Cl**

2.57 (a) C_6H_6 **92.26% C, 7.74% H**
(b) C_2H_5OH **52.14% C, 13.13% H, 34.73% O**
(c) $K_2Cr_2O_7$ **26.58% K, 35.35% Cr, 38.07% O**
(d) XeF_4 **63.34% Xe, 36.66% F**
(e) $CaCO_3$ **40.04% Ca, 12.00% C, 47.96% O**

2.58 (a) $1 \text{ mol } C_6H_6 \times \frac{6 \text{ mol C}}{\text{mol } C_6H_6} \times \frac{12.01 \text{ g C}}{\text{mole C}} \Leftrightarrow \mathbf{72.06 \text{ g C}}$

$\frac{72.06 \text{ g C}}{78.11 \text{ g } C_6H_6}$ or $\frac{78.11 \text{ g } C_6H_6}{72.06 \text{ g C}}$

(b) $1 \text{ mol } C_2H_5OH \times \frac{2 \text{ mol C}}{\text{mol } C_2H_5OH} \times \frac{12.01 \text{ g C}}{\text{mol C}} \Leftrightarrow \mathbf{24.02 \text{ g C}}$

$\frac{24.02 \text{ g C}}{46.07 \text{ g } C_2H_5OH}$ or $\frac{46.07 \text{ g } C_2H_5OH}{24.02 \text{ g C}}$

(c) $1 \text{ mol } K_2Cr_2O_7 \times \frac{2 \text{ mol K}}{\text{mol } K_2Cr_2O_7} \times \frac{39.10 \text{ g K}}{\text{mol K}} \Leftrightarrow \mathbf{78.20 \text{ g K}}$

$\frac{78.20 \text{ g K}}{294.18 \text{ g } K_2Cr_2O_7}$ or $\frac{294.18 \text{ g } K_2Cr_2O_7}{78.20 \text{ g K}}$

(d) $1 \text{ mol } XeF_4 \times \frac{1 \text{ mol Xe}}{\text{mol } XeF_4} \times \frac{131.30 \text{ g Xe}}{\text{mol Xe}} \Leftrightarrow \mathbf{131.30 \text{ g Xe}}$

$\frac{131.30 \text{ g Xe}}{207.29 \text{ g } XeF_4}$ or $\frac{207.29 \text{ g } XeF_4}{131.30 \text{ g Xe}}$

(e) $1 \text{ mol } CaCO_3 \times \frac{1 \text{ mol Ca}}{\text{mol } CaCO_3} \times \frac{40.08 \text{ g Ca}}{\text{mol Ca}} \Leftrightarrow \mathbf{40.08 \text{ g Ca}}$

$\frac{40.08 \text{ g Ca}}{100.09 \text{ g } CaCO_3}$ or $\frac{100.09 \text{ g } CaCO_3}{40.08 \text{ g Ca}}$

2.59 (a) $15.0 \text{ g } Fe_2O_3 \times \frac{1 \text{ mol } Fe_2O_3}{159.70 \text{ g } Fe_2O_3} \times \frac{2 \text{ mol Fe}}{\text{mol } Fe_2O_3} \times \frac{55.85 \text{ g Fe}}{\text{mole Fe}} \Leftrightarrow \mathbf{10.5 \text{ g Fe}}$

(b) $25.0 \text{ g } Al_2(SO_4)_3 \times \frac{1 \text{ mol } Al_2(SO_4)_3}{342.15 \text{ g } Al_2(SO_4)_3} \times \frac{2 \text{ mol Al}}{\text{mol } Al_2(SO_4)_3} \times \frac{26.98 \text{ g Al}}{\text{mol Al}}$

$\Leftrightarrow \mathbf{3.94 \text{ g Al}}$

(c) $16.0 \text{ g } Na_2CO_3 \times \frac{1 \text{ mol } Na_2CO_3}{106.0 \text{ g } Na_2CO_3} \times \frac{2 \text{ mol Na}}{\text{mol } Na_2CO_3} \times \frac{22.99 \text{ g Na}}{\text{mol Na}} \Leftrightarrow \mathbf{6.94 \text{ g Na}}$

(d) $48.0 \text{ g } MgCl_2 \times \frac{1 \text{ mol } MgCl_2}{95.21 \text{ g } MgCl_2} \times \frac{1 \text{ mol Mg}}{\text{mol } MgCl_2} \times \frac{24.30 \text{ g Mg}}{\text{mol Mg}} \Leftrightarrow \mathbf{12.3 \text{ g Mg}}$

2.60 $30.0 \text{ g gly} \times \frac{1 \text{ mol gly}}{75.07 \text{ g gly}} \times \frac{1 \text{ mol N}}{\text{mol gly}} \times \frac{14.01 \text{ g N}}{\text{mol N}} \Leftrightarrow \mathbf{5.60 \text{ g N}}$

2.61 $12.0 \text{ g } NH_3 \times \frac{1 \text{ mol } NH_3}{17.03 \text{ g } NH_3} \times \frac{3 \text{ mol H}}{\text{mol } NH_3} \times \frac{1.008 \text{ g H}}{\text{mol H}} \Leftrightarrow \mathbf{2.13 \text{ g H}}$

2.62 $\frac{7.04 \text{ g P}}{12.5 \text{ g sample}} \times 100 = \mathbf{56.3\% \text{ P}}$ $\quad \frac{5.46 \text{ g S}}{12.5 \text{ g sample}} \times 100 = \mathbf{43.7\% \text{ S}}$

2.63 (a) $9.34 \text{ g } CO_2 \times \frac{1 \text{ mole } CO_2}{44.01 \text{ g } CO_2} \times \frac{1 \text{ mol C}}{\text{mol } CO_2} \times \frac{12.01 \text{ g C}}{\text{mol C}} \Leftrightarrow \mathbf{2.55 \text{ g C}}$

$5.09 \text{ g } H_2O \times \frac{1 \text{ mol } H_2O}{18.02 \text{ g } H_2O} \times \frac{2 \text{ mol H}}{\text{mol } H_2O} \times \frac{1.008 \text{ g H}}{\text{mol H}} \Leftrightarrow \mathbf{0.569 \text{ g H}}$

(b) g sample = g O + g C + g H or 4.25 - 2.55 - 0.569 = **1.13 g O**

(c) (2.55 g C/4.25 g sample) x 100 = **60.0% C, 13.4% H, 26.6% O**

2.64 A structural formula represents the total number of atoms in a molecule and the way the atoms are linked together. A molecular formula specifies the actual number of each kind of atom found in the molecule but not necessarily how they are linked. The empirical formula gives the relative number of atoms of each element present in the molecule but the actual number present may be a multiple of that indicated by the empirical formula.

2.65 (a) $\mathbf{NH_4SO_4}$ (b) $\mathbf{Fe_2O_3}$ (c) $\mathbf{AlCl_3}$ (d) $\mathbf{CH}$ (e) $\mathbf{C_3H_8O_3}$
(f) $\mathbf{CH_2O}$ (g) $\mathbf{Hg_2SO_4}$

2.66 **$C_2H_6O_2$ and CH_3O**

2.67 The simplest formula is calculated from experimentally observed or measured data obtained by a chemical analysis of the compound. One dictionary defined empirical as "founded upon experiment or experience."

2.68 To calculate the empirical formula of a compound, one needs to know the mass of each element present in a sample and the atomic masses of those elements.

2.69 $S_?O_?$ S = 1.40 g x (1 mol S/32.066 g S) = 0.0437 mol
O = 2.10 g x (1 mol O/15.999 g O) = 0.131 mol

$S_{0.0437}O_{0.131}$ or $S_{0.0437/0.0437}O_{0.131/0.0437}$

= S_1O_3 or **SO_3**

2.70 moles C = 0.423 g x (1 mol C/12.011 g C) = 0.0352 mol C

moles Cl = 2.50 g x (1 mol Cl/35.45 g Cl) = 0.0705 mol Cl

moles F = 1.34 g x (1 mol F/19.00 g F) = 0.0705 mol F

$C_{0.0352}Cl_{0.0705}F_{0.0705}$ = **CCl_2F_2**

2.71 moles P = 7.04 g P x (1 mol P/30.97 g P) = 0.227 mol P
moles S = 5.46 g S x (1 mol S/32.07 g S) = 0.170 mol S
$P_{0.227}S_{0.170} = P_{1.34}S_{1.00}$ = **P_4S_3**

2.72 2.55 g C x (1 mol C/12.01 g C) = 0.212 mol C
0.569 g H x (1 mol H/1.008 g H) = 0.564 mol H
1.13 g O x (1 mol O/15.999 g O) = 0.0706 mol O
$C_{0.212/0.0706}H_{0.564/0.0706}O_{0.0706/0.0706}$ = **C_3H_8O**

2.73 (Assume a 100 g sample). 14.5 g C x (1 mol C/12.01 g C) = 1.21 mol C
85.5 g Cl x (1 mol Cl/35.45 g Cl) = 2.41 mol Cl
$C_{1.21/1.21}Cl_{2.41/1.21}$ = **CCl_2**

2.74 75.7 g As x (1 mol As/74.92 g As) = 1.01 mol As
24.3 g O x (1 mol O/15.999 g O) = 1.52 mol O
$As_{1.01/1.01}O_{1.52/1.01}$ = **As_2O_3**

2.75 1.31 g S and 4.22 g - 1.31 g or 2.91 g Cl
1.31 g S x (1mol S/32.07 g S) = 0.0408 mol S
2.91 g Cl x (1 mol Cl/35.45 g Cl) = 0.0821 mol Cl
$S_{0.0408/0.0408}Cl_{0.0821/0.0408}$ = **SCl_2**

2.76 60.8 g Na x (1 mol Na/22.99 g Na) = 2.64 mol Na
28.5 g B x (1 mol B/10.81 g B) = 2.64 mol B
10.5 g H x (1 mol H/1.008 g H) = 10.4 mol H **$NaBH_4$**

2.77 63.2 g C x (1 mol C/12.01 g C) = 5.26 mol C
5.26 g H x (1 mol H/1.008 g H) = 5.22 mol H
31.6 g O x (1 mol O/16.00 g O) = 1.98 mol O
$C_{5.26/1.98}H_{5.22/1.98}O_{1.98/1.98}$ or $C_{2.66}H_{2.64}O_{1.00}$ times 3 = **$C_8H_8O_3$**

2.78 $C_?H_?O_? + ?O_2 \rightarrow CO_2 + H_2O$
From the mass of CO_2 one can calculate the mass of C in the sample. From the mass of H_2O one can calculate the mass of H in the sample. From the mass of the sample and masses of C and H, one can obtain the mass of O in the sample. From these masses, the moles of each and the empirical formula can be calculated.

$$1.030 \text{ g } CO_2 \text{ x } \frac{1 \text{ mol } CO_2}{44.01 \text{ g } CO_2} \text{ x } \frac{1 \text{ mol C}}{\text{mol } CO_2} \text{ x } \frac{12.01 \text{ g C}}{\text{mol C}} = 0.281 \text{ g C}$$

$$0.632 \text{ g } H_2O \text{ x } \frac{1 \text{ mol } H_2O}{18.02 \text{ g } H_2O} \text{ x } \frac{2 \text{ mol H}}{\text{mol } H_2O} \text{ x } \frac{1.008 \text{ g H}}{\text{mol H}} = 0.0707 \text{ g H}$$

0.537 - 0.281 - 0.0707 = 0.185 g O
0.281 g C x (1 mol C/12.01 g C) = **0.0234 mol C**
0.0707 g H x (1 mol H/1.008 g H) = **0.0701 mol H**
0.185 g O x (1 mol O/16.00 g O) = **0.0116 mol O**
$C_{0.0234/0.0116}H_{0.0701/0.0116}O_{0.0116/0.0116}$ = **C_2H_6O**

2.79 (a) $4.072 \text{ g AgCl x } \frac{1 \text{ mol AgCl}}{143.32 \text{ g AgCl}} \text{ x } \frac{1 \text{ mol Cl}}{\text{mol AgCl}} \text{ x } \frac{35.45 \text{ g Cl}}{\text{mol C}} = 1.007 \text{ g Cl}$

(b) From part (a), grams Cl = 1.007 g
Grams Cr = 1.500 - 1.007 = 0.493 g

(c) 1.007 g Cl x (1 mol Cl/35.45 g Cl) = 0.0284 mol Cl
0.493 g Cr x (1 mol Cr/52.00 g Cr) = 0.00948 mol Cr
$Cr_{0.00948/0.00948}Cl_{0.0284/0.00948}$ = **$CrCl_3$**

2.80 From the amount of H_2O and CO_2 produced from the 1.35 g sample, one can calculate the % H and % C. From the amount of NH_3 produced from the 0.735 g sample, one can obtain the % N. From the difference, one can obtain % O. Once the % composition is known, the problem becomes a problem very much like Problems 2.73, 2.74, 2.76 and 2.77.

% H: $0.810 \text{ g } H_2O \text{ x } \frac{1 \text{ mol } H_2O}{18.02 \text{ g } H_2O} \text{ x } \frac{2 \text{ mol H}}{\text{mol } H_2O} \text{ x } \frac{1.01 \text{ g H}}{\text{mol H}} = 0.0908 \text{ g H}$

$\frac{0.0908 \text{ g H}}{1.35 \text{ g sample}} \text{ x } 100 = \mathbf{6.73\% \text{ H}}$

(continued)

2.80 (continued)

$$\%\ C:\ 1.32\ g\ CO_2 \times \frac{1\ mol\ CO_2}{44.01\ g\ CO_2} \times \frac{1\ mol\ C}{mol\ CO_2} \times \frac{12.01\ g\ C}{mol\ C} = 0.360\ g\ C$$

$$\frac{0.360\ g\ C}{1.35\ g\ sample} \times 100 = \mathbf{26.7\%\ C}$$

$$\%\ N:\ 0.284\ g\ NH_3 \times \frac{1\ mol\ NH_3}{17.03\ g\ NH_3} \times \frac{1\ mol\ N}{mol\ NH_3} \times \frac{14.01\ g\ N}{mol\ N} = 0.234\ g\ N$$

$$\frac{0.234\ g\ N}{0.735\ g\ sample} \times 100 = \mathbf{31.8\%\ N}$$

% O: 100.0 - 6.73 - 26.7 - 31.8 = **34.8% O**

6.73 g H x (1 mol H/1.008 g H) = 6.68 mol H
26.7 g C x (1 mol C/12.01 g C) = 2.22 mol C
31.8 g N x (1 mol N/14.01 g N) = 2.27 mol N
34.8 g O x (1 mol O/16.00 g O) = 2.18 mol O
Empirical Formula: $C_{2.22}H_{6.68}N_{2.27}O_{2.18}$ = $\mathbf{CH_3NO}$

2.81 (See the discussion for Problem 2.80.) This problem has the added element, Cl.

(a)

$$\%\ C:\ 0.138\ g\ CO_2 \times \frac{1\ mol\ CO_2}{44.01\ g\ CO_2} \times \frac{1\ mol\ C}{mol\ CO_2} \times \frac{12.01\ g\ C}{mol\ C} = 0.0377\ g\ C$$

$$\frac{0.377\ g\ C}{0.150\ g\ sample} \times 100 = \mathbf{25.1\%\ C}$$

$$\%\ H:\ 0.0566\ g\ H_2O \times \frac{1\ mol\ H_2O}{18.02\ g\ H_2O} \times \frac{2\ mol\ H}{mol\ H_2O} \times \frac{1.01\ g\ H}{mol\ H} = 0.00634\ g\ H$$

$$\frac{0.00634\ g\ H}{0.150\ g\ sample} \times 100 = \mathbf{4.23\%\ H}$$

$$\%\ N:\ 0.0238\ g\ NH_3 \times \frac{1\ mol\ NH_3}{17.03\ g\ NH_3} \times \frac{1\ mol\ N}{mol\ NH_3} \times \frac{14.01\ g\ N}{mol\ N} = 0.0196\ g\ N$$

$$\frac{0.0196\ g\ N}{0.200\ g\ sample} \times 100 = \mathbf{9.79\%\ N}$$

$$\%\ Cl:\ 0.251\ g\ AgCl \times \frac{1\ mol\ AgCl}{143.3\ g\ AgCl} \times \frac{1\ mol\ Cl}{mol\ AgCl} \times \frac{35.45\ g\ Cl}{mol\ Cl} = 0.0621\ g\ Cl$$

$$\frac{0.0621\ g\ Cl}{0.125\ g\ sample} \times 100 = \mathbf{49.7\%\ Cl}$$

% O: = 100.0 - 25.1 - 4.23 - 9.79 - 49.7 = **11.2% O**

(b) Empirical Formula = $C_{25.1/12.01}H_{4.23/1.01}N_{9.79/14.01}Cl_{49.7/35.45}O_{11.2/16.0}$
= $C_{2.09}H_{4.19}N_{0.70}Cl_{1.40}O_{0.70}$ = $\mathbf{C_3H_6NCl_2O}$

2.82 (a) Empirical mass = 135.1, therefore $\mathbf{Na_2S_4O_6}$
(b) Empirical mass = 73.5, therefore $\mathbf{C_6H_4Cl_2}$
(c) Empirical mass= 60.5, therefore $\mathbf{C_6H_3Cl_3}$
(d) Empirical mass = 122.1, therefore $\mathbf{Na_{12}Si_6O_{18}}$
(e) Empirical mass = 102.0, therefore $\mathbf{Na_3P_3O_9}$

2.83 mass C: $0.6871 \text{ g } CO_2 \times \frac{1 \text{ mol } CO_2}{44.01 \text{ g } CO_2} \times \frac{1 \text{ mol C}}{\text{mol } CO_2} \times \frac{12.01 \text{ g C}}{\text{mol C}} = \mathbf{0.1875 \text{ g C}}$

mass H: $0.1874 \text{ g } H_2O \times \frac{1 \text{ mol } H_2O}{18.02 \text{ g } H_2O} \times \frac{2 \text{ mol H}}{\text{mol } H_2O} \times \frac{1.008 \text{ g H}}{\text{mol H}} = \mathbf{0.02097 \text{ g H}}$

mass O = 0.5000 g total - 0.1875 g C - 0.02097 g H = **0.2915 g O**

Empirical Formula = $C_{0.1875/12.01}H_{0.02097/1.008}O_{0.2915/16.00}$ =
$C_{0.01561}H_{0.02080}O_{0.01822} = C_{1.000}H_{1.332}O_{1.167}$
Multiplied by 6 it becomes $\mathbf{C_6H_8O_7}$

Molecular Formula? Empirical formula mass= 192. MM = empirical formula mass. Therefore, the molecular formula is $\mathbf{C_6H_8O_7}$

2.84 The formula mass of the C_8H_8 unit within the brackets is 104.

$\frac{MM}{FM\ C_8H_8} = \frac{10^6}{104} = 1 \times 10^4$ **Therefore, n = 1×10^4 styrene units/molecule**

2.85 For carbon-12, atomic mass = 12.0000
For the isotope of gold, atomic mass = 14.9977 x 12.0000 or 179.972

2.86 C_4H_{10} + excess $O_2 \rightarrow 4CO_2 + 5H_20$

$2.50 \text{ mol butane} \times \frac{4 \text{ mol } CO_2}{\text{mol butane}} \Leftrightarrow 10.0 \text{ mol } CO_2$

C_3H_8 + excess $O_2 \rightarrow 3CO_2 + 4H_2O$

$10 \text{ mol } CO_2 \times \frac{1 \text{ mol propane}}{3 \text{ mol } CO_2} \Leftrightarrow \mathbf{3.33 \text{ mol propane}}$

2.87 $65.0 \text{ g } CuSO_4{\bullet}5H_2O \times \frac{1 \text{ mol } CuSO_4{\bullet}5H_2O}{249.69 \text{ g } CuSO_4{\bullet}5H_2O} \times \frac{9 \text{ mol O}}{1 \text{ mol } CuSO_4{\bullet}5H_2O} \times \frac{16.0 \text{ g O}}{\text{mol O}}$

$= \mathbf{37.5 \text{ g O}}$

2.88 Ba: 41.23 g x (1 mol Ba/137.33 g) = **0.3002 mol Ba**
Cu: 28.62 g x (1 mol Cu/63.546 g) = **0.4504 mol Cu**
Y: 13.35 g x (1 mol Y/88.906 g) = **0.1502 mol Y**
O: 16.81 g x (1 mol O/15.999 g) = **1.051 mol O**

$Ba_{0.3002/0.1502}Cu_{0.4504/0.1502}Y_{0.1502/0.1502}O_{1.051/0.1502} = \mathbf{Ba_2Cu_3YO_7}$

2.89 1 Cr atom ⇔ ? volume

$$\frac{51.996\text{ g Cr}}{\text{mol Cr}} \times \frac{1\text{ mol Cr}}{6.022 \times 10^{23}\text{ atom Cr}} \times 1\text{ atom Cr} \times \frac{1\text{ cm}^3}{7.20\text{ g}} \Leftrightarrow \mathbf{1.20 \times 10^{-23}\ cm^3}$$

2.90 From question 2.89, the volume of an atom = 1.20×10^{-23} cm^3. Assuming that the atom is cubic, the thickness of the atom would be the cube root of 1.20×10^{-23} cm^3 or 2.29×10^{-8} cm per atom. To find the number of atoms thick : 7.5×10^{-5} cm thick/ 2.29×10^{-8} cm per atom. = **3.3×10^3 atoms thick**

2.91 $Ag_xS_y \rightarrow BaSO_4$

$$0.8689\text{ g BaSO}_4 \times \frac{1\text{ mol BaSO}_4}{233.39\text{ g BaSO}_4} \times \frac{1\text{ mol S}}{1\text{ mol BaSO}_4} \times \frac{32.07\text{ g S}}{\text{mol S}} \Leftrightarrow 0.1194\text{ g S}$$

0.9225 g sample - 0.1194 g S ⇔ 0.8031 g Ag

0.1194 g S x (1 mol S/32.07 g S) = 0.003723 mol S

0.8031 g Ag x (1 mol Ag/107.87 g Ag) = 0.007445 mol Ag

$Ag_{0.007445}S_{0.003723}$ = **Ag_2S**

2.92 84.98 g Hg x (1 mol Hg/200.59 g Hg) = 0.4237 mol Hg

(100 - 84.98)g Cl x (1 mol Cl/35.45 g Cl) = 0.4237 mol Cl

Empirical Formula = $Hg_{0.4237}Cl_{0.4237}$ = HgCl

Formula mass of HgCl = 236.04

Molecular mass equals 472. Therefore, molecular formula is **Hg_2Cl_2**

2.93 (a) **2.00 moles** (b) **0.720 moles** (c) **6.00 moles** (d) **3.00 moles**

3 CHEMICAL REACTIONS AND THE MOLE CONCEPT

3.1 The law of conservation of mass

3.2 The coefficients are: (a) **1,2,1,1** (b) **2,1,1,1** (c) **8,3,4,9** (d) **1,1,2** (e) **2,1,2,1**

3.3 The coefficients are: (a) **2,3,1,6** (b) **2,1,1,2,2** (c) **3,2,1,6** (d) **1,3,1,3,3** (e) **1,1,1,1,2**

3.4 The coefficients are: (a) **2,13,8,10 [or 1,13/2,4,5]** (b) **2,15,14,6 [or 1,15/2,7,3]** (c) **1,6,4** (d) **4,11,2,8** (e) **4,5,4,6 [or 2,5/2,2,3]**

3.5 The coefficients are: (a) **1,2,1,1** (b) **1,4,2,1,1** (c) **1,3,1,3** (d) **3,1,1,2** (e) **2,9,4,6,2**

3.6 (a) $0.200 \text{ mol MnO(OH)} \times \frac{1 \text{ mol Zn}}{2 \text{ mol MnO(OH)}} \Leftrightarrow \mathbf{0.100 \text{ mol Zn}}$

(b) $0.150 \text{ mol Zn} \times \frac{2 \text{ mol } H_2O}{1 \text{ mol Zn}} \Leftrightarrow \mathbf{0.300 \text{ mol } H_2O}$

(c) $0.100 \text{ mol } MnO_2 \times \frac{1 \text{ mol } Zn(OH)_2}{2 \text{ mol } MnO_2} \Leftrightarrow \mathbf{0.0500 \text{ mol } Zn(OH)_2}$

(d) $0.600 \text{ mol } MnO_2 \times \frac{2 \text{ mol } H_2O}{2 \text{ mol } MnO_2} \Leftrightarrow \mathbf{0.600 \text{ mol } H_2O}$

3.7 (a) $2.50 \text{ mol } CaC_2 \text{ x } \frac{1 \text{ mol } C_2H_2}{\text{mol } CaC_2} \Leftrightarrow \mathbf{2.50 \text{ mol } C_2H_2}$

(b) $0.500 \text{ mol } CaC_2 \text{ x } \frac{1 \text{ mol } C_2H_2}{\text{mol } CaC_2} \text{ x } \frac{26.0 \text{ g } C_2H_2}{\text{mol } C_2H_2} \Leftrightarrow \mathbf{13.0 \text{ g } C_2H_2}$

(c) $(3.20 \text{ mol } C_2H_2) \text{ x } (2 \text{ mol } H_2O/\text{mol}C_2H_2) \Leftrightarrow \mathbf{6.40 \text{ mol } H_2O}$

(d) $28.0 \text{ g } C_2H_2 \text{ x } \frac{1 \text{ mol } C_2H_2}{26.0 \text{ g } C_2H_2} \text{ x } \frac{1 \text{ mol } Ca(OH)_2}{\text{mol } C_2H_2} \text{ x } \frac{74.1 \text{ g } Ca(OH)_2}{\text{mol } Ca(OH)_2}$

$\Leftrightarrow \mathbf{79.8 \text{ g } Ca(OH)_2}$

3.8 (a) $14.3 \text{ g } ClO_2 \text{ x } \frac{1 \text{ mol } ClO_2}{67.45 \text{ g } ClO_2} \text{ x } \frac{5 \text{ mol } HClO_3}{6 \text{ mol } ClO_2} \Leftrightarrow \mathbf{0.177 \text{ mol } HClO_3}$

(b) $5.74 \text{ g } HCl \text{ x } \frac{1 \text{ mol } HCl}{36.46 \text{ g } HCl} \text{ x } \frac{3 \text{ mol } H_2O}{\text{mol } HCl} \text{ x } \frac{18.02 \text{ g } H_2O}{\text{mol } H_2O} \Leftrightarrow \mathbf{8.51 \text{ g } H_2O}$

3.9 (a) $\mathbf{P_4 + 5O_2 \rightarrow P_4O_{10}}$

(b) $0.500 \text{ mol } O_2 \text{ x } \frac{1 \text{ mol } P_4O_{10}}{5 \text{ mol } O_2} \Leftrightarrow \mathbf{0.100 \text{ mol } P_4O_{10}}$

(c) $50.0 \text{ g } P_4O_{10} \text{ x } \frac{1 \text{ mol } P_4O_{10}}{283.9 \text{ g } P_4O_{10}} \text{ x } \frac{1 \text{ mol } P_4}{\text{mol } P_4O_{10}} \text{ x } \frac{123.9 \text{ g } P_4}{\text{mol } P_4} \Leftrightarrow \mathbf{21.8 \text{ g } P_4}$

(d) $25.0 \text{ g } O_2 \text{ x } \frac{1 \text{ mol } O_2}{32.0 \text{ g } O_2} \text{ x } \frac{1 \text{ mol } P_4}{5 \text{ mol } O_2} \text{ x } \frac{123.9 \text{ g } P_4}{\text{mol } P_4} \Leftrightarrow \mathbf{19.4 \text{ g } P_4}$

3.10 (a) $0.0250 \text{ mol } N_2H_4 \text{ x } \frac{2 \text{ mol } HNO_3}{\text{mol } N_2H_4} \Leftrightarrow \mathbf{0.0500 \text{ mol } HNO_3}$

(b) $1.35 \text{ mol } H_2O \text{ x } \frac{7 \text{ mol } H_2O_2}{8 \text{ mol } H_2O} \Leftrightarrow \mathbf{1.18 \text{ mol } H_2O_2}$

(continued)

3.10 (continued)

(c) $1.87 \text{ mol } HNO_3 \times \frac{8 \text{ mol } H_2O}{2 \text{ mol } HNO_3} \Leftrightarrow \mathbf{7.48 \text{ mol } H_2O}$

(d) $22.0 \text{ g } N_2H_4 \times \frac{1 \text{ mol } N_2H_4}{32.05 \text{ g } N_2H_4} \times \frac{7 \text{ mol } H_2O_2}{\text{mol } N_2H_4} \Leftrightarrow \mathbf{4.80 \text{ mol } H_2O_2}$

(e) $45.8 \text{ g } HNO_3 \times \frac{1 \text{ mol } HNO_3}{63.02 \text{ g } HNO_3} \times \frac{7 \text{ mol } H_2O_2}{2 \text{ mol } HNO_3} \times \frac{34.02 \text{ g } H_2O_2}{\text{mol } H_2O_2}$

$\Leftrightarrow \mathbf{86.5 \text{ g } H_2O_2}$

3.11 (a) $35.0 \text{ mol Fe} \times \frac{3 \text{ mol CO}}{2 \text{ mol Fe}} \Leftrightarrow \mathbf{52.5 \text{ mol CO}}$

(b) $4.50 \text{ mol } CO_2 \times \frac{1 \text{ mol } Fe_2O_3}{3 \text{ mol } CO_2} \Leftrightarrow \mathbf{1.50 \text{ mol } Fe_2O_3}$

(c) $0.570 \text{ mol Fe} \times \frac{1 \text{ mol } Fe_2O_3}{2 \text{ mol Fe}} \times \frac{159.7 \text{ g } Fe_2O_3}{\text{mol } Fe_2O_3} \Leftrightarrow \mathbf{45.5 \text{ g } Fe_2O_3}$

(d) $48.5 \text{ g } Fe_2O_3 \times \frac{1 \text{ mol } Fe_2O_3}{159.7 \text{ g } Fe_2O_3} \times \frac{3 \text{ mol CO}}{\text{mol } Fe_2O_3} \Leftrightarrow \mathbf{0.911 \text{ mol CO}}$

(e) $18.6 \text{ g CO} \times \frac{1 \text{ mol CO}}{28.0 \text{ g CO}} \times \frac{2 \text{ mol Fe}}{3 \text{ mol CO}} \times \frac{55.85 \text{ g Fe}}{\text{mol Fe}} \Leftrightarrow \mathbf{24.7 \text{ g Fe}}$

3.12 (a) $6.50 \text{ mol } TiCl_4 \times \frac{2 \text{ mol } H_2O}{\text{mol } TiCl_4} \Leftrightarrow \mathbf{13.0 \text{ mol } H_2O}$

(b) $8.44 \text{ mol } TiCl_4 \times \frac{4 \text{ mol HCl}}{\text{mol } TiCl_4} \Leftrightarrow \mathbf{33.8 \text{ mol HCl}}$

(c) $14.4 \text{ mol } TiCl_4 \times \frac{1 \text{ mol } TiO_2}{\text{mol } TiCl_4} \times \frac{79.9 \text{ g } TiO_2}{\text{mol } TiO_2} \Leftrightarrow \mathbf{1.15 \times 10^3 \text{ g } TiO_2}$

(d) $85.0 \text{ g } TiCl_4 \times \frac{1 \text{ mol } TiCl_4}{189.7 \text{ g } TiCl_4} \times \frac{4 \text{ mol HCl}}{\text{mol } TiCl_4} \times \frac{36.46 \text{ g HCl}}{\text{mol HCl}} \Leftrightarrow \mathbf{65.3 \text{ g HCl}}$

3.13 $1{,}000 \text{ kg } C_6H_5Cl \times \frac{1 \text{ kmol } C_6H_5Cl}{112.6 \text{ kg } C_6H_5Cl} \times \frac{1 \text{ kmol DDT}}{2 \text{ kmol } C_6H_5Cl} \times \frac{354.5 \text{ kg DDT}}{\text{kmol DDT}}$

$\Leftrightarrow$ **1.574 x 10^3 kg DDT**

3.14 $\frac{5 \text{ grains}}{\text{tablet}} \times 2 \text{ tablets} \times \frac{1 \text{ g}}{15.4 \text{ grains}} \times \frac{1 \text{ mol aspirin}}{180.2 \text{ g}} \times \frac{1 \text{ mol } C_7H_6O_3}{\text{mol aspirin}}$

$\times \frac{138.1 \text{ g } C_7H_6O_3}{\text{mol } C_7H_6O_3}$ $\Leftrightarrow$ **0.5 g $C_7H_6O_3$**

3.15 (a) **$(CH_3)_2NNH_2 + 2N_2O_4 \rightarrow 4H_2O + 2CO_2 + 3N_2$**

(b) $50.0 \text{ kg } (CH_3)_2NNH_2 \times \frac{1 \text{ kmol } (CH_3)_2NNH_2}{60.10 \text{ kg } (CH_3)_2NNH_2}$

$\times \frac{2 \text{ kmol } N_2O_4}{1 \text{ kmol } (CH_3)_2NNH_2} \times \frac{92.02 \text{ kg } N_2O_4}{\text{kmol } N_2O_4}$ $\Leftrightarrow$ **153 kg N_2O_4**

3.16 $500 \text{ g sugar} \times \frac{1 \text{ mol sugar}}{180.2 \text{ g sugar}} \times \frac{2 \text{ mol } C_2H_5OH}{\text{mol sugar}} \times \frac{46.08 \text{ g } C_2H_5OH}{\text{mol } C_2H_5OH}$

$\Leftrightarrow$ **256 g C_2H_5OH**

3.17 (a) $20.0 \text{ g Pb} \times \frac{1 \text{ mol Pb}}{207.2 \text{ g Pb}} \times \frac{1 \text{ mol white lead}}{6 \text{ mol Pb}} \times \frac{775.6 \text{ g white lead}}{\text{mol white lead}}$

$\Leftrightarrow$ **12.5 g white lead**

(b) $14.0 \text{ g } O_2 \times \frac{1 \text{ mol } O_2}{32.0 \text{ g } O_2} \times \frac{2 \text{ mol } CO_2}{3 \text{ mol } O_2} \times \frac{44.01 \text{ g } CO_2}{\text{mol } CO_2}$ $\Leftrightarrow$ **12.8 g CO_2**

3.18 (a) **16.0 pounds** (b) **23.0 pounds** (c) **160 pounds**
(d) **46.0 pounds** (e) **55.8 pounds**

3.19 $2.40 \text{ ton-mol } CaCl_2 \times \frac{1 \text{ ton-mol Ca}}{1 \text{ ton-mol } CaCl_2} \times \frac{40.08 \text{ tons Ca}}{1 \text{ ton-mol Ca}} \Leftrightarrow$ **96.2 tons Ca**

3.20 $25.0 \text{ tons } Ca_3(PO_4)_2 \times \frac{1 \text{ ton-mol } Ca_3(PO_4)_2}{310.2 \text{ tons } Ca_3(PO_4)_2} \times \frac{2 \text{ ton-mol } H_2SO_4}{\text{ton-mol } Ca_3(PO_4)_2}$

$$\times \frac{98.08 \text{ tons } H_2SO_4}{\text{ton-mol } H_2SO_4} \Leftrightarrow \mathbf{15.8 \text{ tons } H_2SO_4}$$

3.21 $650 \text{ lb } H_2 \times \frac{1 \text{ lb-mol } H_2}{2.02 \text{ lb } H_2} \times \frac{2 \text{ lb-mol } NH_3}{3 \text{ lb-mol } H_2} \times \frac{17.04 \text{ lb } NH_3}{\text{lb-mol } NH_3}$

$$\Leftrightarrow \mathbf{3.66 \times 10^3 \text{ lb } NH_3}$$

3.22 That reactant that is completely consumed before any of the remaining reactants are used up is the **limiting reactant**. First calculate the number of moles of each reactant present. Then compare their ratios with that for the balanced chemical equation. From this deduce which reactant will be depleted first.

3.23 (a) $0.40 \text{ mol Fe} \times \frac{2 \text{ mol HCl}}{\text{mol Fe}} \Leftrightarrow 0.80$ mol HCl required to react with 0.40 mol Fe

Since only 0.75 mol HCl is available, HCl is the limiting reactant.

(b) $0.75 \text{ mol HCl} \times \frac{1 \text{ mol } H_2}{2 \text{ mol HCl}} \Leftrightarrow$ **0.38 mol H_2**

(c) Moles of Fe required to react with 0.75 mole of HCl is:

$$0.75 \text{ mol HCl} \times \frac{1 \text{ mol Fe}}{2 \text{ mol HCl}} \Leftrightarrow 0.38 \text{ mol Fe}$$

There is 0.40 mol of Fe present. Therefore, 0.40 - 0.38 or **0.02 mol of Fe will remain after all the HCl has reacted.**

3.24 $6\,ClO_2 + 3H_2O \rightarrow 5HClO_3 + HCl$

Find the limiting reactant:

$$4.25 \text{ g } ClO_2 \times \frac{1 \text{ mol } ClO_2}{67.45 \text{ g } ClO_2} \times \frac{3 \text{ mol } H_2O}{6 \text{ mol } ClO_2} \times \frac{18.02 \text{ g } H_2O}{\text{mol } H_2O} \Leftrightarrow 0.568 \text{ g } H_2O$$

Since there were 0.853 g of H_2O available, the ClO_2 will be the limiting reactant. Find the mass of $HClO_3$ produced by 4.25 g ClO_2 and excess H_2O:

$$4.25 \text{ g } ClO_2 \times \frac{1 \text{ mol } ClO_2}{67.45 \text{ g } ClO_2} \times \frac{5 \text{ mol } HClO_3}{6 \text{ mol } ClO_2} \times \frac{84.46 \text{ g } HClO_3}{\text{mol } HClO_3}$$

$$\Leftrightarrow \mathbf{4.43 \text{ g } HClO_3}$$

3.25 (a) Limiting reactant? Moles Al? $20.0 \text{ g Al} \times \frac{1 \text{ mol}}{26.98 \text{ g}} = 0.741 \text{ mol Al}$

Moles H_2SO_4? $115 \text{ g } H_2SO_4 \times \frac{1 \text{ mol}}{98.08 \text{ g}} = 1.17 \text{ mol } H_2SO_4$

Select either reactant as the limiting reactant and determine if a sufficient quantity of the other is present. Let's select the H_2SO_4. How many moles of Al are needed to react with it?

$$1.17 \text{ mol } H_2SO_4 \times \frac{2 \text{ mol Al}}{3 \text{ mol } H_2SO_4} \Leftrightarrow 0.780 \text{ mole Al}$$

The Al present is less than that required to react with all the H_2SO_4. **Therefore, Al is the limiting reactant.**

(b) $0.741 \text{ mol Al} \times \frac{3 \text{ mol } H_2}{2 \text{ mol Al}} \Leftrightarrow \mathbf{1.11 \text{ mol } H_2}$

(c) $0.741 \text{ mol Al} \times \frac{1 \text{ mol } Al_2(SO_4)_3}{2 \text{ mol Al}} \times \frac{342.1 \text{ g } Al_2(SO_4)_3}{\text{mol } Al_2(SO_4)_3}$

$$\Leftrightarrow \mathbf{127 \text{ g } Al_2(SO_4)_3}$$

(d) $0.741 \text{ mol Al} \times \frac{3 \text{ mol } H_2SO_4}{2 \text{ mol Al}} \times \frac{98.08 \text{ g } H_2SO_4}{\text{mol } H_2SO_4} \Leftrightarrow 109 \text{ g } H_2SO_4 \text{ needed}$

Excess = 115 g - 109 g = **6 g H_2SO_4**

3.26 (a) $35.0 \text{ g } C_2H_2 \text{ x } \frac{1 \text{ mol } C_2H_2}{26.04 \text{ g } C_2H_2} = 1.34 \text{ mol } C_2H_2$

$51.0 \text{ g HCl x } \frac{1 \text{ mol HCl}}{36.46 \text{ g HCl}} = 1.40 \text{ mol HCl}$

Since they react in a 1:1 mole ratio, **the C_2H_2 is the limiting reactant.**

(b) $1.34 \text{ mol } C_2H_2 \text{ x } \frac{1 \text{ mol } C_2H_3Cl}{\text{mol } C_2H_2} \text{ x } \frac{62.50 \text{ g } C_2H_3Cl}{\text{mol } C_2H_3Cl} \Leftrightarrow \mathbf{83.8 \text{ g } C_2H_3Cl}$

(c) 1.40 - 1.34 = 0.06 mole of excess HCl

$0.06 \text{ mole HCl x } \frac{36.46 \text{ g HCl}}{\text{mol HCl}} = \mathbf{2 \text{ g HCl}}$

3.27 First find the limiting reactant. $150 \text{ g } CCl_4 \text{ x } \frac{1 \text{ mol } CCl_4}{153.8 \text{ g } CCl_4} = 0.975 \text{ mol } CCl_4$

$100 \text{ g } SbF_3 \text{ x } \frac{1 \text{ mol } SbF_3}{178.8 \text{ g } SbF_3} = 0.559 \text{ mol } SbF_3$

The needed mole ratio is 3 mol CCl_4 to 2 mol SbF_3. Is 0.975 mol of CCl_4 to 0.559 mol of SbF_3 larger or smaller than the needed 3:2 ratio?

$\frac{0.975}{0.559} = 1.74$ Larger! Therefore, the **SbF_3 is the limiting reactant.**

(a) $0.559 \text{ mol } SbF_3 \text{ x } \frac{3 \text{ mol } CCl_2F_2}{2 \text{ mol } SbF_3} \text{ x } \frac{120.9 \text{ g } CCl_2F_2}{\text{mol } CCl_2F_2} \Leftrightarrow \mathbf{101 \text{ g } CCl_2F_2}$

(b) Moles of CCl_4 needed to react with 0.559 mole of SbF_3 is:

$0.559 \text{ mol } SbF_3 \text{ x } \frac{3 \text{ mol } CCl_4}{2 \text{ mol } SbF_3} \Leftrightarrow 0.838 \text{ mol } CCl_4$

$(0.975 \text{ mol } CCl_4 - 0.838 \text{ mol } CCl_4) \text{ x } \frac{153.8 \text{ g } CCl_4}{\text{mol } CCl_4} = \mathbf{21.1 \text{ g } CCl_4 \text{ excess}}$

3.28 Calculate the limiting reactant first.

$$0.950\text{ g Ag x }\frac{1\text{ mol Ag}}{107.9\text{ g Ag}} = 8.80\text{ x }10^{-3}\text{ mol Ag}$$

$$0.140\text{ g }H_2S\text{ x }\frac{1\text{ mol }H_2S}{34.08\text{ g }H_2S} = 4.11\text{ x }10^{-3}\text{ mol }H_2S$$

$$0.0800\text{ g }O_2\text{ x }\frac{1\text{ mol }O_2}{32.00\text{ g }O_2} = 2.50\text{ x }10^{-3}\text{ mol }O_2$$

Needed mole ratio is 4:2:1. If H_2S is the limiting reactant, 8.22 x 10^{-3} mol of Ag and 2.06 x 10^{-3} mol of O_2 are needed. Therefore, we see that H_2S is the limiting reactant and that Ag and O_2 are present in excess. From the moles of H_2S, one can calculate the maximum mass of Ag_2S that can be obtained.

$$4.11\text{ x }10^{-3}\text{ mol }H_2S\text{ x }\frac{2\text{ mol }Ag_2S}{2\text{ mol }H_2S}\text{ x }\frac{247.8\text{ g }Ag_2S}{\text{mol }Ag_2S} \Leftrightarrow \mathbf{1.02\text{ g }Ag_2S}$$

3.29 (a) $0.430\text{ mol }COCl_2\text{ x }\frac{2\text{ mol HCl}}{1\text{ mol }COCl_2} \Leftrightarrow \mathbf{0.860\text{ mol HCl}}$

(b) $11.0\text{ g }CO_2\text{ x }\frac{1\text{ mol }CO_2}{44.01\text{ g }CO_2}\text{ x }\frac{2\text{ mol HCl}}{\text{mol }CO_2}\text{ x }\frac{36.45\text{ g HCl}}{\text{mol HCl}} \Leftrightarrow \mathbf{18.2\text{ g HCl}}$

(c) Needed mole ratio is 1:1. Therefore, the $COCl_2$ is the limiting reactant.

$$0.200\text{ mol }COCl_2\text{ x }\frac{2\text{ mol HCl}}{\text{mol }COCl_2} \Leftrightarrow \mathbf{0.400\text{ mol HCl}}$$

3.30 $40.0\text{ g }Br_2\text{ x }\frac{1\text{ mol }Br_2}{159.8\text{ g }Br_2} = 0.250\text{ mol }Br_2$ before any reaction.

What is present after the first reaction is complete but before any of the second takes place?

$$5.00\text{ g }C_2H_2\text{ x }\frac{1\text{ mol }C_2H_2}{26.04\text{ g }C_2H_2} = 0.192\text{ mol }C_2H_2\text{ to start with.}$$

(continued)

3.30 (continued)

$$1.92 \text{ mol } C_2H_2 \text{ x } \frac{1 \text{ mol } C_2H_2Br_2}{\text{mol } C_2H_2} \Leftrightarrow 0.192 \text{ mol } C_2H_2Br_2 \text{ ,}$$

after the first reaction is complete.

Excess Br_2 after first reaction is: 0.250 - 0.192 = 0.058 mol Br_2. Therefore, after the first reaction, there would be present 0.192 mol $C_2H_2Br_2$ and 0.058 mol Br_2. The Br_2 will be the limiting reactant for the second reaction. The maximum mass of $C_2H_2Br_4$ that can be produced is:

$$0.058 \text{ mol } Br_2 \text{ x } \frac{1 \text{ mol } C_2H_2Br_4}{\text{mol } Br_2} \text{ x } \frac{345.6 \text{ g } C_2H_2Br_4}{\text{mol } C_2H_2Br_4} = \mathbf{20 \text{ g } C_2H_2Br_4}$$

The mass of unreacted $C_2H_2Br_2$ is:

$$(0.192 \text{ mol} - 0.058 \text{ mol}) \text{ x } \frac{185.8 \text{ g } C_2H_2Br_2}{\text{mol } C_2H_2Br_2} = \mathbf{24.9 \text{ g } C_2H_2Br_2}$$

3.31 The theoretical yield is the maximum amount of product that could be produced from a given quantity of reactant if the reaction gave only that product, with no side reactions. The percent yield is a comparison of the yield that is actually obtained to the theoretical yield; it is the efficiency of the reaction. The actual yield is the amount of product that you actually obtain in a given experiment when the reaction is carried out.

3.32 (a) $$14.6 \text{ g } SbF_3 \text{ x } \frac{1 \text{ mol } SbF_3}{178.8 \text{ g } SbF_3} \text{ x } \frac{3 \text{ mol } CCl_2F_2}{2 \text{ mol } SbF_3} \text{ x } \frac{120.9 \text{ g } CCl_2F_2}{\text{mol } CCl_2F_2}$$

$$\Leftrightarrow 14.8 \text{ g } CCl_2F_2$$

Theoretical yield = **14.8 g CCl_2F_2**

(b) Actual yield was given as 8.62 g CCl_2F_2

(c) Percentage yield = (8.62/14.8) x 100% = **58.2%**

3.33 (a) $$6.40 \text{ g } CH_3OH \text{ x } \frac{1 \text{ mol } CH_3OH}{32.0 \text{ g } CH_3OH} \text{ x } \frac{2 \text{ mol } CO_2}{2 \text{ mol } CH_3OH} \text{ x } \frac{44.0 \text{ g } CO_2}{\text{mol } CO_2}$$

$\Leftrightarrow$ **8.80 g CO_2** = theoretical yield of CO_2

(b) Actual yield given as 6.12 g CO_2

(c) Percentage yield = (6.12/8.80) x 100% = **69.5%**

3.34 (a) $15.0 \text{ g } C_6H_6 \times \frac{1 \text{ mol } C_6H_5Br}{78.12 \text{ g } C_6H_6} \times \frac{157.0 \text{ g } C_6H_5Br}{\text{mol } C_6H_5Br} \Leftrightarrow \mathbf{30.1 \text{ g } C_6H_5Br}$

(b) $2.50 \text{ g } C_6H_4Br_2 \times \frac{1 \text{ mol } C_6H_4Br_2}{235.9 \text{ g } C_6H_4Br_2} \times \frac{1 \text{ mol } C_6H_6}{\text{mol } C_6H_4Br_2} \times \frac{78.12 \text{ g } C_6H_6}{\text{mol } C_6H_6}$

$\Leftrightarrow \mathbf{0.828 \text{ g } C_6H_6}$

(c) $(15.0 \text{ g } C_6H_6 - 0.828 \text{ g } C_6H_6) \times \frac{1 \text{ mol } C_6H_6}{78.12 \text{ g } C_6H_6} \times \frac{1 \text{ mol } C_6H_5Br}{\text{mol } C_6H_6}$

$\times \frac{157.0 \text{ g } C_6H_5Br}{\text{mol } C_6H_5Br} \Leftrightarrow \mathbf{28.5 \text{ g } C_6H_5Br}$

(d) Percentage yield = (28.5/30.1) x 100% = **94.7%**

3.35 (a) Mol CH_3Cl + mol CH_2Cl_2 + mol $CHCl_3$ + mol CCl_4 must equal mol of CH_4 at the start if the total amount of C is to be maintained.

mol CH_4 ? $\frac{20.8 \text{ g } CH_4}{16.04 \text{ g/mol}} = 1.30 \text{ mol } CH_4$

mol CH_3Cl ? $\frac{5.0 \text{ g } CH_3Cl}{50.5 \text{ g/mol}} = 0.099 \text{ mol } CH_3Cl$

mol CH_2Cl_2 ? $\frac{25.5 \text{ g } CH_2Cl_2}{84.93 \text{ g/mol}} = 0.300 \text{ mol } CH_2Cl_2$

mol $CHCl_3$? $\frac{59.0 \text{ g } CHCl_3}{119.37 \text{ g/mol}} = 0.494 \text{ mol } CHCl_3$

mol CCl_4 ? 1.30 - 0.099 - 0.30 - 0.494 = 0.41 mol CCl_4

mass CCl_4 ? 0.41 mol CCl_4 x 153.8 g/mol = **63 g CCl_4**

(continued)

3.35 (continued)

(b) If all the available CH_4 had been converted to CCl_4, the theoretical yield would be:

$$1.30 \text{ mol } CH_4 \times \frac{1 \text{ mol } CCl_4}{\text{mol } CH_4} \times \frac{153.8 \text{ g } CCl_4}{\text{mol } CCl_4} \Leftrightarrow \mathbf{2.00 \times 10^2 \text{ g } CCl_4}$$

(c) Percentage yield = (63/200) x 100% = **32%**

(d) $CH_4 + Cl_2 \rightarrow CH_3Cl + HCl$,
$CH_4 + 2Cl_2 \rightarrow CH_2Cl_2 + 2HCl$
$CH_4 + 3Cl_2 \rightarrow CHCl_3 + 3HCl$,
$CH_4 + 4Cl_2 \rightarrow CCl_4 + 4HCl$

(0.099 mol CH_3Cl x 1) + (0.300 mol CH_2Cl_2 x 2) + (0.494 mol $CHCl_3$ x 3) + (0.41 mol CCl_4 x 4) = 3.82 mol Cl_2

3.82 mol Cl_2 x 70.91 g/mol = **271 g Cl_2**

3.36 $\frac{30.0 \text{ g}}{80.0 \text{ g/mol}}$ = 0.375 mol starting material

Step # 1 0.375 x 50% = 0.188 mol

Step # 2 0.188 x 50% = 0.0940 mol

Step # 3 0.0940 x 50% = 0.0470 mol

Step # 4 0.0470 x 50% = 0.0235 mol

Step # 5 0.0235 x 50% = 0.0118 mol

Step # 6 0.0118 x 50% = 0.00590 mol

0.00590 mol x 100 g/mol = **0.590 g** (assuming that 50% has more than 2 significant figures)

If 30.0 g yields 0.590 g, then how much would be required to produce 10.0 g?
10.0 g product x (30.0 g reactant/0.590 g product) = **508 g reactant**

3.37 Solutions are used to carry out chemical reactions because the reactions will take place much more rapidly when the reactants are in a dissolved state.

3.38 (a) A **precipitate** in a chemical reaction is a solid that is formed in a solution as a result of the chemical reaction.
(b) A **solution** is a homogeneous mixture that has uniform properties throughout.
(c) The **solvent** is the component whose physical state doesn't change when the solution is formed. It is usually the major component.
(d) **Solute(s)** is (are) the substance(s) that is (are) dissolved in the solvent to form a solution.
(e) **Concentration** is the term used to describe the relative amounts of solute and solvent.
(f) A **concentrated** solution has a relatively high concentration of solute.
(g) A **dilute** solution has a relatively low concentration of solute.

3.39 Molar concentration = number of moles of solute/total volume of the solution in liters.

3.40 Place 180 g of $C_6H_{12}O_6$ in a 1.00-liter flask. Dissolve the sugar in less than 1.00 liter of water; then dilute to a total volume of 1.00 liter.

3.41 0.20 M Na_3PO_4 means; $\dfrac{0.20 \text{ mol } Na_3PO_4}{1.0 \text{ L solution}}$ and/or $\dfrac{1{,}000 \text{ mL solution}}{0.20 \text{ mol } Na_3PO_4}$

3.42 (a) 0.250 mol/0.400 L soln. = 0.625 mol/L soln. = **0.625 M NaCl**

(b) 1.45 mol/0.345 L = **4.20 M sucrose**

(c) $\dfrac{195 \text{ g } H_2SO_4}{0.875 \text{ L soln.}} \times \dfrac{1 \text{ mol}}{98.08 \text{ g}} = \mathbf{2.27 \text{ M } H_2SO_4}$

(d) $\dfrac{80.0 \text{ g KOH}}{0.200 \text{ L}} \times \dfrac{1 \text{ mol}}{56.1 \text{ g}} = \mathbf{7.13 \text{ M KOH}}$

3.43 (a) 1.35 mol NH_4Cl/2.45 L soln. = 0.551 mol/Lsoln. = **0.551 M NH_4Cl**

(b) 0.422 mol $AgNO_3$/0.742 L soln. = **0.569 M $AgNO_3$**

(c) 3.00×10^{-3} mol KCl/0.0100 L soln. = **0.300 M KCl**

(d) $\dfrac{4.80 \times 10^{-2} \text{ g } NaHCO_3}{0.0250 \text{ L}} \times \dfrac{1 \text{ mol}}{84.01 \text{ g}} = \dfrac{0.0229 \text{ mol } NaHCO_3}{\text{L soln.}}$

$= \mathbf{0.0229 \text{ M } NaHCO_3}$

3.44 (a) $\frac{0.250 \text{ mmol}}{\text{mL}} \times \frac{10^{-3} \text{ mol}}{\text{mmol}} \times \frac{10^3 \text{ mL}}{\text{L}} = \mathbf{0.250 \ mol/L}$

(b) **0.250 M**

(c) **M = mol/L = mmol/mL**

3.45 (a) $\frac{0.150 \text{ mol}}{\text{L}} \times 0.250 \text{ L} = \mathbf{0.0375 \ mol \ Li_2CO_3}$

(b) $\frac{0.150 \text{ mol } Li_2CO_3}{\text{L}} \times 0.630 \text{ L soln.} \times \frac{73.89 \text{ g } Li_2CO_3}{\text{mol } Li_2CO_3} = \mathbf{6.98 \ g \ Li_2CO_3}$

(c) $\frac{0.0100 \text{ mol } Li_2CO_3}{0.150 \text{ mol } Li_2CO_3\text{/L soln.}} = 0.0667 \text{ L} = \mathbf{66.7 \ mL \ solution}$

(d) $0.0800 \text{ g } Li_2CO_3 \times \frac{1 \text{ mol } Li_2CO_3}{73.89 \text{ g } Li_2CO_3} \times \frac{1 \text{ L}}{0.150 \text{ mol } Li_2CO_3}$

$\times \frac{10^3 \text{ mL}}{\text{L}} = \mathbf{7.22 \ mL \ solution}$

3.46 (a) $0.100 \text{ mol KOH} \times \frac{1 \text{ L soln.}}{0.375 \text{ mol KOH}} \times \frac{10^3 \text{ mL}}{\text{L}} = \mathbf{267 \ mL \ soln.}$

(b) $45.0 \text{ mL soln.} \times \frac{0.375 \text{ mol KOH}}{\text{L soln.}} \times \frac{1 \text{ L}}{10^3 \text{ mL}} = \mathbf{0.0169 \ mol \ KOH}$

(c) $10.0 \text{ g KOH} \times \frac{1 \text{ mol KOH}}{56.11 \text{ g KOH}} \times \frac{1 \text{ L soln.}}{0.375 \text{ mol KOH}} \times \frac{10^3 \text{ mL}}{\text{L}} = \mathbf{475 \ mL \ soln.}$

(d) $1 \text{ mL soln.} \times \frac{1 \text{ L}}{10^3 \text{ mL}} \times \frac{0.375 \text{ mol KOH}}{\text{L soln.}} \times \frac{56.11 \text{ g KOH}}{1 \text{ mol KOH}} = \mathbf{0.0210 \ g \ KOH}$

3.47 $\frac{0.250 \text{ mol}}{\text{L}} \times \frac{2.00 \text{ L}}{1} \times \frac{158.2 \text{ g } Ca(C_2H_3O_2)_2}{\text{mol}} = \mathbf{79.1 \ g \ Ca(C_2H_3O_2)_2}$

3.48 $250.0\text{ mL} \times \frac{3.000 \times 10^{-2}\text{ mmol}}{\text{mL}} \times \frac{10^{-3}\text{ mol}}{\text{mmol}} \times \frac{101.11\text{ g } KNO_3}{\text{mol}} = \mathbf{0.7583\ g\ KNO_3}$

3.49 $\frac{0.150\text{ mol } MgSO_4}{\text{L}} \times 0.500\text{ L} \times \frac{1\text{ mol } MgSO_4{\bullet}7H_2O}{\text{mol } MgSO_4} \times \frac{246.5\text{ g } MgSO_4{\bullet}7H_2O}{1\text{ mol } MgSO_4{\bullet}7H_2O}$

$\Leftrightarrow \mathbf{18.5\ g\ MgSO_4{\bullet}7H_2O}$

3.50 The number of moles of solute in solution does not change as a solution is diluted.

3.51 Add the more dense, concentrated reagent slowly to the water.

3.52 $M_iV_i = M_fV_f$; (V_i = initial volume in mL);

$(18.0\ \mathbf{M})\ V_i = (5.00\ \mathbf{M})\ (V_i + 100)\text{mL}$; $18.0\ V_i = 5.00\ V_i + 500$;

$13.0\ V_i = 500\text{ mL}$; $V_i = \mathbf{38.5\ mL}$

3.53 Concentrated NH_3 is 15 **M** (Table 3.1)

$15\ \mathbf{M} \times V_i = 0.500\ \mathbf{M} \times 250\text{ mL}$

$V_i = \frac{0.500\ \mathbf{M} \times 250\text{ mL}}{15\ \mathbf{M}}$

$V_i = \mathbf{8.3\ mL}$

3.54 $18.0\ \mathbf{M} \times V_i = 3.00\ \mathbf{M} \times 400\text{ mL}$

$V_i = \frac{3.00\ \mathbf{M} \times 400\text{ mL}}{18.0\ \mathbf{M}} = \mathbf{66.7\ mL}$

3.55 $0.500\ \mathbf{M} \times 100\text{ mL} = 0.200\ \mathbf{M} \times V_f$

$V_f = \frac{0.500\ \mathbf{M} \times 100\text{ mL}}{0.200\ \mathbf{M}} = \mathbf{2.50 \times 10^2\ mL}$

3.56 1.00 **M** x 85.0 mL = 0.650 **M** x V_f

$$V_f = \frac{1.00\ \mathbf{M} \times 85.0\ \text{mL}}{0.650\ \mathbf{M}}$$

V_f = 131 mL

V of H_2O added is 131 mL - 85.0 mL = **46 mL**

3.57 Known values: **M** = moles/L or **M** x L = moles

Let ? = L of 1.00 **M** HCl added

$$0.600\ \mathbf{M} = \frac{(0.500\ \mathbf{M} \times 0.0500\ \text{L}) + (1.00\ \mathbf{M} \times ?\ \text{L})}{(0.0500\ \text{L} + ?\ \text{L})}$$

0.0300 + 0.600 ? = 0.0250 + ?

0.40 ? = 0.0050

? = 0.0125 L or **12 mL**

3.58 (a) $3.50\ \text{g Al} \times \frac{1\ \text{mol Al}}{26.98\ \text{g Al}} \times \frac{3\ \text{mol}\ H_2SO_4}{2\ \text{mol Al}} \times \frac{1\ \text{L solution}}{0.200\ \text{mol}\ H_2SO_4} \times \frac{10^3\ \text{mL}}{\text{L}}$

⇔ **973 mL**

(b) $\frac{0.200\ \text{mol}\ H_2SO_4}{\text{L}} \times 0.400\ \text{L} \times \frac{3\ \text{mol}\ H_2}{3\ \text{mol}\ H_2SO_4}$ ⇔ **8.00×10^{-2} mol H_2**

3.59 (a) $\frac{0.250\ \text{mol NaBr}}{\text{L}} \times 0.300\ \text{L} = 0.0750$ mol NaBr

(0.400 mol $AgNO_3$/L) x 0.200 L = 0.0800 mol $AgNO_3$

Since they react in a 1:1 mole ratio, the **NaBr is the limiting reactant.**

(b) $0.0750\ \text{mol NaBr} \times \frac{1\ \text{mol AgBr}}{\text{mol NaBr}} \times \frac{187.8\ \text{g AgBr}}{\text{mol AgBr}}$ ⇔ **14.1 g AgBr**

3.60 (a) $\frac{0.400 \text{ mol NaOH}}{\text{L}} \text{ x } 0.0250 \text{ L} = 0.0100 \text{ mol NaOH}$

$$0.0100 \text{ mol NaOH x } \frac{1 \text{ mol } H_2SO_4}{2 \text{ mol NaOH}} \Leftrightarrow 0.00500 \text{ mol } H_2SO_4$$

Use $\mathbf{M} = \frac{\text{moles}}{\text{L}}$ $\quad 0.200\ \mathbf{M} = \frac{0.00500 \text{ mol}}{?\text{ L}}$

? L = 0.00500/0.200 = 0.0250 L = **25.0 mL**

(b) $\frac{0.270 \text{ mol } H_2SO_4}{\text{L}} \text{ x } 0.0500 \text{ L} = 0.0135 \text{ mol } H_2SO_4$

$$0.0135 \text{ mol } H_2SO_4 \text{ x } \frac{2 \text{ mol NaOH}}{\text{mol } H_2SO_4} \Leftrightarrow 0.0270 \text{ mol NaOH}$$

$$0.100\ \mathbf{M} = \frac{0.0270}{?\text{ L}}$$

? L = 0.0270/0.100 = 0.270 L = **270 mL**

(c) (0.300 mol NaOH/L) x 0.0400 L = 0.0120 mol NaOH

(0.350 mol H_2SO_4/L) x 0.0150 L = 0.00525 mol H_2SO_4

Needed mole ratio is 2:1. Therefore, the limiting reactant is H_2SO_4.

$$0.00525 \text{ mol } H_2SO_4 \text{ x } \frac{1 \text{ mol } Na_2SO_4}{\text{mol } H_2SO_4} \Leftrightarrow \mathbf{0.00525 \text{ mol } Na_2SO_4}$$

3.61 Let's use the definition of molarity as mmol solute divided by mL solution as a working equation. $\mathbf{M} = \frac{\text{mmol}}{\text{mL}}$ can be used as needed.

(a) $\frac{0.200 \text{ mmol } MgCl_2}{\text{mL}} \text{ x } 75.0 \text{ mL} = 15.0 \text{ mmol } MgCl_2$

$$15.0 \text{ mmol } MgCl_2 \text{ x } \frac{2 \text{ mmol NaOH}}{\text{mmol } MgCl_2} \Leftrightarrow 30.0 \text{ mmol NaOH}$$

(continued)

3.61 (continued)

$$0.300\ M = \frac{30.0\ mmol}{?\ mL} \qquad ?\ mL = 30.0/0.300 = \mathbf{100\ mL}$$

(b) $$\frac{0.600\ mmol\ MgCl_2}{mL} \times 50.0\ mL = 30.0\ mmol\ MgCl_2$$

$$30.0\ mmol\ MgCl_2 \times \frac{1\ mmol\ Mg(OH)_2}{mmol\ MgCl_2} \Leftrightarrow 30.0\ mmol\ Mg(OH)_2$$

$$30.0\ mmol\ Mg(OH)_2 \times \frac{1\ mol}{10^3\ mmol} \times \frac{58.32\ g\ Mg(OH)_2}{mol\ Mg(OH)_2} = \mathbf{1.75\ g\ Mg(OH)_2}$$

(c) $$\frac{0.200\ mmol\ MgCl_2}{mL} \times 30.0\ mL = 6.00\ mmol\ MgCl_2$$

$$\frac{0.140\ mmol\ NaOH}{mL} \times 100\ mL = 14.0\ mmol\ NaOH$$

The needed mole - or mmol - ratio is 1 of $MgCl_2$ to 2 of NaOH. In this reaction the $MgCl_2$ is the limiting reactant.

$$6.00\ mmol\ MgCl_2 \times \frac{1\ mmol\ Mg(OH)_2}{mmol\ MgCl_2} \times \frac{10^{-3}\ mol}{mmol} \times \frac{58.32\ g\ Mg(OH)_2}{mol\ Mg(OH)_2}$$

$$\Leftrightarrow \mathbf{0.350\ g\ Mg(OH)_2}$$

3.62 $$0.2867\ g\ AgCl \times \frac{1\ mol\ AgCl}{143.32\ g\ AgCl} \times \frac{1\ mol\ AgNO_3}{mol\ AgCl} \Leftrightarrow 2.000 \times 10^{-3}\ mol\ AgNO_3$$

$$M = \frac{2.000 \times 10^{-3}\ mol\ AgNO_3}{0.02000\ L} = \mathbf{0.1000\ M\ AgNO_3}$$

3.63 $$15.0\text{ mL NaOH} \times \frac{0.750\text{ mol NaOH}}{\text{L}} \times \frac{1\text{ L}}{1000\text{ mL}} = 0.0112\text{ mol NaOH}$$

$$0.0112\text{ mol NaOH} \times \frac{1\text{ mol }H_2SO_4}{2\text{ mol NaOH}} \Leftrightarrow 5.60 \times 10^{-3}\text{ mol }H_2SO_4$$

$$\frac{5.60 \times 10^{-3}\text{ mol }H_2SO_4}{0.0250\text{ L}} = 0.224\ \mathbf{M}\ H_2SO_4\text{ used in reaction}$$

$$1.40\ \mathbf{M} \times 250\text{ mL} = 0.224\ \mathbf{M} \times V_f$$

$$V_f = \frac{1.40\ \mathbf{M} \times 250\text{ mL}}{0.224\ \mathbf{M}} = \mathbf{1.56 \times 10^3}\text{ mL}$$

3.64 (a) $\mathbf{2Al + 6HCl \rightarrow 3H_2 + 2AlCl_3}$

(b) $$0.300\text{ mol Al} \times \frac{3\text{ mol }H_2}{2\text{ mol Al}} \Leftrightarrow \mathbf{0.450\text{ mol }H_2}$$

(c) $$0.200\text{ mol }H_2 \times \frac{2\text{ mol }AlCl_3}{3\text{ mol }H_2} \Leftrightarrow \mathbf{0.133\text{ mol }AlCl_3}$$

(d) $$0.600\text{ mol HCl} \times \frac{2\text{ mol Al}}{6\text{ mol HCl}} \Leftrightarrow \mathbf{0.200\text{ mol Al}}$$

(e) $$9.13\text{ g HCl} \times \frac{1\text{ mol HCl}}{36.46\text{ g HCl}} \times \frac{2\text{ mol Al}}{6\text{ mol HCl}} \times \frac{26.98\text{ g Al}}{\text{mol Al}} \Leftrightarrow \mathbf{2.25\text{ g Al}}$$

3.65 $$0.400\text{ L} \times \frac{0.200\text{ mol}}{\text{L}} = 0.0800\text{ mol (solution \# 1)}$$

$$0.800\text{ L} \times \frac{0.600\text{ mol}}{\text{L}} = 0.480\text{ mol (solution \# 2)}$$

(Assume that the volumes are additive)

$$\text{Final solution is: } \frac{0.0800\text{ mol} + 0.480\text{ mol}}{0.400\text{ L} + 0.800\text{ L}} = \mathbf{0.467\ M}$$

3.66 (a) $\mathbf{2Al(s) + 6HCl(aq) \rightarrow 3H_2(g) + 2AlCl_3(aq)}$

(b) $0.300\ L \times \frac{0.150\ mol\ HCl}{L} \times \frac{2\ mol\ Al}{6\ mol\ HCl} \times \frac{26.98\ g\ Al}{mol\ Al} \Leftrightarrow \mathbf{0.405\ g\ Al}$

(c) $\frac{0.300\ L}{0.300\ L} \times \frac{0.150\ mol\ HCl}{L} \times \frac{2\ mol\ AlCl_3}{6\ mol\ HCl}$

$\Leftrightarrow \frac{0.0500\ mol\ AlCl_3}{L} = \mathbf{0.0500\ M\ AlCl_3}$

3.67 (a) If the Mg is the limiting reactant:
$6.00\ g\ Mg \times \frac{1\ mol\ Mg}{24.305\ g\ Mg} \times \frac{1\ mol\ H_2}{mol\ Mg} \times \frac{2.02\ g\ H_2}{mol\ H_2} \Leftrightarrow 0.499\ g\ H_2$

If the HCl is the limiting reactant:
$0.350\ L \times \frac{0.800\ mol\ HCl}{L} \times \frac{1\ mol\ H_2}{2\ mol\ HCl} \times \frac{2.02\ g\ H_2}{mol\ H_2} = 0.283\ g\ H_2$

Since the amount of HCl will produce less H_2, the HCl solution is the limiting reactant and the answer is the **0.283 g H_2.**

(b) First calculate theoretical yield.
$0.350\ L \times \frac{0.800\ mol\ HCl}{L} \times \frac{1\ mol\ MgCl_2}{2\ mol\ HCl} \times \frac{95.2\ g\ MgCl_2}{mol\ MgCl_2}$

$\Leftrightarrow 13.3\ g\ MgCl_2$ (theoretical yield)

Percentage yield = (8.64/13.3 g) x 100% = **65.0%**

3.68 $10.0 \times 10^3\ g\ Pb \times \frac{1\ mol\ Pb}{207.2\ g} \times \frac{2\ mol\ first\ product}{2\ mol\ Pb} = 48.26\ mol\ product$

Actual yield = 48.26 mol x 0.858 = 41.4 mol

$41.4\ mol\ first\ product \times \frac{1\ mol\ white\ lead}{6\ mol\ first\ product} \times \frac{775.6\ g\ white\ lead}{mol\ white\ lead}$

$= 5.35 \times 10^3\ g\ white\ lead$

Actual yield of step 2 = 5.35 kg x 0.723 = **3.87 kg white lead**

3.69 Limiting reactant? HCl? $0.235\ L \times 0.600\ mol/L = 0.141\ mol\ HCl$

Na_2CO_3? $0.0940\ L \times 0.750\ mol/L = 0.0705\ mol\ Na_2CO_3$

Both are the limiting reactants since they are mixed in a ratio equal to that indicated by the balanced equation. Either could be used to calculate the answer.

$$0.141\ \text{mol HCl} \times \frac{2\ \text{mol NaCl}}{2\ \text{mol HCl}} \times \frac{1}{(.235\ \text{L} + 0.0940\ \text{L})} \Leftrightarrow \frac{0.429\ \text{mol NaCl}}{\text{L soln.}}$$

$$= \mathbf{0.429\ M\ NaCl}$$

4 THE PERIODIC TABLE AND SOME PROPERTIES OF THE ELEMENTS

4.1 The ability to deform when hammered is called malleability. A blacksmith relies on the malleability of iron when forging a horseshoe.

4.2 Ductility is the ability of a metal to stretch when pulled from opposite directions. This property is used in the manufacture of wire.

4.3 Three properties of metals, other than malleability and ductility, are: (1) metallic luster, (2) good conductors of heat and (3) good conductors of electricity.

4.4 Sodium reacts rapidly with oxygen in the air and moisture. Iron reacts slowly with air and moisture. Gold does not react with air and moisture. Iron is not used to make jewelry because it reacts with air and moisture, also known as tarnishing or rusting.

4.5 $2Na + 2H_2O \rightarrow 2NaOH + H_2$

4.6 Gold and copper

4.7 Since they are good conductors of heat, metals feel hot when left in the sun. As your hand absorbs heat from the metal, heat travels quickly from the neighboring parts of the object to replace the heat your hand absorbed, thus providing more heat for your hand to absorb.

4.8 The plating of electrical contacts

4.9 Tungsten has the highest melting point of any element which accounts for its use as the filament in electrical light bulbs. Mercury has the lowest melting point of any metal. Mercury is the fluid used in some thermometers.

4.10 Oxygen (O_2), nitrogen (N_2), hydrogen (H_2), fluorine (F_2), chlorine (Cl_2), bromine (Br_2), and Iodine (I_2) are diatomic molecules. O_2, N_2, H_2, F_2 and Cl_2 are gases. Br_2 is a liquid. I_2 is a solid.

4.11 Graphite and diamond. Both are made up of carbon. Both lack luster, are nonmalleable and nonductile. Graphite is soft and opaque while diamond is transparent and very hard.

4.12 Yes. Copper has the luster of a metal while sulfur appears to not possess metallic luster. The copper also appears to be a wire which would indicate that it is ductile.

4.13 Oxygen and nitrogen

4.14 Fluorine

4.15 Metalloids look somewhat like metals but are darker in color. They conduct electricity but not nearly as well as metals. Metalloids are much more like nonmetals than metals.

4.16 Mendeleev arranged his periodic table so that the elements were in order of increasing atomic mass and elements with similar properties were in columns or groups.

4.17 Mendeleev left spaces for yet undiscovered elements because there were not known elements with the properties that fit into the pattern.

4.18 In Mendeleev's table elements in any particular column had to have similar properties.

4.19 Co and Ni, Th and Pa, and U and Np

4.20 The Noble Gases were missing because none of them had been discovered.

4.21 One coulomb is equal to the amount of charge that moves past a given point in a wire when an electric current of 1 ampere flows for 1 second.
1 C = 1A x 1s

The charge on a mole of electrons =

$$\frac{1.602 \times 10^{-19}\ \text{C}}{e^-} \times \frac{6.022 \times 10^{23}\ e^-}{\text{mol}\ e^-} = \frac{\mathbf{9.647 \times 10^{4}\ C}}{\mathbf{mol\ e^-}}$$

4.22 (a) $\frac{9.11 \times 10^{-28}\text{ g}}{e^-} \times \frac{6.022 \times 10^{23}\ e^-}{1.00 \text{ mol } e^-} = \mathbf{5.49 \times 10^{-4}\ g/mol\ e^-}$

(b) $\frac{5.49 \times 10^{-4}\text{ g/mol } e^-}{1.0079 \text{ g/mol H}} \times 100\% = \mathbf{5.45 \times 10^{-2}\ \%}$

4.23 The number of protons in the nuclei of the atoms (and e^-'s in neutral atoms)

4.24 A different number of protons in their nuclei; 7 for N and 8 for O

4.25 $\frac{+4.8 \times 10^{-19}\text{C}}{1.6 \times 10^{-19}\text{ C}/e^-\text{ lost}} = +3 \text{ or } \mathbf{Al^{3+}}$

4.26 -3.2×10^{-19} C/(1.6×10^{-19} C/e^- gained) = -2 or $\mathbf{Se^{2-}}$

4.27 Density = m/V and mass = 1.67×10^{-24} g
$V = (4/3)\pi r^3 = (4/3) \times 3.142 \times (0.500 \times 10^{-13}\text{ cm})^3$
$= (4/3) \times 3.142 \times 1.25 \times 10^{-40}\text{ cm}^3 = 5.24 \times 10^{-40}\text{ cm}^3$
Density = 1.67×10^{-24} g/5.24×10^{-40} cm^3 = $\mathbf{3.19 \times 10^{15}\ g/cm^3}$

4.28 $V = (4/3)\ \pi r^3$ vol. of nucleus $= (4/3)\ \pi \left(\frac{1 \times 10^{-13}}{2}\right)^3 \text{cm}^3$

$= 5 \times 10^{-40}\text{ cm}^3$

vol. of atom $= (4/3)\ \pi \left(\frac{2 \times 10^{-8}}{2}\right)^3 \text{cm}^3 = 4 \times 10^{-24}\text{ cm}^3$

% occupied by nucleus $= \frac{5 \times 10^{-40}}{4 \times 10^{-24}} \times 100\% = \mathbf{1 \times 10^{-14}\ \%}$

4.29 Mass of the earth in grams is:

$6.59 \times 10^{21}\text{ tons} \times \frac{2000\text{ lb}}{\text{ton}} \times \frac{454\text{ g}}{\text{lb}} = 5.98 \times 10^{27}\text{ g}$

(continued)

4.29 (continued)
Use Density = 3.19×10^{15} g/cm^3 from problem 4.27

$$\frac{5.98 \times 10^{27}\ \text{g}}{3.19 \times 10^{15}\ \text{g/cm}^3} = 1.87 \times 10^{12}\ \text{cm}^3$$

Use $V = (4/3)\ \pi\ r^3$ to obtain the radius. $1.87 \times 10^{12} = (4/3)\ \pi\ r^3$
$r = 7.64 \times 10^3$ cm
Diameter would be: $2 \times 7.64 \times 10^3$ cm or 1.53×10^4 cm or **153 meters** or **502 ft** or **0.0951 mile**

4.30 Neutrons are found in the nuclei of atoms. The lightest isotope of hydrogen, hydrogen-1, does not have any neutrons.

4.31 (a) A proton has a mass of 1.007276 u, or approximately 1 u, and a charge of 1+.
(b) A neutron has a mass of 1.008665 u, or approximately 1 u, and a charge of 0.
(c) An electron has a mass of 0.0005486 u, or approximately 0 u, and a charge of 1-

4.32 The observed atomic weights of elements are obtained as an average of the masses contributed by each isotope of an element. In the case of some elements, such as Cl and Cu, the average is far from being a whole number.

4.33 The mass number is simply the total count of protons plus neutrons and is not quite equal to the atomic mass of an atom which is the atom's actual mass.

4.34 ^{132}Cs has 55 protons, 77 neutrons and 55 electrons.
$^{115}Cd^{2+}$ has 48 protons, 67 neutrons and 46 electrons.
^{194}Tl has 81 protons, 113 neutrons and 81 electrons.
$^{105}Ag^{+}$ has 47 protons, 58 neutrons and 46 electrons.
$^{78}Se^{2-}$ has 34 protons, 44 neutrons and 36 electrons.

4.35 ^{131}Ba has 56 protons, 75 neutrons and 56 electrons.
$^{109}Cd^{2+}$ has 48 protons, 61 neutrons and 46 electrons.
$^{36}Cl^{-}$ has 17 protons, 19 neutrons and 18 electrons.
^{63}Ni has 28 protons, 35 neutrons and 28 electrons.
^{107}Tm has 69 protons, 101 neutrons and 69 electrons.

4.36 (a) **^{55}Fe** (b) **^{86}Rb** (c) **^{204}Tl** (d) **^{170}Lu** (e) **^{169}Yb**

4.37 (a) **29** (b) **49** (c) **123** (d) **99** (e) **99**

4.38 The mass of 47.82% of a mole of ^{151}Eu is:
(150.9 g/mol) x 0.4782 mol = 72.16
The mass of 52.18% of a mole of ^{153}Eu is:
(152.9 g/mol) x 0.5218 mol = 79.78
Total weight of 1 mole of naturally-occurring Eu is: 72.16 + 79.78 = **151.9**.

4.39 ^{10}B (10.01294 u/atom) x 0.196 atom = 1.96 u
^{11}B (11.00931 u/atom) x 0.804 atom = 8.85 u
Average mass of one atom = **10.81 u**
Atomic mass = **10.81 g/mol**

4.40 ^{204}Pb (203.973 g/mol) x 0.0148 mol = 3.02
^{206}Pb (205.9745 g/mol) x 0.236 mol = 48.6
^{207}Pb (206.9759 g/mol) x 0.226 mol = 46.8
^{208}Pb (207.9766 g/mol) x 0.523 mol = 109.
Average mass of one mole = **207 g**

4.41 (34.96885 x ?% ^{35}Cl) +[36.96590 x (100 - ?% ^{35}Cl)] = 35.453 x 100%
(34.96885 x ?% ^{35}Cl) +(3696.590 - 36.96590 x ?% ^{35}Cl) = 3545.3
(36.96590 - 34.96885) ?% ^{35}Cl = 3696.590 - 3545.3
1.99705 ?% ^{35}Cl = 151.3
?% ^{35}Cl = **75.76%** % ^{37}Cl = (100.00 - 75.76)% = **24.24%**

4.42 (106.9041 x ?% ^{107}Ag) + 108.9047 (100 - ?% ^{107}Ag) = 107.868 x 100%
(106.9041 x ?% ^{107}Ag + 10890.47 - 108.9047 x ?% ^{107}Ag) = 10786.8
(2.0006 x ?)% ^{107}Ag = 103.7
?% ^{107}Ag =**51.83%**
% ^{109}Ag = 100.00 - 51.83 = **48.17%**

4.43

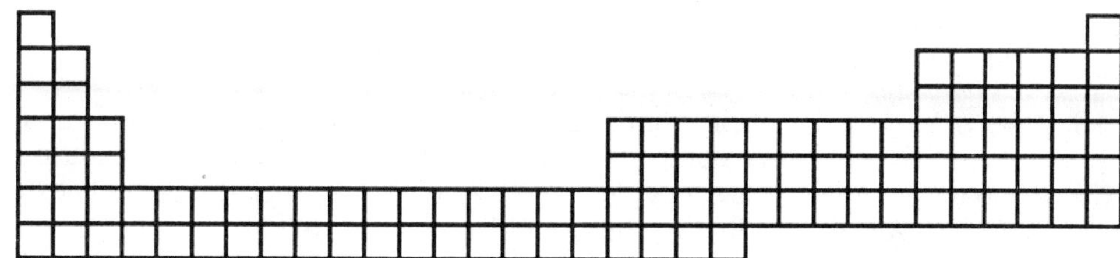

4.44 The vertical columns in the periodic table are called **groups**. The rows in the periodic table are called **periods**.

4.45 Mg, Se and Br

4.46 Ru, W and Ag

4.47 Elements 58 through 71 and 90 through 103

4.48 F

4.49 F_2, Cl_2, Br_2 and I_2

4.50 K

4.51 Ba

4.52 Ta, Nd and Cs

4.53 B, Si, Ge, As, Sb, Te, Po and At (Al is not a metalloid)

4.54 The metallic character of the elements decreases from left to right across a period of the periodic table from metals to metalloids to nonmetals. From the top to the bottom of a group from Group IA to Group VA the elements increase in metallic character.

4.55

	Metal	Nonmetal	Metalloids
Period 4	K, Ca, Sc, Ti V, Cr, Mn, Fe, Co Ni, Cu, Zn, Ga	Se, Br, Kr	Ge, As
Group VA	Bi	N, P	As, Sb

4.56 K. It should be similar to Na; both are members of Group IA.

4.57 $RaCl_2$

4.58 (b) iron, (c) chromium, (d) copper, (f) silver and (g) gold are transition metals

4.59 A combination reaction is one in which two or more substances combine to form a single product.

4.60 We write only empirical formulas for compounds such as CaO, NaCl, and BaF_2 because they are ionic compounds and we cannot say that any particular positive ion belongs to any particular negative ion. We write the empirical formula to show the smallest ratio of ions that represents a neutral combination.

4.61 (a) Sr^{2+} (b) Na^+ (c) S^{2-} (d) Al^{3+} (e) Br^-

4.62 (a) K^+ (b) N^{3-} (c) Mg^{2+} (d) O^{2-} (e) Ca^{2+}

4.63 (a) Fe^{2+} and Fe^{3+} (b) Cu^{+} and Cu^{2+} (c) Sn^{2+} and Sn^{4+}
(d) Zn^{2+} (e) Cr^{2+} and Cr^{3+}

4.64 See Table 4.3 (a) $CrCl_2$, $CrCl_3$, CrS and Cr_2S_3
(b) $MnCl_2$, $MnCl_3$, MnS and Mn_2S_3
(c) $FeCl_2$, $FeCl_3$, FeS and Fe_2S_3
(d) $CoCl_2$, $CoCl_3$, CoS and Co_2S_3
(e) $NiCl_2$ and NiS
(f) CuCl, $CuCl_2$, Cu_2S and CuS
(g) AgCl and Ag_2S

4.65 See Table 4.3 (a) AuBr, $AuBr_3$, Au_2O and Au_2O_3
(b) $ZnBr_2$ and ZnO
(c) $CdBr_2$ and CdO
(d) AgBr and Ag_2O
(e) $SnBr_2$, $SnBr_4$, SnO and SnO_2
(f) $PbBr_2$, $PbBr_4$, PbO and PbO_2
(g) $BiBr_3$ and Bi_2O_3

4.66 (a) ammonium ion (b) carbonate ion (c) chromate ion
(d) sulfite ion (e) acetate ion

4.67 (a) CN^- (b) ClO_4^- (c) MnO_4^- (d) NO_3^- (e) PO_4^{3-}
(f) OH^- (g) $C_2O_4^{2-}$ (h) $Cr_2O_7^{2-}$ (i) SO_4^{2-} (j) HCO_3^-
(k) SO_3^{2-} (l) NO_2^-

4.68 (a) Na_2CO_3 (b) $Ca(ClO_3)_2$ (c) SrS (d) $CrCl_3$ (e) $Ti(ClO_4)_4$

4.69 (a) $CrCO_3$, $Cr_2(CO_3)_3$, $CrCrO_4$, $Cr_2(CrO_4)_3$, $CrSO_3$, $Cr_2(SO_3)_3$, $Cr(C_2H_3O_2)_2$, $Cr(C_2H_3O_2)_3$

(b) $MnCO_3$, $Mn_2(CO_3)_3$, $MnCrO_4$, $Mn_2(CrO_4)_3$, $MnSO_3$, $Mn_2(SO_3)_3$, $Mn(C_2H_3O_2)_2$, $Mn(C_2H_3O_2)_3$

(c) $FeCO_3$, $Fe_2(CO_3)_3$, $FeCrO_4$, $Fe_2(CrO_4)_3$, $FeSO_3$, $Fe_2(SO_3)_3$, $Fe(C_2H_3O_2)_2$, $Fe(C_2H_3O_2)_3$

(d) through (n) Repeat using the other 11 metals.

4.70 (a) $Fe_2(HPO_4)_3$ (b) K_3N (c) $Ni(NO_3)_2$ (d) $Cu(C_2H_3O_2)_2$ (e) $BaSO_3$

4.71 Compounds formed between two nonmetals are not held together by an attraction between ions but rather by the sharing of electrons. In BaF_2 the force between positive Ba^{2+} ions and negative F^- ions is the attraction between ions of opposite charge. In H_2 the force is a sharing of electrons.

4.72 (a) PH_3 (b) H_2S (c) HBr
(d) SiH_4 (e) SbH_3

4.73 PbO and PbO_2

4.74 PCl_3 and PCl_5

4.75 (a) CF_4 (b) NF_3 (c) OF_2
(d) AsF_3 (e) ClF

4.76 Ionic compounds usually have melting points much higher than the melting points of molecular compounds.

4.77 Ionic compounds are more brittle than molecular compounds because the forces in ionic compounds are such that even a small slippage of one part of the solid can cause attractions to change to repulsions. The repulsive forces cause the crystal to break. Molecular solids, on the other hand, are much more flexible since there are no strong attractions or repulsions between molecules.

4.78 (b) AlF_3 and (d) CaF_2 will conduct electricity when melted since they are ionic.

4.79 (a) CaF_2 (b) $AlCl_3$ (c) TiO_2 (d) NaH
In each case, the ionic compound will have the higher melting point.

4.80 The modern definitions of oxidation are loss of electrons and increase in oxidation number. Reduction is the gain of electrons and decrease in oxidation number. The increase or decrease in oxidation number is the preferred definition.

4.81 Oxidation number and oxidation state are used interchangeably.

4.82 In the reaction $2Ca + O_2 \rightarrow 2CaO$, the Ca ($Ca \rightarrow Ca^{2+} + 2e^-$) is being oxidized and the oxygen is being reduced ($O_2 + 4e^- \rightarrow 2O^{2-}$). Therefore, the reaction is a redox reaction.

4.83 The actual rules are given in the textbook.

4.84 The oxidizing agent is reduced and the reducing agent is oxidized.

4.85 (a) K, +1; Cl, +3; O, -2 (b) Ba, +2; Mn, +6; O, -2
(c) Fe, +8/3; O, -2 (d) O, +1; F, -1
(e) I, +5; F, -1 (f) H, +1; O, -2; Cl, +1
(g) Ca, +2; S, +6; O, -2 (h) Cr, +3; S, +6; O, -2
(i) O, 0 (j) Hg, +1; Cl, -1

4.86 H, +1; S, +6; O,-2; C, +4; Br,-1;
O, +2; F, -1; K,+1; O, -1/2;
Cr, +3; Cl, -1; Mn, +7; O, -2;
K, +1; Mn, +7; O, -2; H, +1; C, +3; O, -2:
K, +1; Cl, +5; O, -2; Li, +1; N, +5; O, -2

4.87 (a) **oxidation** (b) **reduction** (c) **oxidation**
(d) **oxidation** (e) **reduction**

4.88

	Oxidized	Reduced	Oxidizing Agent	Reducing Agent
(a)	H_3AsO_3	HNO_3	HNO_3	H_3AsO_3
(b)	NaI	HOCl	HOCl	NaI
(c)	$H_2C_2O_4$	$KMnO_4$	$KMnO_4$	$H_2C_2O_4$
(d)	Al	H_2SO_4	H_2SO_4	Al
(e)	HCl	$K_2Cr_2O_7$	$K_2Cr_2O_7$	HCl

4.89

	Oxidized	Reduced	Oxidizing Agent	Reducing Agent
(a)	NaI	$NaIO_3$	$NaIO_3$	NaI
(b)	Cu	HNO_3	HNO_3	Cu
(c)	Cu	HNO_3	HNO_3	Cu
(d)	Cu	H_2SO_4	H_2SO_4	Cu
(e)	SO_2	HNO_3	HNO_3	SO_2

4.90 (a) NaBr sodium bromide
(b) CaO calcium oxide
(c) $FeCl_3$ ferric chloride; iron(III) chloride
(d) $CuCO_3$ cupric carbonate; copper(II) carbonate
(e) CBr_4 carbon tetrabromide
(f) P_4O_6 tetraphosphorus hexoxide
(g) $AsCl_5$ arsenic pentachloride
(h) $Mn(HCO_3)_2$ manganous hydrogen carbonate; manganous bicarbonate; manganese(II) hydrogen carbonate; manganese(II) bicarbonate
(i) $NaMnO_4$ sodium permanganate
(j) O_2F_2 dioxygen difluoride

4.91 (a) $Al(NO_3)_3$ (b) $FeSO_4$ (c) $NH_4H_2PO_4$ (d) IF_5
(e) PCl_3 (f) N_2O_4 (g) $KMnO_4$ (h) $Mg(OH)_2$
(i) H_2Se (j) NaH

4.92 (a) chromium(III) oxide (b) magnesium dihydrogen phosphate
(c) copper(II) nitrate (d) calcium sulfate
(e) barium hydroxide (f) aluminum phosphate
(g) magnesium nitride (h) lead(II) oxalate
(i) ammonium carbonate (j) potassium dichromate

4.93 (a) TiO_2 (b) $SiCl_4$ (c) $CaSe$
(d) KNO_3 (e) $Al_2(SO_4)_3$ (f) $Ni(HCO_3)_2$
(g) $NaHSO_4$ (h) $(NH_4)_2Cr_2O_7$ (i) $Ca(C_2H_3O_2)_2$
(j) $Sr(OH)_2$

4.94 SnF_2

4.95 (a) strontium chloride (b) calcium nitrate (c) copper(II) sulfide
(d) tin(II) phosphate (e) nickel(II) chlorate (f) zinc acetate
(g) bromic acid (h) mercury(II) bromide (i) cobalt(II) sulfate
(j) potassium dihydrogen arsenate

4.96 (a) chlorine trifluoride (b) dinitrogen pentoxide
(c) diammonium hydrogen phosphate (d) selenium tetrachloride
(e) tetrasulfur tetranitride (f) hydrogen telluride
(g) hydrotelluric acid (h) magnesium phosphide
(i) tin(IV) sulfide (j) oxygen difluoride

4.97 (a) $Pb(C_2H_3O_2)_4$ (b) Na_2Se (c) $Ba_3(PO_4)_2$
(d) HI (e) HI(aq) (f) PBr_3
(g) $Ca(OCl)_2$ [or $Ca(ClO)_2$] (h) $Ag_2C_2O_4$ (i) H_2CrO_4
(j) SiF_4

4.98 (a) SnF_2 (b) AlP (c) $HBrO_3$ (d) $Sr(H_2PO_4)_2$ (e) V_2O_3
(f) $SbCl_5$ (g) NO (h) CuBr (i) $CuBr_2$ (j) NH_4HSO_3

4.99 (a) S_3O_9 (b) ICl_5 (c) $Cr_2(SO_4)_3$ (d) $Fe_2(SO_4)_3$
(e) PbS (f) HIO_4 (g) LiOI [or LiIO] (h) $Hg(NO_3)_2$
(i) $Au_2(SO_4)_3$ (j) Bi_2O_5

4.100 (a) As_4O_{10} (b) N_2 (c) S_2N_2 (d) S_5N_6 (e) $Mn(CN)_3$
(f) HCN(aq)

4.101 (a) $Fe_2(SO_4)_3$ (b) $FeCl_2$ (c) $Hg_2(NO_3)_2$
(d) CuCl (e) $SnCl_4$ (f) $Co(OH)_2$
(g) $AuCl_3$ (h) $Cr(C_2H_3O_2)_3$

4.102 (a) chromium(II) carbonate, chromium(III) carbonate, chromium(II) chromate, chromium(III) chromate, chromium(II) sulfite, chromium(III) sulfite, chromium(II) acetate, chromium(III) acetate
[Answers to parts (b) through (n) are not provided.]

5 CHEMICAL REACTIONS IN AQUEOUS SOLUTION

5.1 Solutions are usually employed for carrying out chemical reactions because the homogeneous nature of solutions allows dissolved substances to intermingle freely. Thus, the reactions involving solutions are permitted to occur very rapidly.

5.2 Solvent: substance present in the greatest proportion in a solution
Solutes: all substances in solution other than the solvent
Concentrated: relatively large amounts of solute present in a small amount of solvent
Dilute: only a small amount of solute in a large amount of solvent
Saturated: at a given temperature, the maximum amount of solute that the solvent can hold while in contact with undissolved solute
Supersaturated: at a given temperature, the solvent contains more solute than ordinarily required for saturation
Unsaturated: the solvent contains less than the maximum amount of solute at a given temperature while in contact with undissolved solute

5.3 Solubility: the amount of solute required to produce a saturated solution with a given amount of solvent at a particular temperature

5.4 The solubility of a solute in a particular solvent changes with temperature.

5.5 Yes, for a solute which has a limited solubility in a particular solvent, a saturated solution will have a relatively small proportion of solute in large amounts of solvent.

5.6 If a crystal of solute is added to a supersaturated solution, additional solute crystallizes on the "seed" crystal until the concentration drops to the point of saturation.

5.7 To prepare a supersaturated solution of sugar in water, heat the water and add enough sugar to form a saturated solution at the elevated temperature. When the solution is allowed to cool, a supersaturated solution is the result if no crystals are present for the excess to crystallize onto.

5.8 If the solution is at 25°C, it would be described as supersaturated. At some higher temperature it would be saturated and above that temperature unsaturated. Because of the low solubility of $PbSO_4$ in water, all of these solutions would be considered to be dilute solutions.

5.9 An electrolyte is a substance which when dissolved in a solvent produces ions which make the solution able to conduct electricity. A nonelectrolyte is a substance which, when dissolved in a solvent, does not produce ions; therefore, it does not give an electrically conducting solution.

5.10 We can distinguish between strong and weak electrolytes with the aid of the apparatus shown in Figure 5.2. A strong electrolyte will cause the bulb to glow brightly and a weak electrolyte will cause the bulb to glow dimly. Ions are present in solutions of electrolytes but not in solutions of nonelectrolytes.

5.11 $KCl(aq) \rightarrow K^+(aq) + Cl^-(aq)$

$(NH_4)_2SO_4(aq) \rightarrow 2NH_4^+(aq) + SO_4^{2-}(aq)$

$Na_3PO_4(aq) \rightarrow 3Na^+(aq) + PO_4^{3-}(aq)$

$NaOH(aq) \rightarrow Na^+(aq) + OH^-(aq)$

$HCl(aq) \rightarrow H^+(aq) + Cl^-(aq)$

5.12 H^+ is the short form of H_3O^+. We often leave out the H_2O which is merely a carrier for the H^+ ion.

5.13 A dynamic equilibrium is one in which opposing processes occur at equal rates, so there is no net change in the system, e.g., ions react to form molecules while molecules react to form ions.

5.14 $CdSO_4(aq) \rightleftharpoons Cd^{2+}(aq) + SO_4^{2-}(aq)$

5.15 $H_2O \rightleftharpoons H^+ + OH^-$ (or, $2H_2O \rightleftharpoons H_3O^+ + OH^-$)

5.16 By position of equilibrium, we mean the relative proportions of reactants and products in a chemical system when the system is at equilibrium.

5.17 The position of the equilibrium lies mainly to the left in favor of undissociated H_2O in the equation for the dissociation of water. For HCl the position of equilibrium lies mainly to the right, virtually 100% in favor of the ionic products.

5.18 The rate of the reverse reaction would decrease because the concentrations of $C_2H_3O_2^-$ and H_3O^+ would be less and there would be fewer collisions between them. The rate of the forward reaction would also decrease, but by a lesser amount, because the concentration of $HC_2H_3O_2$ would be less and the concentration of H_2O would be more (only slightly more for dilute solutions). The percentage ionization increases as the solution is made more dilute.

5.19 Precipitate

5.20 Metathesis is the kind of reaction in which the cations and anions have changed partners. This is also known as double replacement.

5.21 A spectator ion is an ion that does not change during a reaction.

5.22 By filtering a mixture the precipitate can be separated from the reaction mixture.

5.23 In a molecular equation all reactants and products are written as if they were molecules. An ionic equation more accurately represents a reaction as it actually occurs in solution by showing all soluble ionic substances as being dissociated. A net ionic equation only represents the net chemical change that occurs. The spectator ions are not shown in a net ionic equation.

5.24 The net ionic equation focuses attention on the species that participate in the changes occurring in the solution and emphasizes that any substances that produce the same species in solution will react in the same way.

5.25 Ionic equation:

$Ca^{2+}(aq) + 2Cl^-(aq) + 2K^+(aq) + CO_3^{2-}(aq) \rightarrow CaCO_3(s) + 2K^+(aq) + 2Cl^-(aq)$

Net ionic equation: $Ca^{2+}(aq) + CO_3^{2-}(aq) \rightarrow CaCO_3(s)$

5.26 (a) $Cu^{2+} + 2Cl^- + Pb^{2+} + 2NO_3^- \rightarrow Cu^{2+} + 2NO_3^- + PbCl_2(s)$

Net ionic: $Pb^{2+}(aq) + 2Cl^-(aq) \rightarrow PbCl_2(s)$

(b) $Fe^{2+} + SO_4^{2-} + 2Na^+ + 2OH^- \rightarrow Fe(OH)_2(s) + 2Na^+ + SO_4^{2-}$

Net ionic: $Fe^{2+}(aq) + 2OH^-(aq) \rightarrow Fe(OH)_2(s)$

(c) $Zn^{2+} + SO_4^{2-} + Ba^{2+} + 2Cl^- \rightarrow Zn^{2+} + 2Cl^- + BaSO_4(s)$

Net ionic: $Ba^{2+}(aq) + SO_4^{2-}(aq) \rightarrow BaSO_4(s)$

(d) $2Ag^+ + 2NO_3^- + 2K^+ + SO_4^{2-} \rightarrow Ag_2SO_4(s) + 2K^+ + 2NO_3^-$

Net ionic: $2Ag^+(aq) + SO_4^{2-}(aq) \rightarrow Ag_2SO_4(s)$

(e) $2NH_4^+ + CO_3^{2-} + Ca^{2+} + 2Cl^- \rightarrow 2NH_4^+ + 2Cl^- + CaCO_3(s)$

Net ionic: $Ca^{2+}(aq) + CO_3^{2-}(aq) \rightarrow CaCO_3(s)$

5.27 (a) $Cu^{2+} + 2NO_3^- + 2Na^+ + 2OH^- \rightarrow Cu(OH)_2(s) + 2Na^+ + 2NO_3^-$
$Cu^{2+}(aq) + 2OH^-(aq) \rightarrow Cu(OH)_2(s)$ (net ionic)
(b) $3Ba^{2+} + 6Cl^- + 2Al^{3+} + 3SO_4^{2-} \rightarrow 3BaSO_4(s) + 2Al^{3+} + 6Cl^-$
$Ba^{2+}(aq) + SO_4^{2-}(aq) \rightarrow BaSO_4(s)$ (net ionic)
(c) $Hg_2^{2+} + 2NO_3^- + 2H^+ + 2Cl^- \rightarrow Hg_2Cl_2(s) + 2H^+ + 2NO_3^-$
$Hg_2^{2+}(aq) + 2Cl^-(aq) \rightarrow Hg_2Cl_2(s)$ (net ionic)
(d) $2Bi^{3+} + 6NO_3^- + 6Na^+ + 3S^{2-} \rightarrow Bi_2S_3(s) + 6Na^+ + 6NO_3^-$
$2Bi^{3+}(aq) + 3S^{2-}(aq) \rightarrow Bi_2S_3(s)$ (net ionic)
(e) $Ca^{2+} + 2Cl^- + 2Na^+ + SO_4^{2-} \rightarrow CaSO_4(s) + 2Na^+ + 2Cl^-$
$Ca^{2+}(aq) + SO_4^{2-}(aq) \rightarrow CaSO_4(s)$ (net ionic)

5.28 Solutions of acids have a sour taste and solutions of bases have a bitter taste. Another property is their effect on indicators, e.g., a basic solution turns the dye litmus blue and an acidic solution turns it pink. An acid reacts with a base to produce a salt and water; a base reacts with an acid to produce a salt and water (neutralization).

5.29 An acid is any substance that increases the concentration of hydronium ions (hydrogen ions) by reaction with water. A base is any substance that increases the concentration of hydroxide ions in aqueous solutions.

5.30 Use an indicator; e.g., dip a strip of pink litmus paper into the solution; if the litmus paper turns blue, the solution is basic.

5.31 A strong acid is essentially 100% ionized in solution while only a small fraction of a weak acid ionizes in solution.

5.32

Hydrochloric acid	HCl	$HCl + H_2O \rightarrow H_3O^+ + Cl^-$
Hydrobromic acid	HBr	$HBr + H_2O \rightarrow H_3O^+ + Br^-$
Hydroiodic acid	HI	$HI + H_2O \rightarrow H_3O^+ + I^-$
Chloric acid	$HClO_3$	$HClO_3 + H_2O \rightarrow H_3O^+ + ClO_3^-$
Perchloric acid	$HClO_4$	$HClO_4 + H_2O \rightarrow H_3O^+ + ClO_4^-$
Periodic acid	HIO_4	$HIO_4 + H_2O \rightarrow H_3O^+ + IO_4^-$
Nitric acid	HNO_3	$HNO_3 + H_2O \rightarrow H_3O^+ + NO_3^-$
Sulfuric acid	H_2SO_4	$H_2SO_4 + H_2O \rightarrow H_3O^+ + HSO_4^-$

(note: only the first dissociation of H_2SO_4 is strong)

5.33 (a) $H_2SO_3 \rightleftharpoons H^+ + HSO_3^-$

$HSO_3^- \rightleftharpoons H^+ + SO_3^{2-}$

(b) $H_3AsO_4 \rightleftharpoons H^+ + H_2AsO_4^-$

$H_2AsO_4^- \rightleftharpoons H^+ + HAsO_4^{2-}$

$HAsO_4^{2-} \rightleftharpoons H^+ + AsO_4^{3-}$

5.34 (a) A monoprotic acid is able to furnish only one hydrogen ion per molecule of acid.
(b) A diprotic acid is able to furnish two hydrogen ions per molecule of acid.
(c) A triprotic acid is able to furnish three hydrogen ions per molecule of acid.
(d) A polyprotic acid is an acid which is able to furnish more than one hydrogen ion per molecule of acid; includes diprotic and triprotic acids.

5.35 Acid anhydride - nonmetal oxides that react with H_2O to yield acid solutions.

e.g., $SO_2 + H_2O \rightleftharpoons H_2SO_3$

Basic anhydride - metal oxides that react with H_2O to give corresponding hydroxides. e.g., $CaO + H_2O \rightarrow Ca(OH)_2$

5.36 (a) acidic (b) basic (c) acidic (d) acidic (e) basic

5.37 $H_2O + N_2O_5 \rightarrow 2HNO_3$

5.38 Phosphoric acid; H_3PO_4: $P_4O_{10} + 6H_2O \rightarrow 4H_3PO_4$

5.39 A base turns litmus blue. Therefore, the element would be classified as a metal because its oxide yields a base on reaction with water while if it were a nonmetal its oxide would be an acidic anhydride.

5.40 The two kinds of bases are ionic hydroxides and molecular substances that react with water to produce OH^-.

5.41 $O^{2-} + H_2O \rightarrow 2OH^-$

5.42 potassium hydroxide, KOH

5.43 $NH_3(aq) + H_2O \rightleftharpoons NH_4^+(aq) + OH^-(aq)$ Ammonia is a weak base.

5.44 $N_2H_4(aq) + H_2O \rightleftharpoons N_2H_5^+(aq) + OH^-(aq)$

5.45 $H_3O^+ + OH^- \rightarrow 2H_2O$ or $H^+ + OH^- \rightarrow H_2O$

5.46 (a) $KOH + HCl \rightarrow KCl + H_2O$

(b) $NaOH + HC_2H_3O_2 \rightarrow NaC_2H_3O_2 + H_2O$

(c) $NH_3(aq) + HCl \rightarrow NH_4Cl(aq)$

(d) $CuO + 2HBr \rightarrow CuBr_2 + H_2O$

(e) $Fe_2O_3 + 3H_2SO_4 \rightarrow Fe_2(SO_4)_3 + 3H_2O$

5.47 Acid salts are the products of partial neutralization of a polyprotic acid. Examples include $NaHCO_3$, $NaHSO_4$, and Na_2HPO_4.

KH_2PO_4	potassium dihydrogen phosphate
K_2HPO_4	dipotassium hydrogen phosphate
K_3PO_4	tripotassium phosphate (or potassium phosphate)

5.48 The three "driving forces" for metathesis reactions are: (a) formation of a precipitate, (b) formation of a weak electrolyte and (c) formation of a gas.

5.49

Soluble	Insoluble
KCl	$PbSO_4$
$(NH_4)_2SO_4$	$Mn(OH)_2$
$AgNO_3$	$FePO_4$
$Zn(ClO_4)_2$	$CaCO_3$
$Ba(C_2H_3O_2)_2$	NiO

5.50

Soluble	Insoluble
KNO_3	$NiCO_3$
$FeCl_2$	Hg_2Cl_2
$(NH_4)_2HPO_4$	$Al(OH)_3$
CuI_2	PbI_2
$SrBr_2$	CoS

5.51 (a) Ionic: $Al(OH)_3(s) + 3H^+(aq) + 3Cl^-(aq) \rightarrow Al^{3+}(aq) + 3Cl^-(aq) + 3H_2O$

Net ionic: $Al(OH)_3(s) + 3H^+(aq) \rightarrow Al^{3+}(aq) + 3H_2O$

(b) Ionic: $CuCO_3(s) + 2H^+(aq) + SO_4^{2-}(aq) \rightarrow Cu^{2+}(aq) + SO_4^{2-}(aq) + H_2O + CO_2(g)$

Net ionic: $CuCO_3(s) + 2H^+(aq) \rightarrow Cu^{2+}(aq) + H_2O + CO_2(g)$

(c) Ionic: $Cr_2(CO_3)_3(s) + 6H^+(aq) + 6NO_3^-(aq)$
$\rightarrow 2Cr^{3+}(aq) + 6NO_3^-(aq) + 3H_2O + 3CO_2(g)$

Net ionic: $Cr_2(CO_3)_3(s) + 6H^+(aq) \rightarrow 2Cr^{3+}(aq) + 3H_2O + 3CO_2(g)$

5.52 (a) $Ag^+(aq) + Br^-(aq) \rightarrow AgBr(s)$
(b) $CoCO_3(s) + 2H^+(aq) \rightarrow Co^{2+}(aq) + CO_2(g) + H_2O$
(c) $C_2H_3O_2^-(aq) + H^+(aq) \rightarrow HC_2H_3O_2(aq)$
(d) $Pb^{2+}(aq) + SO_4^{2-}(aq) \rightarrow PbSO_4(s)$
(e) $H_2S(aq) + Cu^{2+}(aq) \rightarrow 2H^+(aq) + CuS(s)$
(f) $NH_4^+(aq) + OH^-(aq) \rightarrow NH_3(g) + H_2O$

5.53 (a) $CoS(s) + 2H^+(aq) \rightarrow H_2S(g) + Co^{2+}(aq)$
(b) $PbCO_3(s) + 2H^+(aq) \rightarrow H_2O + CO_2(g) + Pb^{2+}(aq)$
(c) $PbCO_3(s) + 2H^+(aq) + SO_4^{2-}(aq) \rightarrow PbSO_4(s) + H_2O + CO_2(g)$
(d) $Sn^{2+}(aq) + 2OH^-(aq) \rightarrow Sn(OH)_2(s)$
(e) $Ag_2O(s) + 2H^+(aq) + 2Cl^-(aq) \rightarrow 2AgCl(s) + H_2O$
(f) (This reaction does not have a driving force.)

5.54 (a) $Na_2SO_4(aq) + BaCl_2(aq) \rightarrow BaSO_4(s) + 2NaCl(aq)$
$2Na^+(aq) + SO_4^{2-}(aq) + Ba^{2+}(aq) + 2Cl^-(aq) \rightarrow BaSO_4(s) + 2Na^+(aq) + 2Cl^-(aq)$
$Ba^{2+}(aq) + SO_4^{2-}(aq) \rightarrow BaSO_4(s)$

(b) $Ca(NO_3)_2(aq) + (NH_4)_2CO_3(aq) \rightarrow CaCO_3(s) + 2NH_4NO_3(aq)$
$Ca^{2+}(aq) + 2NO_3^-(aq) + 2NH_4^+(aq) + CO_3^{2-}(aq)$
$\rightarrow CaCO_3(s) + 2NH_4^+(aq) + 2NO_3^-(aq)$
$Ca^{2+}(aq) + CO_3^{2-}(aq) \rightarrow CaCO_3(s)$

(c) $NaC_2H_3O_2(aq) + HNO_3(aq) \rightarrow NaNO_3(aq) + HC_2H_3O_2(aq)$
$Na^+(aq) + C_2H_3O_2^-(aq) + H^+(aq) + NO_3^-(aq) \rightarrow Na^+(aq) + NO_3^-(aq) + HC_2H_3O_2(aq)$
$H^+(aq) + C_2H_3O_2^-(aq) \rightarrow HC_2H_3O_2(aq)$

(d) $2NaOH(aq) + CuCl_2(aq) \rightarrow 2NaCl(aq) + Cu(OH)_2(s)$
$2Na^+(aq) + 2OH^-(aq) + Cu^{2+}(aq) + 2Cl^-(aq) \rightarrow 2Na^+(aq) + 2Cl^-(aq) + Cu(OH)_2(s)$
$Cu^{2+}(aq) + 2OH^-(aq) \rightarrow Cu(OH)_2(s)$

(e) $(NH_4)_2CO_3(aq) + 2HNO_3(aq) \rightarrow 2NH_4NO_3(aq) + H_2O + CO_2(g)$
$2NH_4^+(aq) + CO_3^{2-}(aq) + 2H^+(aq) + 2NO_3^-(aq)$
$\rightarrow 2NH_4^+(aq) + 2NO_3^-(aq) + H_2O + CO_2(g)$
$2H^+(aq) + CO_3^{2-}(aq) \rightarrow H_2O + CO_2(g)$

5.55 (a) no reaction between reactants
(b) no reaction between reactants
(c) $K_2S(aq) + Ni(C_2H_3O_2)_2(aq) \rightarrow 2KC_2H_3O_2(aq) + NiS(s)$
$2K^+(aq) + S^{2-}(aq) + Ni^{2+}(aq) + 2C_2H_3O_2^-(aq) \rightarrow 2K^+(aq) + 2C_2H_3O_2^-(aq) + NiS(s)$
$Ni^{2+}(aq) + S^{2-}(aq) \rightarrow NiS(s)$
(d) $MgSO_4(aq) + 2LiOH(aq) \rightarrow Li_2SO_4(aq) + Mg(OH)_2(s)$
$Mg^{2+}(aq) + SO_4^{2-}(aq) + 2Li^+(aq) + 2OH^-(aq) \rightarrow 2Li^+(aq) + SO_4^{2-}(aq) + Mg(OH)_2(s)$
$Mg^{2+}(aq) + 2OH^-(aq) \rightarrow Mg(OH)_2(s)$
(e) $AgC_2H_3O_2(aq) + KCl(aq) \rightarrow AgCl(s) + KC_2H_3O_2(aq)$
$Ag^+(aq) + C_2H_3O_2^-(aq) + K^+(aq) + Cl^-(aq) \rightarrow AgCl(s) + K^+(aq) + C_2H_3O_2^-(aq)$
$Ag^+(aq) + Cl^-(aq) \rightarrow AgCl(s)$

5.56 (a) $AgBr(s) + KI(aq) \rightarrow AgI(s) + KBr(aq)$
$AgBr(s) + K^+(aq) + I^-(aq) \rightarrow AgI(s) + K^+(aq) + Br^-(aq)$
$AgBr(s) + I^-(aq) \rightarrow AgI(s) + Br^-(aq)$
(b) $SO_2(aq) + H_2O + BaCl_2(aq) \rightarrow BaSO_3(s) + 2HCl(aq)$
$SO_2(aq) + H_2O + Ba^{2+}(aq) + 2Cl^-(aq) \rightarrow BaSO_3(s) + 2H^+(aq) + 2Cl^-(aq)$
$SO_2(aq) + H_2O + Ba^{2+}(aq) \rightarrow BaSO_3(s) + 2H^+(aq)$
(c) $Na_2C_2O_4(aq) + 2HCl(aq) \rightarrow 2NaCl(aq) + H_2C_2O_4(aq)$
$2Na^+(aq) + C_2O_4^{2-}(aq) + 2H^+(aq) + 2Cl^-(aq) \rightarrow 2Na^+ + 2Cl^- + H_2C_2O_4(aq)$
$2H^+(aq) + C_2O_4^{2-}(aq) \rightarrow H_2C_2O_4(aq)$
(d) $K_2SO_3(aq) + 2HCl(aq) \rightarrow 2KCl(aq) + H_2SO_3(aq)$
plus $H_2SO_3(aq) \rightarrow H_2O(\ell) + SO_2(g)$ to yield an overall equation of:
$K_2SO_3(aq) + 2HCl(aq) \rightarrow 2KCl(aq) + H_2O + SO_2(g)$
$2K^+(aq) + SO_3^{2-}(aq) + 2H^+(aq) + 2Cl^-(aq) \rightarrow 2K^+(aq) + 2Cl^-(aq) + H_2O + SO_2(g)$
$SO_3^{2-}(aq) + 2H^+(aq) \rightarrow H_2O + SO_2(g)$
(e) $BaCO_3(s) + H_2SO_4(aq) \rightarrow BaSO_4(s) + H_2O + CO_2(g)$
$BaCO_3(s) + 2H^+(aq) + SO_4^{2-}(aq) \rightarrow BaSO_4(s) + H_2O + CO_2(g)$
The net ionic equation is the same as the ionic equation.

5.57 $H_2O + CO_2(g) \rightarrow H_2CO_3(aq)$
$H_2CO_3(aq) + 2NaOH(aq) \text{ (excess)} \rightarrow Na_2CO_3(aq) + 2H_2O$

5.58 The following are only the answers. The necessary half-reactions and steps are not shown. The states of the reactants and products are also not shown.

(a) $8H^+ + 2NO_3^- + 3Cu \rightarrow 2NO + 3Cu^{2+} + 4H_2O$

(b) $10H^+ + NO_3^- + 4Zn \rightarrow NH_4^+ + 4Zn^{2+} + 3H_2O$

(c) $2Cr + 6H^+ \rightarrow 2Cr^{3+} + 3H_2$

(d) $8H^+ + Cr_2O_7^{2-} + 3H_3AsO_3 \rightarrow 2Cr^{3+} + 4H_2O + 3H_3AsO_4$

(e) $10H^+ + SO_4^{2-} + 8I^- \rightarrow 4I_2 + H_2S + 4H_2O$

(f) $4H_2O + 8Ag^+ + AsH_3 \rightarrow H_3AsO_4 + 8Ag + 8H^+$

(g) $H_2O + S_2O_8^{2-} + HNO_2 \rightarrow NO_3^- + 2SO_4^{2-} + 3H^+$

(h) $4H^+ + MnO_2 + 2Br^- \rightarrow Mn^{2+} + Br_2 + 2H_2O$

(i) $2S_2O_3^{2-} + I_2 \rightarrow 2I^- + S_4O_6^{2-}$

(j) $IO_3^- + 3HSO_3^- \rightarrow I^- + 3SO_4^{2-} + 3H^+$

5.59 (See the note at the beginning of the answers to question 5.58).

(a) $8H^+ + Cr_2O_7^{2-} + 3CH_3CH_2OH \rightarrow 2Cr^{3+} + 3CH_3CHO + 7H_2O$

(b) $4H^+ + PbO_2 + 2Cl^- \rightarrow Pb^{2+} + Cl_2 + 2H_2O$

(c) $14H^+ + 2Mn^{2+} + 5BiO_3^- \rightarrow 2MnO_4^- + 5Bi^{3+} + 7H_2O$

(d) $3H_2O + ClO_3^- + 3HAsO_2 \rightarrow 3H_3AsO_4 + Cl^-$

(e) $2H_2O + PH_3 + 2I_2 \rightarrow H_3PO_2 + 4I^- + 4H^+$

(f) $16H^+ + 2MnO_4^- + 10S_2O_3^{2-} \rightarrow 5S_4O_6^{2-} + 2Mn^{2+} + 8H_2O$

(g) $4H^+ + 2Mn^{2+} + 5PbO_2 \rightarrow 2MnO_4^- + 5Pb^{2+} + 2H_2O$

(h) $2H^+ + As_2O_3 + 2NO_3^- + 2H_2O \rightarrow 2H_3AsO_4 + N_2O_3$

(i) $8H_2O + 2P + 5Cu^{2+} \rightarrow 5Cu + 2H_2PO_4^- + 12H^+$

(j) $6H^+ + 2MnO_4^- + 5H_2S \rightarrow 2Mn^{2+} + 5S + 8H_2O$

5.60 (See the note at the beginning of the answers to question 5.58).

(a) $H_2O + CN^- + AsO_4^{3-} \rightarrow AsO_2^- + CNO^- + 2OH^-$

(b) $2CrO_2^- + 3HO_2^- \rightarrow 2CrO_4^{2-} + H_2O + OH^-$

(c) $7OH^- + 4Zn + NO_3^- + 6H_2O \rightarrow 4Zn(OH)_4^{2-} + NH_3$

(d) $4OH^- + Cu(NH_3)_4^{2+} + S_2O_4^{2-} \rightarrow 2SO_3^{2-} + Cu + 4NH_3 + 2H_2O$

(e) $N_2H_4 + 2Mn(OH)_3 \rightarrow 2Mn(OH)_2 + 2NH_2OH$

(f) $4OH^- + 2MnO_4^- + 3C_2O_4^{2-} \rightarrow 2MnO_2 + 6CO_3^{2-} + 2H_2O$

(g) $6OH^- + 7ClO_3^- + 3N_2H_4 \rightarrow 6NO_3^- + 7Cl^- + 9H_2O$

5.61 (See the note at the beginning of the answers to question 5.58).

(a) $3H_2O + P_4 + 3OH^- \rightarrow PH_3 + 3H_2PO_2^-$

(b) $12H^+ + 12Cu + 12Cl^- + As_4O_6 \rightarrow 12CuCl + 4As + 6H_2O$

(c) $9H_2O + 5IPO_4 \rightarrow I_2 + 3IO_3^- + 5H_2PO_4^- + 8H^+$

(d) $3NO_2 + H_2O \rightarrow 2NO_3^- + NO + 2H^+$

(e) $6OH^- + 3Br_2 \rightarrow 5Br^- + BrO_3^- + 3H_2O$

(f) $4HSO_2NH_2 + 6NO_3^- \rightarrow 4SO_4^{2-} + 2H^+ + 5N_2O + 5H_2O$

(g) $4H^+ + 2ClO_3^- + 2Cl^- \rightarrow 2ClO_2 + Cl_2 + 2H_2O$

(h) $2OH^- + 2ClO_2 \rightarrow ClO_2^- + ClO_3^- + H_2O$

(i) $6OH^- + 3Se \rightarrow 2Se^{2-} + SeO_3^{2-} + 3H_2O$

(j) $3H_2O + 5ICl \rightarrow 2I_2 + IO_3^- + 5Cl^- + 6H^+$

(k) $4OH^- + 2FNO_3 \rightarrow O_2 + 2F^- + 2NO_3^- + 2H_2O$

(l) $2H_2O + 4Fe(OH)_2 + O_2 \rightarrow 4Fe(OH)_3$

5.62 Chlorine is generally not used in the laboratory as an oxidizing agent because it is a poisonous gas and requires special precautions to be used safely.

5.63 (a) CrO_4^{2-} --- yellow
(b) $Cr_2O_7^{2-}$ --red-orange
(c) MnO_4^- --- purple

5.64 (a) $3HSO_3^-(aq) + 5H^+ + Cr_2O_7^{2-}(aq) \rightarrow 2Cr^{3+}(s) + 4H_2O + 3SO_4^{2-}$

(b) $2S_2O_3^{2-}(aq) + I_2(aq) \rightarrow 2I^-(aq) + S_4O_6^{2-}(aq)$

(c) $S_2O_3^{2-}(aq) + 4Cl_2(aq) + 5H_2O \rightarrow 8Cl^-(aq) + 2SO_4^{2-}(aq) + 10H^+(aq)$

5.65 $3HSO_3^-(aq) + Cr_2O_7^{2-}(aq) + 5H^+(aq) \rightarrow 3SO_4^{2-}(aq) + 2Cr^{3+}(aq) + 4H_2O$
or (if very acidic)
$3H_2SO_3(aq) + Cr_2O_7^{2-}(aq) + 2H^+(aq) \rightarrow 3SO_4^{2-}(aq) + 2Cr^{3+}(aq) + 4H_2O$

5.66 $3SO_3^{2-}(aq) + 2MnO_4^-(aq) + H_2O \rightarrow 3SO_4^{2-}(aq) + 2MnO_2(s) + 2OH^-(aq)$

5.67 $3SO_3^{2-}(aq) + 2CrO_4^{2-}(aq) + H_2O \rightarrow 3SO_4^{2-}(aq) + 2CrO_2^-(aq) + 2OH^-(aq)$

5.68 The percent by mass is the number of grams of solute per 100 g of solution.

5.69 Parts per million is the number of grams or volumes of solute per million (10^6) grams or volumes of solution.

5.70 Parts per billion is the number of grams or volumes of solute per billion (10^9) grams or volumes of solution.

5.71 (a) $\frac{0.001\text{ g F}^-}{1{,}000\text{ g soln.}} \times 100 = \mathbf{1 \times 10^{-4}\ \%\ F^-\ by\ mass}$

(b) $\frac{0.001\text{ g F}^-}{1{,}000\text{ g soln.}} \times 1{,}000{,}000 = \mathbf{1\ part\ F^-\ per\ million}$

(c) $\frac{0.001\text{ g F}^-}{1{,}000\text{ g soln.}} \times 1{,}000{,}000{,}000 = \mathbf{1 \times 10^3\ parts\ F^-\ per\ billion}$

5.72 $\frac{2.1 \times 10^{-5}\text{ mol Hg}}{25.0\text{ g sample}} \times \frac{200.59\text{ g Hg}}{\text{mol Hg}} \times 1{,}000{,}000 = 1.7 \times 10^2\text{ ppm}$

1.7×10^2 ppm is greater than the allowed 0.50 ppm. Therefore, the shipment **must be confiscated.**

5.73 (a) $\frac{1.50\text{ mol NaCl}}{2.00\text{ L soln.}} = \mathbf{0.750\ M}$

(b) **0.992 M**

(c) **0.556 M**

(d) $\frac{85.5\text{ g HNO}_3}{1.00\text{ L soln.}} \times \frac{1\text{ mol HNO}_3}{63.02\text{ g HNO}_3} = \mathbf{1.36\ M}$

(e) $\frac{44.5\text{ g NH}_4\text{C}_2\text{H}_3\text{O}_2}{600\text{ mL soln.}} \times \frac{1{,}000\text{ mL}}{\text{L}} \times \frac{1\text{ mol NH}_4\text{C}_2\text{H}_3\text{O}_2}{77.08\text{ NH}_4\text{C}_2\text{H}_3\text{O}_2} = \mathbf{0.962\ M}$

5.74 (a) $\frac{0.100 \text{ moles solute}}{\text{L soln.}} \times 0.250 \text{ L soln.} = \mathbf{0.0250 \text{ mol KCl}}$

(b) **2.31 moles $HClO_4$**

(c) **2.50×10^{-4} mole $HC_2H_3O_2$**

5.75 $\frac{0.150 \text{ mol } Na_2CO_3}{\text{L soln.}} \times 0.300 \text{ L soln.} \times \frac{106.0 \text{ g } Na_2CO_3}{\text{mol}} = \mathbf{4.77 \text{ g } Na_2CO_3}$

5.76 $\frac{0.300 \text{ mol } OH^-}{\text{L soln.}} \times \frac{1 \text{ mol } Ba(OH)_2}{2 \text{ mol } OH^-} \times \frac{171.3 \text{ g } Ba(OH)_2}{\text{mol } Ba(OH)_2} \times 0.250 \text{ L soln.}$

$\Leftrightarrow \mathbf{6.42 \text{ g } Ba(OH)_2}$

5.77 In pure nitric acid, nitric acid is both the solute and the solution. Therefore, a density of 1.513 g/mL can be expressed as 1.513 g solute/mL solution.

$\frac{1.513 \text{ g } HNO_3}{\text{mL soln.}} \times \frac{1{,}000 \text{ mL soln.}}{\text{L soln.}} \times \frac{1 \text{ mole } HNO_3}{63.012 \text{ g } HNO_3} = \mathbf{24.01 \text{ M}}$

5.78 $\frac{273.8 \text{ g salt}}{1.000 \text{ L soln.}} \times \frac{1 \text{ L}}{1{,}000 \text{ mL}} \times \frac{100 \text{ g soln.}}{22.0 \text{ g salt}} = \frac{1.24 \text{ g soln.}}{\text{mL soln.}}$

Density = 1.24 g/mL

$\frac{273.8 \text{ g } MgSO_4}{1.000 \text{ L soln}} \times \frac{1 \text{ mol } MgSO_4}{120.36 \text{ } MgSO_4} = 2.275 \text{ mol } MgSO_4/\text{L} = \mathbf{2.275 \text{ M}}$

5.79 (a) **0.100 M Li^+** and **0.100 M Cl^-**

(b) **0.250 M Ca^{2+}** and **0.500 M Cl^-**

(c) **2.40 M NH_4^+** and **1.20 M SO_4^{2-}**

(d) **0.600 M Na^+** and ~ **0.600 M HSO_4^-**

(e) **0.800 M Fe^{3+}** and **1.20 M SO_4^{2-}**

5.80 (a) **0.0250 M Ba^{2+}** and **0.0500 M OH^-**

(b) **0.300 M Cd^{2+}** and **0.600 M NO_3^-**

(c) **0.800 M Na^+** and **0.400 M HPO_4^{2-}**

(d) **0.200 M Cr^{3+}** and **0.300 M SO_4^{2-}**

(e) **0.0450 M Hg_2^{2+}** and **0.0900 M NO_3^-**

5.81 $\frac{0.100 \text{ mol } SO_4^{2-}}{L} \times \frac{1 \text{ mol } Na_2SO_4}{1 \text{ mol } SO_4^{2-}} \Leftrightarrow \frac{0.100 \text{ mol } Na_2SO_4}{L} \Leftrightarrow$ **$0.100\ M\ Na_2SO_4$**

5.82 $\frac{0.160 \text{ mol } Cl^-}{L} \times \frac{1 \text{ mol } FeCl_3}{3 \text{ mol } Cl^-} \Leftrightarrow \frac{0.0533 \text{ mol } FeCl_3}{L} =$ **$0.0533\ M\ FeCl_3$**

5.83 **0.0700 M** (See problem 5.81 or 5.82 for an example of the method for obtaining this answer).

5.84 (a) **0.0100 mol Na^+** and **0.0100 mol Cl^-**
(b) **0.00480 mol Ca^{2+}** and **0.00960 mol Cl^-**
(c) **0.0351 mol Na^+** and **0.0176 mol SO_4^{2-}**
(d) **0.221 mol NH_4^+** and **0.111 mol SO_4^{2-}**
(e) **0.0375 mol Al^{3+}** and **0.0562 mol SO_4^{2-}**

5.85 M.M. of $CuSO_4 \cdot 5H_2O$ = 249.68 The concentration of the solution is:

$$\frac{10.45 \text{ g salt}}{150.0 \text{ mL soln.}} \times \frac{1{,}000 \text{ mL}}{L} \times \frac{1 \text{ mol salt}}{249.68 \text{ g}} = 0.2790 \text{ M salt}$$

The salt solution will contain: **0.2790 M Cu^{2+}** and **0.2790 M SO_4^{2-}**

5.86 $2NaOH(aq) + H_2SO_4(aq) \rightarrow Na_2SO_4(aq) + 2H_2O$

$$5.00 \times 10^{-3} \text{ mol } H_2SO_4 \times \frac{2 \text{ mol NaOH}}{\text{mol } H_2SO_4} \times \frac{1 \text{ L NaOH soln.}}{0.100 \text{ mol NaOH}}$$

$\Leftrightarrow$ 0.100 L NaOH soln. or **100 mL**

5.87 $3.22 \text{ g Cu} \times \frac{1 \text{ mol Cu}}{63.546 \text{ g Cu}} \times \frac{8 \text{ mol } HNO_3}{3 \text{ mol Cu}} \times \frac{1 \text{ L soln.}}{1.250 \text{ mol } HNO_3} \Leftrightarrow$

0.1081 L soln. = **108 mL** (based upon H^+ provided by the HNO_3)

$$3.22 \text{ g Cu} \times \frac{1 \text{ mol Cu}}{63.546 \text{ g Cu}} \times \frac{2 \text{ mol } HNO_3}{3 \text{ mol Cu}} \times \frac{1 \text{ L soln.}}{1.250 \text{ mol } HNO_3} \Leftrightarrow$$

0.0270 L soln. = **27.0 mL** (based upon NO_3^- provided by the HNO_3)

5.88 (a) Molecular equation: $CuCO_3(s) + 2HClO_4 \rightarrow H_2O + CO_2(g) + Cu(ClO_4)_2$
Net ionic eq.: $2H^+(aq) + CuCO_3(s) \rightarrow CO_2(g) + H_2O(l) + Cu^{2+}(aq)$

(b) To show the formation of $Cu(ClO_4)_2$ the molecular equation must be used. Solid $Cu(ClO_4)_2$ can be obtained by evaporation.

$$5.25 \text{ g } Cu(ClO_4)_2 \times \frac{1 \text{ mol } Cu(ClO_4)_2}{262.5 \text{ g}} \times \frac{2 \text{ mol } HClO_4}{1 \text{ mol } Cu(ClO_4)_2} \times \frac{1 \text{ L soln.}}{1.35 \text{ mol } HClO_4} \times \frac{1{,}000 \text{ mL}}{\text{L}} \Leftrightarrow \mathbf{29.6 \text{ mL } HClO_4}$$

(c) $$5.25 \text{ g } Cu(ClO_4)_2 \times \frac{1 \text{ mol } Cu(ClO_4)_2}{262.5 \text{ g}} \times \frac{1 \text{ mol } CuCO_3}{1 \text{ mol } Cu(ClO_4)_2} \times \frac{123.6 \text{ g } CuCO_3}{\text{mol } CuCO_3} \Leftrightarrow \mathbf{2.47 \text{ g } CuCO_3}$$

5.89 (a) $H_3PO_4(aq) + 3NaOH(aq) \rightarrow Na_3PO_4(aq) + 3H_2O$

$$\frac{0.170 \text{ mol } H_3PO_4}{\text{L soln.}} \times 0.500 \text{ L soln.} \times \frac{3 \text{ mol NaOH}}{\text{mol } H_3PO_4} \times \frac{1 \text{ L NaOH soln.}}{0.300 \text{ mol NaOH}} \Leftrightarrow 0.850 \text{ L NaOH soln.} = \mathbf{850 \text{ mL}}$$

(b) $H_3PO_4(aq) + 2NaOH(aq) \rightarrow Na_2HPO_4(aq) + 2H_2O$

$$\frac{0.170 \text{ mol } H_3PO_4}{\text{L soln.}} \times 0.500 \text{ L soln.} \times \frac{2 \text{ mol NaOH}}{\text{mol } H_3PO_4} \times \frac{1 \text{ L NaOH soln.}}{0.300 \text{ mol NaOH}} \Leftrightarrow 0.567 \text{ L NaOH soln.} = \mathbf{567 \text{ mL}}$$

(c) $H_3PO_4(aq) + NaOH(aq) \rightarrow NaH_2PO_4(aq) + H_2O$

$$\frac{0.170}{1} \times \frac{.500}{1} \times \frac{1}{1} \times \frac{1}{0.300} = 0.283 \text{ L} = \mathbf{283 \text{ mL}}$$

5.90 $BaCl_2(aq) + H_2SO_4(aq) \rightarrow 2HCl(aq) + BaSO_4(s)$

$$\frac{0.200 \text{ mol } H_2SO_4}{\text{L}} \times 0.0250 \text{ L} \times \frac{1 \text{ mol } BaCl_2}{\text{mol } H_2SO_4} \times \frac{1 \text{ L } BaCl_2 \text{ soln.}}{0.100 \text{ mol } BaCl_2} \Leftrightarrow 0.0500 \text{ L } BaCl_2 \text{ soln.} = \mathbf{50.0 \text{ mL } BaCl_2 \text{ soln.}}$$

5.91 $3BaCl_2(aq) + Fe_2(SO_4)_3(aq) \rightarrow 3BaSO_4(s) + 2FeCl_3(aq)$

$$\frac{0.200 \text{ mol } Fe_2(SO_4)_3}{L} \times 0.0250 \text{ L} \times \frac{3 \text{ mol } BaCl_2}{\text{mol } Fe_2(SO_4)_3} \times \frac{1 \text{ L } BaCl_2 \text{ soln.}}{0.100 \text{ mol } BaCl_2}$$

$$\times \frac{1{,}000 \text{ mL}}{L} \Leftrightarrow \mathbf{150 \text{ mL } BaCl_2 \text{ soln.}}$$

5.92 $1\ NaOH + 1$ monoprotic acid $\rightarrow$ salt $+ 1H_2O$

$$\frac{1 \text{ L NaOH soln.}}{0.100 \text{ mol NaOH}} \times \frac{1}{0.0200 \text{ L NaOH soln.}} \times \frac{1 \text{ mol NaOH}}{\text{mol benzoic acid}} \times$$

0.244 g benzoic acid $\Leftrightarrow$ **122 g benzoic acid/mol benzoic acid**

5.93 (a) Molecular equation: $AgNO_3(aq) + NaCl(aq) \rightarrow NaNO_3(aq) + AgCl(s)$
Net ionic equation: $\mathbf{Ag^+(aq) + Cl^-(aq) \rightarrow AgCl(s)}$

(b) 20.0 mL of 0.200 M $AgNO_3$ contains:
$0.0200 \text{ L} \times \frac{0.200 \text{ mol}}{L}$ or 0.00400 mole of Ag^+

30.0 mL of 0.200 M NaCl contains:
$0.0300 \text{ L} \times \frac{0.200 \text{ mol}}{L}$ or 0.00600 mole of Cl^-

From this, one can see that the Ag^+ is the limiting reactant and that only **0.00400 mole of AgCl** can be precipitated.

(c) $0.00400 \text{ mol AgCl} \times \frac{143.3 \text{ g AgCl}}{\text{mol AgCl}} = \mathbf{0.573 \text{ g AgCl}}$

(d) The amount of each ion before reaction is: $Ag^+ = 0.00400$ moles, $NO_3^- = 0.00400$ moles, $Na^+ = 0.00600$ moles, and $Cl^- = 0.00600$ moles. The precipitation process will remove 0.00400 moles of Ag^+ and 0.00400 moles of Cl^- leaving in solution 0.0 moles Ag^+, 0.00200 moles Cl^-, 0.00400 moles NO_3^- and 0.00600 moles Na^+. The concentration of each ion will be the number of moles of ion in the final solution divided by the total volume of solution.

$Ag^+ = \mathbf{0\ M}$ $\quad Cl^- = \frac{0.00200 \text{ mol}}{0.0500 \text{ L}} = \mathbf{0.0400\ M}$

$NO_3^- = \frac{0.00400 \text{ mol}}{0.0500 \text{ L}} = \mathbf{0.0800\ M}$ $\quad Na^+ = \frac{0.00600 \text{ mol}}{0.0500 \text{ L}} = \mathbf{0.120\ M}$

5.94 $AgNO_3(aq) + HCl \rightarrow AgCl(s) + HNO_3(aq)$
If the limiting reactant is HCl,

$$\frac{0.050 \text{ mol HCl}}{\text{L}} \text{ x } 0.0250 \text{ L x } \frac{1 \text{ mol AgCl}}{1 \text{ mol HCl}} \text{ x } \frac{143 \text{ g AgCl}}{\text{mol}} \Leftrightarrow \mathbf{0.18\ g\ AgCl}$$

If the limiting reactant is $AgNO_3$,

$$\frac{0.50 \text{ mol AgNO}_3}{\text{L}} \text{ x } 0.100 \text{ L x } \frac{1 \text{ mol AgCl}}{1 \text{ mol AgNO}_3} \text{ x } \frac{143 \text{ g AgCl}}{\text{mol}} \Leftrightarrow \mathbf{7.2\ g\ AgCl}$$

The theoretical yield will be 0.18 g AgCl.

5.95 (a) $$\frac{0.0500 \text{ mol NaOH}}{\text{L NaOH soln.}} \text{ x } 0.0172 \text{ L NaOH soln. x } \frac{1 \text{ mol cap. acid}}{1 \text{ mol NaOH}}$$

$$\Leftrightarrow 0.000860 \text{ mol caproic acid}$$

$$\frac{0.100 \text{ g cap. acid}}{0.000860 \text{ mol}} = \mathbf{116\ g/mol = M.M.}$$

(b) C_3H_6O empirical formula mass = 58.1

From its molecular mass and its empirical formula mass, its molecular formula must be twice its empirical formula. $\mathbf{C_6H_{12}O_2}$

5.96 $Ba(OH)_2(aq) + 2HCl(aq) \rightarrow Ba^{2+}(aq) + 2Cl^-(aq) + 2H_2O$

$$\frac{0.273 \text{ mol Ba(OH)}_2}{\text{L}} \text{ x } 0.380 = 0.104 \text{ mol Ba(OH)}_2$$

$$\frac{0.520 \text{ mol HCl}}{\text{L}} \text{ x } 0.500 \text{ L} = 0.260 \text{ mol HCl}$$

(a) HCl is in excess. Therefore, the mixture will be **acidic.**

(b) To calculate the excess of HCl:

$$0.104 \text{ mol Ba(OH)}_2 \text{ x } \frac{2 \text{ mol HCl}}{1 \text{ mol Ba(OH)}_2} \Leftrightarrow 0.208 \text{ mol HCl (reacted)}$$

$$\frac{0.260 \text{ mol HCl total} - 0.208 \text{ mol reacted}}{0.380 \text{ L} + 0.500 \text{ L}} = 0.059 \text{ M HCl} = \mathbf{0.059\ M\ H^+}$$

5.97 (a) $\mathbf{3Ba^{2+}(aq) + 6OH^{-}(aq) + 2Al^{3+}(aq) + 3SO_4^{2-}(aq) \rightarrow 2Al(OH)_3(s) + 3BaSO_4(s)}$

(b) $0.270 \text{ mol } Ba^{2+} L^{-1} \times 0.0400 \text{ L} = 0.0108 \text{ mol } Ba^{2+}$

$$0.0108 \text{ mol } Ba^{2+} \times \frac{2 \text{ mol } OH^-}{\text{mol } Ba^{2+}} \Leftrightarrow 0.0216 \text{ mol } OH^-$$

$$\frac{0.330 \text{ mol } Al_2(SO_4)_3}{\text{L}} \times \frac{2 \text{ mol } Al^{3+}}{\text{mol } Al_2(SO_4)_3} \times 0.0250 \text{ L} \Leftrightarrow 0.0165 \text{ mol } Al^{3+}$$

$$\frac{0.330 \text{ mol } Al_2(SO_4)_3}{\text{L}} \times \frac{3 \text{ mol } SO_4^{2-}}{\text{mol } Al_2(SO_4)_3} \times 0.0250 \text{ L} \Leftrightarrow 0.0248 \text{ mol } SO_4^{2-}$$

Ba^{2+} is the limiting reactant for the formation of $BaSO_4$.

$$0.0108 \text{ mol } Ba^{2+} \times \frac{3 \text{ mol } BaSO_4}{3 \text{ mol } Ba^{2+}} \times \frac{233.4 \text{ g } BaSO_4}{\text{mol } BaSO_4} \Leftrightarrow 2.52 \text{ g of } BaSO_4 \text{ ppt.}$$

OH^- is the limiting reactant for the formation of $Al(OH)_3$

$$0.0216 \text{ mol } OH^- \times \frac{2 \text{ mol } Al(OH)_3}{6 \text{ mol } OH^-} \times \frac{78.0 \text{ g } Al(OH)_3}{\text{mol } Al(OH)_3} \Leftrightarrow 0.562 \text{ g of } Al(OH)_3 \text{ ppt.}$$

Total weight of ppt. is 2.52 + 0.562 = **3.08 g**

(c) **~0 M Ba^{2+}, ~0 M OH^-**

$$0.0108 \text{ mol } Ba^{2+} \times \frac{1 \text{ mol } SO_4^{2-}}{\text{mol } Ba^{2+}} \Leftrightarrow 0.0108 \text{ mol } SO_4^{2-} \text{ reacted}$$

$$\frac{0.0248 \text{ mol } SO_4^{2-} \text{ total} - 0.0108 \text{ mol } SO_4^{2-} \text{ ppt.}}{0.0650 \text{ L}} = \mathbf{0.215 \text{ M } SO_4^{2-}}$$

$$0.0216 \text{ mol } OH^- \times \frac{\text{mol } Al^{3+}}{3 \text{ mol } OH^-} \Leftrightarrow 0.00720 \text{ mol } Al^{3+} \text{ reacted}$$

$$\frac{0.0165 \text{ mol } Al^{3+} - 0.00720 \text{ mol } Al^{3+} \text{ ppt.}}{0.0650 \text{ L}} = \mathbf{0.143 \text{ M } Al^{3+}}$$

5.98 Chemical analysis is the experimental determination of chemical composition. Chemical analysis is used by many companies and laboratories to analyze a wide variety of samples.

5.99 $0.694 \text{ g AgCl} \times \frac{1 \text{ mol AgCl}}{143.4 \text{ g}} \times \frac{1 \text{ mol Cl}}{\text{mol AgCl}} \Leftrightarrow 0.00484 \text{ mol Cl}$

$0.00484 \text{ mol Cl} \times \frac{35.45 \text{ g Cl}}{\text{mol}} = 0.172 \text{ g Cl}$

g Ti = 0.249 g of sample - 0.172 g Cl = 0.077 g Ti

$0.077 \text{ g Ti} \times \frac{1 \text{ mol Ti}}{47.9 \text{ g}} = 0.0016 \text{ mol Ti}$

Formula is: $Ti_{0.0016}Cl_{0.00484}$ or $Ti_{0.0016/0.0016}Cl_{0.00484/0.0016}$ or **$TiCl_3$**

5.100 $PbCO_3(s) + 2HNO_3(aq) \rightarrow H_2O + CO_2(g) + Pb^{2+}(aq) + 2NO_3^-(aq)$

$Pb^{2+}(aq) + Na_2SO_4(aq) \rightarrow PbSO_4(s) + 2Na^+(aq)$

$1.081 \text{ g } PbSO_4 \times \frac{1 \text{ mol } PbSO_4}{303.3 \text{ g } PbSO_4} \times \frac{1 \text{ mol } PbCO_3}{1 \text{ mol } PbSO_4} \times \frac{267.2 \text{ g } PbCO_3}{\text{mol } PbCO_3}$

$\Leftrightarrow 0.9523 \text{ g } PbCO_3$

$\frac{0.9523 \text{ g } PbCO_3}{1.526 \text{ g sample}} \times 100.0\% = \mathbf{62.40\%\ PbCO_3}$

5.101 $2AgCl(s)(\text{excess}) + CuBr_2(aq) \rightarrow 2AgBr(s) + CuCl_2(aq) + AgCl(s)$

$1.800 \text{ g AgCl(initial)} \times \frac{1 \text{ mol AgCl}}{143.32 \text{ g AgCl}} = 0.01256 \text{ mol AgCl(initial)}$

0.01256 mol AgCl(initial) = mol AgBr + mol AgCl(excess)
mol AgBr = X
mol AgCl(final) = 0.01256 - X

(continued)

mole AgBr . Xg/m +

5.101 (continued)

2.052 g sample = (X mol AgBr x 187.77 g/mol) + [(0.01256 - X) mol AgCl x 143.32 g/mol]

2.052 = 187.77X + 1.800 - 143.32X X = 0.00567 mol AgBr

$$0.00567 \text{ mol AgBr} \times \frac{1 \text{ mol } CuBr_2}{2 \text{ mol AgBr}} \times \frac{223.35 \text{ g } CuBr_2}{\text{mol } CuBr_2} \Leftrightarrow 0.633 \text{ g } CuBr_2$$

$$\frac{0.633 \text{ g } CuBr_2}{1.850} \times 100\% = \mathbf{34.2\%}$$

5.102 (a) A buret is a long tube fitted at one end with a valve (called a stopcock) and is precisely graduated in milliliters and tenths of milliliters. It is used to deliver known quantities of a liquid or solution.
(b) Titration is an analytical procedure that allows us to measure the amount of one solution needed to react exactly with the contents of another solution.
(c) The titrant is the solution delivered via the buret.
(d) The end point in a titration is the point that delivery of the titrant is stopped and is usually signaled by the color change of an indicator.

5.103 An indicator signals when the reaction is complete.
(a) colorless (b) pink

5.104 $H_2SO_4(aq) + 2NaOH(aq) \rightarrow Na_2SO_4(aq) + 2H_2O$

$$\frac{0.1500 \text{ mol NaOH}}{\text{L NaOH soln.}} \times 0.02130 \text{ L NaOH soln.} \times \frac{1 \text{ mol } H_2SO_4}{2 \text{ mol NaOH}} \times$$

$$\frac{1}{0.01500 \text{ L } H_2SO_4 \text{ soln.}} \Leftrightarrow \frac{0.1065 \text{ mol } H_2SO_4}{\text{L } H_2SO_4 \text{ soln.}} = \mathbf{0.1065\ M\ H_2SO_4}$$

5.105 $CaCO_3 \rightarrow CaO \rightarrow Ca(OH)_2$

$Ca(OH)_2(aq) + 2HCl(aq) \rightarrow CaCl_2(aq) + 2H_2O$

$$\frac{0.120 \text{ mol HCl}}{\text{L}} \times 0.03725 \text{ L} \times \frac{1 \text{ mol } Ca(OH)_2}{2 \text{ mol HCl}} \Leftrightarrow 0.00224 \text{ mol } Ca(OH)_2$$

$0.00224 \text{ mol } Ca(OH)_2 = 0.00224 \text{ mol } CaCO_3$

$$\frac{0.00224 \text{ mol } CaCO_3 \times 100.1 \text{ g mol}^{-1}}{1.030 \text{ g sample}} \times 100\% = \mathbf{21.8\%}$$

5.106 $\frac{0.0500 \text{ mol NaOH}}{\text{L}} \times 0.0204 \text{ L} \times \frac{1 \text{ mol acids}}{1 \text{ mol NaOH}} \Leftrightarrow 0.00102 \text{ mol acids}$

Let X = moles L.A.; then 0.00102 - X = moles C.A.
0.1000 g = (X mol L.A. x 90.09 g/mol)+{[(0.00102 - X) mol C.A.] (116.18 g/mol)}
0.1000 = 90.09X + 0.1185 - 116.18X
0.0185 = 26.09X
X = 0.000709
Mass L.A. = 0.000709 mol x 90.09 g/mol = **0.064 g Lactic Acid**
Mass C.A. = (0.00102 - 0.000709 mol) (116.18 g/mol) = **0.036 g Caproic Acid**

5.107 (a) $\frac{0.05000 \text{ mol HCl}}{\text{L}} \times 0.0500 \text{ L} = 0.00250 \text{ mol HCl total}$

$\frac{0.0600 \text{ mol NaOH}}{\text{L}} \times 0.03057 \text{ L} \times \frac{1 \text{ mol HCl neut.}}{1 \text{ mol NaOH}}$

$\Leftrightarrow 0.00183$ mol HCl neut. by NaOH

0.00250 mol HCl total = 0.00183 mol HCl neut. by NaOH + X mol HCl neut. by NH_3

X = 0.00067 mol HCl neutralized by NH_3

(b) $0.00067 \text{ mol HCl} \times \frac{1 \text{ mol } NH_3}{\text{mol HCl}} \times \frac{1 \text{ mol N}}{\text{mol } NH_3} \times \frac{14.01 \text{ g N}}{\text{mol N}} \Leftrightarrow$ **0.0094 g N**

(c) $\frac{0.0094 \text{ g N}}{0.0500 \text{ g sample}} \times 100\% =$ **19% N in sample**

$\frac{14.01 \text{ g N in gly}}{75.08 \text{ g gly}} \times 100\% =$ **18.66% N in gly**

Glycine and the sample have the same percent nitrogen to two significant figures.

5.108 For acids and bases, an equivalent is the amount of substance that supplies or reacts with one mole of H^+. For redox, an equivalent is the amount of substance that gains or loses one mole of electrons.

5.109 The number of equivalents of A that react is exactly equal to the number of equivalents of B that react in any given reaction.

5.110 $0.200 \text{ mol Ba(OH)}_2 \times \frac{2 \text{ eq Ba(OH)}_2}{\text{mol Ba(OH)}_2} =$ **0.400 eq $Ba(OH)_2$**

5.111 $5.00 \text{ eq } H_3PO_4 \times \frac{1 \text{ mol } H_3PO_4}{3 \text{ eq } H_3PO_4} = \mathbf{1.67 \text{ mol } H_3PO_4}$

5.112 $0.140 \text{ mol } H_3AsO_4 \times \frac{2 \text{ eq } H_3AsO_4}{\text{mol } H_3AsO_4} = \mathbf{0.280 \text{ eq } H_3AsO_4}$

5.113 (a) $MnSO_4 \rightarrow Mn_2O_3$ or $Mn^{2+} \rightarrow Mn^{3+}$ (1 electron change)

F.M. of $MnSO_4$ = 151.00 $\quad$ eq wt $= \frac{151.00 \text{ g}}{\text{mol}} \times \frac{1 \text{ mol}}{\text{eq}} = \mathbf{151.00 \text{ g/eq}}$

(b) $Mn^{2+} \rightarrow Mn^{4+}$ $\quad$ eq wt $= \frac{151.00 \text{ g}}{\text{mol}} \times \frac{1 \text{ mol}}{2 \text{ eq}} = \mathbf{75.50 \text{ g/eq}}$

(c) $Mn^{2+} \rightarrow Mn^{6+}$ $\quad$ eq wt $= \frac{151.00 \text{ g}}{\text{mol}} \times \frac{1 \text{ mol}}{4 \text{ eq}} = \mathbf{37.75 \text{ g/eq}}$

(d) $Mn^{2+} \rightarrow Mn^{7+}$ $\quad$ eq wt $= \frac{151.00 \text{ g}}{\text{mol}} \times \frac{1 \text{ mol}}{5 \text{ eq}} = \mathbf{30.20 \text{ g/eq}}$

5.114 $Cr^{6+} \rightarrow Cr^{3+}$

$0.400 \text{ eq } Na_2CrO_4 \times \frac{1 \text{ mol}}{3 \text{ eq}} \times \frac{161.98 \text{ g}}{\text{mol}} = \mathbf{21.6 \text{ g } Na_2CrO_4}$

5.115 $Mn^{2+} \rightarrow Mn^{7+}$

$\frac{0.100 \text{ eq}}{\text{L}} \times 0.300 \text{ L} \times \frac{1 \text{ mol}}{5 \text{ eq}} \times \frac{259.1 \text{ g}}{\text{mol}} = \mathbf{1.55 \text{ g } MnSO_4 \cdot 6H_2O}$

5.116 $Bi^{5+} \rightarrow Bi^{3+}$ ($-2e^-$) and $Mn^{2+} \rightarrow Mn^{7+}$ ($+5e^-$)

eq wt of $NaBiO_3 = \frac{280.0 \text{ g/mole}}{2 \text{ eq/mole}} = 140 \text{ g/eq}$

eq wt of $Mn(NO_3)_2 = \frac{178.9 \text{ g/mole}}{5 \text{ eq/mole}} = 35.79 \text{ g/eq}$

$$0.500 \text{ g } Mn(NO_3)_2 \times \frac{1 \text{ eq } Mn(NO_3)_2}{35.79 \text{ } Mn(NO_3)_2} \times \frac{1 \text{ eq } NaBiO_3}{1 \text{ eq } Mn(NO_3)_2} \times \frac{140.0 \text{ g } NaBiO_3}{1 \text{ eq } NaBiO_3}$$

$\Leftrightarrow \mathbf{1.96 \text{ g } NaBiO_3}$

5.117 (a) $\frac{98.00 \text{ g } H_3PO_4}{\text{mol}} \times \frac{1 \text{ mol}}{2 \text{ eq}} = \mathbf{49.00 \text{ g } H_3PO_4/eq}$

(b) $\frac{100.5 \text{ g } HClO_4}{\text{mol}} \times \frac{1 \text{ mol}}{\text{eq}} = \mathbf{100.5 \text{ g } HClO_4/eq}$

(c) $I^{5+} \rightarrow I^-$ ($-6e^-$) $\frac{197.89 \text{ g } NaIO_3}{\text{mol}} \times \frac{1 \text{ mol}}{6 \text{ eq}} = \mathbf{32.98 \text{ g } NaIO_3/eq}$

(d) $I^{5+} \rightarrow I^0$ ($-5e^-$) $\frac{197.89 \text{ g } NaIO_3}{\text{mol}} \times \frac{1 \text{ mol}}{5 \text{ eq}} = \mathbf{39.58 \text{ g } NaIO_3/eq}$

(e) $\frac{78.01 \text{ g } Al(OH)_3}{\text{mol}} \times \frac{1 \text{ mol}}{3 \text{ eq}} = \mathbf{26.00 \text{ g } Al(OH)_3/eq}$

5.118 (a) $\frac{22.0 \text{ g } Sr(OH)_2}{0.800 \text{ L soln.}} \times \frac{1 \text{ mol } Sr(OH)_2}{121.64 \text{ g}} \times \frac{2 \text{ eq } Sr(OH)_2}{\text{mol}} = 0.452 \text{ eq/L} = \mathbf{0.452 \text{ N}}$

(b) $\frac{0.25 \text{ mol}}{\text{L}} \times \frac{2 \text{ eq}}{\text{mol}} = 0.50 \text{ eq/L} = \mathbf{0.50 \text{ N}}$

(c) $\frac{0.150 \text{ mol}}{\text{L}} \times \frac{2 \text{ eq}}{\text{mol}} = 0.300 \text{ eq/L} = \mathbf{0.300 \text{ N}}$

(d) $Cr_2O_7^{2-} \rightarrow 2Cr^{3+} + 6e^-$

$\frac{41.7 \text{ g } K_2Cr_2O_7}{0.600 \text{ L soln.}} \times \frac{1 \text{ mol } K_2Cr_2O_7}{294.2 \text{ g}} \times \frac{6 \text{ eq}}{\text{mol}} = 1.42 \text{ eq/L} = \mathbf{1.42 \text{ N}}$

(e) $Na_2O \rightarrow 2NaOH$ 2 eq/mol Na_2O

$\frac{25.0 \text{ g } Na_2O}{1.50 \text{ L soln.}} \times \frac{1 \text{ mol } Na_2O}{62.0 \text{ g}} \times \frac{2 \text{ eq}}{\text{mol}} = 0.538 \text{ eq/L} = \mathbf{0.538 \text{ N}}$

(f) $\frac{0.135 \text{ eq}}{0.400 \text{ L soln.}} = 0.338 \text{ eq/mol} = \mathbf{0.338 \text{ N}}$

5.119 $\frac{0.850 \text{ eq Ba(OH)}_2}{\text{L}} \times 0.129 \text{ L} \times \frac{1 \text{ eq acid}}{1 \text{ eq Ba(OH)}_2} = 0.110$ equivalents acid

$\frac{4.93 \text{ g acid}}{0.110 \text{ eq acid}} =$ **44.8 g/eq**

5.120 Use $V_AN_A = V_BN_B$;

50.0 mL $H_2C_2O_4$ x 0.250 N = 45.0 mL $KMnO_4$ x ?N

?N = 0.278 N $KMnO_4$ for $MnO_4^- \rightarrow Mn^{2+}$

(0.278 eq/L) x (1 mol/5 eq) = 0.0556 mol $KMnO_4$/L

For $K_2C_2O_4$, in which $KMnO_4 \rightarrow MnO_2$ - $3e^-$

(0.250 eq $K_2C_2O_4$/L) x 0.0250 L x (1 eq $KMnO_4$/eq $K_2C_2O_4$) x

(1 mol $KMnO_4$/3 eq) x [1 L $KMnO_4$/0.0556 mol (from above)] ⇔
0.0375 L or **37.5 mL**

5.121 Use $V_AN_A = V_BN_B$ and $V_{dil}N_{dil} = V_{conc}N_{conc}$

For neutralization: 41.0 mL(B) x 0.255 N(B) = 5.00 mL(A) x ?N(A)

N(A) = 2.09 N HCl (the dil. HCl)

From dilution: 50.0 mL x 2.09 N = 10.0 mL x ?N

N = 10.45 To 3 significant figures the answer is **10.5 N HCl or 10.5 M HCl**

5.122 Use $V_AN_A = V_BN_B$

V($K_2Cr_2O_7$) x 0.500 N = 120 mL x 0.850 N

Vol. of $K_2Cr_2O_7$ soln. = 204 mL

5.123 Use $V_{conc}N_{conc} = V_{dil}N_{dil}$

85.0 mL x 1.00 = V_{dil} x 0.650 N

V_{dil} = 131 mL

H_2O added = 131 mL - 85.0 mL = **46 mL**

5.124 First find the concentration of the dilute H_2SO_4 solution
Use $V_AN_A = V_BN_B$

$25.0\text{ mL} \times N(H_2SO_4) = 15.0\text{ mL} \times 0.750\text{ N}$

$0.450\text{ N } H_2SO_4$

Then calculate how this solution could be prepared from 1.40 M H_2SO_4.

$$\frac{1.40\text{ mol } H_2SO_4}{L} \times \frac{2\text{ eq } H_2SO_4}{mol} = 2.80\text{ N } H_2SO_4$$

$V_{conc}N_{conc} = V_{dil}N_{dil}$

$250\text{ mL} \times 2.80N = V_{dil} \times 0.450N$

$V_{dil} = 1{,}560\text{ mL}$ or **1.56 L**

5.125 $\dfrac{825\text{ g solute}}{10^6\text{ g solution}} \times 100 = 8.25 \times 10^{-2}\% = \mathbf{0.0825\%\ by\ mass}$

$$\frac{825\text{ g benzene}}{1.00 \times 10^6\text{ g solution}} \times \frac{1{,}000\text{ g}}{1{,}000\text{ mL}} \times \frac{1{,}000\text{ mL}}{L} \times \frac{1\text{ mol benzene}}{78.11\text{ g}} = \mathbf{0.0106\ M}$$

5.126 The difference in mass of 0.4120 g and 0.4881 g is brought about by the following reaction and the change in mass of the silver solids.
$AgCl(s) + Br^-(aq) \rightarrow AgBr(s) + Cl^-(aq)$
The moles NaCl in original sample = moles AgCl reacting = moles AgBr formed.
Therefore: 0.4881 g - 0.4120 g = mass AgBr - mass AgCl

0.0761 = (X mol AgBr formed x 187.78 g AgBr/mol)
-(X mol AgCl reacted x 143.32 g AgCl/mol)
0.0761 = 187.78X - 143.32X = 44.46X
$X = 0.00171 = 1.71 \times 10^{-3}$ moles AgCl present in sample

$$\frac{1.71 \times 10^{-3} \times 58.44\text{ g/mol}}{0.2000\text{ g sample}} \times 100\% = \mathbf{50.0\%\ NaCl\ by\ mass}$$

$AgBr(s) + I^-(aq) \rightarrow AgI(s) + Br^-(aq)$

0.5868 g - 0.4881 g = mass AgI - mass AgBr

(continued)

5.126 (continued)

0.0987 = (mol AgI x 234.77 g/mol) - (X mol AgBr x 187.77 g/mol)

0.0987 = 234.8X - 187.8X = 47.0X

X = 0.00210 mol AgBr in intermediate sample

0.00210 mol AgBr - 0.00171 mol AgCl converted in previous step = 0.00039 mol AgBr formed from NaBr = 0.00039 mol NaBr in original.

$$\frac{0.00039 \text{ mol NaBr x } 102.89 \text{ g/mol}}{0.2000 \text{ g sample}} \text{ x } 100\% = \mathbf{20\% \text{ NaBr by mass}}$$

To calculate amount of NaI in original sample:

0.2000 g = (0.00171 mol NaCl x 58.44 g/mol)

+ 0.00039 mol NaBr x 102.89 g/mol) + mass NaI

mass NaI = 0.2000 - 0.100 - 0.040 = 0.060 g

$$\frac{0.060 \text{ g NaI}}{0.2000 \text{ g sample}} \text{ x } 100\% = \mathbf{30\% \text{ NaI by mass}}$$

5.127 $5Fe^{2+} + MnO_4^- + 8H^+ \rightarrow 5Fe^{3+} + Mn^{2+} + 4H_2O$

$$\frac{0.00400 \text{ mol } MnO_4^-}{L} \text{ x } 0.0158 \text{ L x } \frac{5 \text{ mol } Fe^{2+}}{\text{mol } MnO_4^-}$$

$$\text{x } \frac{1 \text{ mol } FeSO_4}{\text{mol } Fe^{2+}} \text{ x } \frac{151.9 \text{ g } FeSO_4}{\text{mol } FeSO_4} \Leftrightarrow 0.0480 \text{ g } FeSO_4$$

$$\frac{0.0480 \text{ g } FeSO_4}{0.1000 \text{ g sample}} \text{ x } 100\% = \mathbf{48.0\% \; FeSO_4 \text{ by mass}}$$

5.128 $2Cr^{3+} + 3SO_4^{2-} + 6Na^+ + 6OH^- \rightarrow 2Cr(OH)_3(s) + 6Na^+ + 3SO_4^{2-}$

$2Cr(OH)_3(s) + 6HNO_3(aq) \rightarrow 2Cr^{3+}(aq) + 6H_2O + 6NO_3^-(aq)$

From the sequential reactions shown above, one can see that 1 mole of $Cr_2(SO_4)_3(s)$ in the first reaction will eventually react with 6 moles of nitric acid.

$$0.500\text{ g } Cr_2(SO_4)_3 \times \frac{1\text{ mol } Cr_2(SO_4)_3}{392.2\text{ g } Cr_2(SO_4)_3} \times \frac{6\text{ mol } HNO_3}{\text{mol } Cr_2(SO_4)_3} \times$$

$$\frac{1.00\text{ L } HNO_3}{0.400\text{ mol } HNO_3} \Leftrightarrow 0.0191\text{ L } HNO_3\text{ soln. or } \mathbf{19.1\text{ mL } HNO_3\text{ soln.}}$$

5.129 $$\frac{1.00\text{ mol NaOH}}{\text{L}} \times 19.6 \times 10^{-3}\text{ L} \times \frac{1\text{ mol HCl excess}}{1\text{ mol NaOH}} \Leftrightarrow 0.0196\text{ mol HCl excess}$$

$$\frac{1.00\text{ mol HCl}}{\text{L}} \times 0.100\text{ L HCl} = 0.100\text{ mol HCl total}$$

0.100 mol HCl total - 0.0196 mol HCl excess = 0.0804 mol HCl reacted

$$0.0804\text{ mol HCl react.} \times \frac{1\text{ mol (CaO + MgO)}}{2\text{ mol HCl}} \Leftrightarrow 0.0402\text{ mol (CaO + MgO)}$$

Let X = moles CaO and 0.0402 - X = moles of MgO.

Then:
2.000 g = (X) (56.08 g/mol) + (0.0402 - X) (40.31)
2.000 = 56.08X + 1.620 - 40.31X
0.380 = 15.77X

X = 0.024 moles CaO = moles $CaCO_3$
0.016 = moles MgO = moles $MgCO_3$

$$\%\ CaCO_3 = \frac{0.024\text{ mol } CaCO_3 \times 100.1\text{ g/mol}}{(0.024\text{ mol } CaCO_3 \times 100.1\text{ g/mol}) + (0.016\text{ mol } MgCO_3 \times 84.31\text{ g/mol}}$$

$$\times 100\% = \frac{2.40 \times 100}{2.40 + 1.3} = \mathbf{65\%\ CaCO_3}$$

$$\%\ MgCO_3 = \frac{1.3}{2.40 \times 1.3} \times 100\% = \mathbf{35\%\ MgCO_3}$$

5.130 $CaCO_3 \rightarrow CaO \rightarrow Ca(OH)_2$

$Ca(OH)_2(aq) + 2HCl(aq) \rightarrow CaCl_2(aq) + 2H_2O$

$$\frac{0.1000 \text{ mol HCl}}{1.000 \text{ L solution}} \text{ x } 0.03030 \text{ L x } \frac{1 \text{ mol Ca(OH)}_2}{2 \text{ mol HCl}} \Leftrightarrow$$

$0.001515 \text{ mol Ca(OH)}_2 \Leftrightarrow 0.001515 \text{ mol CaO} \Leftrightarrow 0.001515 \text{ mol CaCO}_3$

0.001515 mol CaO x 56.08 g/mol = 0.08496 g CaO

0.2000 g sample - 0.08496 g CaO = 0.1150 g other material

0.001515 mol $CaCO_3$ x 100.1 g/mol = 0.1517 g $CaCO_3$

$$\frac{0.1517 \text{ g CaCO}_3}{0.1517 \text{ g CaCO}_3 + 0.1150 \text{ g others}} \text{ x } 100\% = \mathbf{56.88\% \text{ CaCO}_3 \text{ by mass}}$$

5.131 $\frac{0.0200 \text{ mol NaOH}}{\text{L}} \text{ x } 0.0152 \text{ L x } \frac{1 \text{ mol acid}}{2 \text{ mol NaOH}} \text{ x } \frac{176.1 \text{ g acid}}{\text{mol}} \Leftrightarrow 0.0268 \text{ g acid}$

$$\frac{0.0268 \text{ g}}{0.1000 \text{ g sample}} \text{ x } 100\% = \mathbf{26.8\%}$$

5.132 (a) $1 \text{ mol HIO}_3 \text{ x } \frac{1 \text{ eq HIO}_3}{\text{mol}} = \mathbf{1 \text{ eq HIO}_3}$ (as an acid)

(b) $2HIO_3 + 10H^+ + 10e^- \rightarrow I_2 + 6H_2O$

$1 \text{ mol HIO}_3 \text{ x } \frac{10 \text{ eq HIO}_3}{2 \text{ mol HIO}_3} = \mathbf{5 \text{ eq HIO}_3}$ (when being reduced to I_2)

5.133 (a) $\mathbf{MnO_4^- + 5Fe^{2+} + 8H^+ \rightarrow Mn^{2+} + 5Fe^{3+} + 4H_2O}$

(b) $\frac{0.0281 \text{ mol KMnO}_4}{\text{L}} \times 0.03942 \text{ L} \times \frac{1 \text{ mol MnO}_4^-}{\text{mol KMnO}_4} \times \frac{5 \text{ mol Fe}^{2+}}{\text{mol MnO}_4^-} \times$

$\frac{1 \text{ mol Fe}_3\text{O}_4}{3 \text{ mol Fe}^{2+}} \times \frac{231.5 \text{ g Fe}_3\text{O}_4}{\text{mol Fe}_3\text{O}_4} \Leftrightarrow 0.427 \text{ g Fe}_3\text{O}_4$

$\frac{0.427 \text{ g}}{1.362 \text{ g sample}} \times 100\% = \mathbf{31.4\%\ of\ Fe_3O_4\ by\ mass}$

5.134 (a) $5H_2C_2O_4(aq) + 2MnO_4^-(aq) + 6H^+(aq) \rightarrow 10CO_2 + 2Mn^{2+}(aq) + 8H_2O$

$\frac{0.2000 \text{ mol KMnO}_4}{\text{L}} \times 0.01964 \text{ L} \times \frac{1 \text{ mol MnO}_4^-}{\text{mol KMnO}_4} \times \frac{5 \text{ mol H}_2\text{C}_2\text{O}_2}{2 \text{ mol MnO}_4^-} \times$

$\frac{1 \text{ mol CaC}_2\text{O}_4}{1 \text{ mol H}_2\text{C}_2\text{O}_2} \Leftrightarrow \mathbf{0.009820\ mol\ CaC_2O_4\ precipitated}$

(b) $0.009820 \text{ mol CaC}_2\text{O}_4 \times \frac{1 \text{ mol CaCl}_2}{\text{mol CaC}_2\text{O}_4} \times \frac{110.98 \text{ g CaCl}_2}{\text{mol}} \Leftrightarrow 1.090 \text{ g CaCl}_2$

$\frac{1.090 \text{ g CaCl}_2}{2.385 \text{ g sample}} \times 100\% = \mathbf{45.70\%\ CaCl_2\ by\ mass}$

6 ENERGY AND ENERGY CHANGES: THERMOCHEMISTRY

6.1 Energy is usually defined as the capacity to do work. Matter can have energy as kinetic energy and as potential energy.

6.2 As shown by the equation K. E. = $1/2\ mv^2$, when the speed of an object is doubled its kinetic energy is increased by a factor of four.

6.3 If the potential energy of an object decreases as it moves away from another object, repulsive forces must exist between the two objects.

6.4 The law of conservation of energy states that energy can be neither created nor destroyed. It can only be transformed from one kind to another.

6.5 Chemical energy is the term used to describe the potential energy that chemicals have because of the attractions and repulsions between their subatomic particles.

6.6 Since the charges on the electron and the proton of the hydrogen atom have opposite signs, when they are moved apart the potential energy increases as would be the case in the separation of any two objects that have attractive forces.

6.7 A ball thrown into the air will have maximum kinetic energy at release. Its kinetic energy decreases as the ball rises and slows but its potential energy increases as it reaches greater height. At its maximum height, it stops thus having zero kinetic energy but maximum potential energy. As it falls the process is reversed--potential energy is decreased and kinetic energy is increased.

6.8 Temperature is a property that we can measure in the laboratory and is related to the speed at which the molecules of a substance are moving. Specifically, the higher the temperature the greater the speed at which the molecules are moving. Speed squared is directly proportional to the absolute temperature (Kelvin temperature).

6.9 The potential energy possessed by atoms and other atomic-sized particles is due to the attractions and repulsions of the electrically charged particles (nuclei and electrons).

6.10 When increasing the temperature from 100 °C to 200 °C, the increase in the absolute temperature is a factor of (200 + 273)/(100 + 273) or 1.27 to 1.00. Knowing that $T \propto$ average K.E., an increase in absolute temperature by a factor of 1.27 (from 100 °C to 200 °C) will increase the K.E. by a factor of 1.27. Increasing the temperature from 27 °C (300 K) to 327 °C (600 K) will double the average kinetic energy.

6.11 Heat energy is the same as kinetic energy. Heat flows from a hot object to a cool object through the transfer of kinetic energy as fast moving particles collide with slower moving particles and impart some of their energy to the less energetic particles.

6.12 When something hot is placed in contact with something cool, heat is transferred from the hot to the cool material until both are the same temperature. The total amount of heat is unchanged; therefore, that gained by the cooler body must equal that lost by the hot object.

6.13 (a) It is exothermic since heat is given up.
(b) The $2Mg + O_2$ must have larger potential energy than the product $2MgO$ in order for the observed exothermic nature of the reaction to take place.
(c) As potential energy is converted to heat, the kinetic energy of the particles is increasing with the increase in temperature.

6.14 During the formation of the ammonium nitrate solution, the potential energy increases as the kinetic energy decreases in this endothermic reaction.

6.15 The joule is defined as the energy corresponding to the energy possessed by an object with a mass of 2 kg traveling at the velocity of 1 meter per second. 4.184 joules equals 1 calorie. 4.184 kilojoules equals 1 kilocalorie.

6.16 The erg is a small energy unit used in physics and equals 1×10^{-7} J
(a) $1\ J = 1 \times 10^{7}$ ergs
(b) 1 calorie = 4.184×10^{7} ergs

6.17 (a) 345 J x (1 cal/4.184 J) = **82.5 cal**
(b) 546 cal x (4.184 J/cal) = **2.28×10^{3} J**
(c) 234 kJ x (1 kcal/4.184 kJ) = **55.9 kcal**
(d) 1.257 kcal x (4.184 kJ/1 kcal) = **5.259 kJ**

6.18 (a) $\frac{1}{2}$ x 4500 kg x $\left(\frac{1.79}{\text{s}}\right)^2$ x $\left(\frac{1\text{ J}}{\text{kg m}^2/\text{s}^2}\right)$ = **7.21 x 10^3 J**

(if 4500 has 3 significant figures)

(b) 7.21 x 10^3 J x (1 cal/4.184 J) = **1.72 x 10^3 cal**

6.19 Knowing that 1 J = 1 kg x m^2/s^2, one would need to convert 145 lb x $(15.3\text{ mi/hr})^2$ to kg x m^2/s^2 before it can be equated to joules.

$$\text{K.E} = \frac{1}{2} \times 145\text{ lb} \times \left(\frac{15.3\text{ mi}}{\text{hr}}\right)^2 \times \frac{1\text{ kg}}{2.205\text{ lb}} \times \left(\frac{5280\text{ ft}}{\text{mi}}\right)^2 \times \left(\frac{12\text{ in.}}{\text{ft}}\right)^2 \times \left(\frac{1\text{ m}}{39.37\text{ in.}}\right)^2$$

$$\times \left(\frac{1\text{ hr}}{60\text{ min}}\right)^2 \times \left(\frac{1\text{ min}}{60\text{ sec}}\right)^2 = 1.54 \times 10^3\text{ kg m}^2/\text{s}^2 = \mathbf{1.54 \times 10^3\ J}$$

6.20 K.E. = 1/2 x 225 kg x (8.00 x 10^4 m/hr)^2 x (1.00 hr/3600 s)^2 = 5.56 x 10^4 kg m^2/s^2 = **5.56 x 10^4 J** 5.56 x 10^4 J x 1 kJ/10^3 J = 5.56 x 10^1 kJ = **55.6 kJ**

6.21 $$\text{K.E} = \frac{1}{2} \times 2.40\text{ tons} \times \left(\frac{35.0\text{ mi}}{\text{hr}}\right)^2 \times \frac{2000\text{ lb}}{\text{ton}} \times \frac{1\text{ kg}}{2.20\text{ lb}} \times \left(\frac{5280\text{ ft}}{\text{mi}}\right)^2 \times \left(\frac{12\text{ in.}}{\text{ft}}\right)^2$$

$$\times \left(\frac{1\text{ m}}{39.37\text{ in.}}\right)^2 \times \left(\frac{1\text{ hr}}{3600\text{ sec}}\right)^2 \times \frac{1\text{ J}}{1\text{ kg} \times \text{m}^2/\text{s}^2} = 2.67 \times 10^5\text{ J}$$

$$2.67 \times 10^5\text{ J} \times \frac{\text{g °C}}{4.18\text{ J}} \times \frac{1}{12.0\text{°C}} = 5.32 \times 10^3\text{ g or } \mathbf{5.32\ kg}$$

6.22 1/2 x 1.5 x 10^3 kg x (60 m/s)^2 x 1 J/kg m^2 s^{-2} = **2.7 x 10^6 J**

6.23 Force x distance = (515 kg m/s^2) x 40.0 m
= 2.06 x 10^4 kg m^2/s^2
(2.06 x 10^4 kg m^2/sec^2) x (1 J/kg m^2 s^{-2}) = **2.06 x 10^4 J**

6.24 Heat capacity is defined as the amount of heat needed to change the temperature of something 1 °C. (Heat Capacity = J/°C) Specific heat is defined as the amount of heat needed to raise the temperature of 1 g of a substance 1 °C.
(Specific Heat = J/g °C)

6.25 The large specific heat of water is responsible for the moderating effects the oceans have on weather.

6.26 The specific heat of water is: **4.18 J g^{-1} °C^{-1}** or **1.00 cal g^{-1} °C^{-1}**. The heat capacity of 1.87 x 10^3 g of water would be: 1.87 x 10^3 g x 4.18 J g^{-1} °C^{-1} =

7.82 x 10^3 J °C^{-1}.

6.27 The heat capacity of water per mol of water or its molar heat capacity is: (18.0 g mol^{-1}) x (4.18 J g^{-1} °C^{-1}) = **75.2 J mol^{-1} °C^{-1}**

6.28 (a) 40.0 J x (g °C /4.18 J) x (1/150 g) = **0.0638°C increase**

(b) 35.0 cal x (4.18 J/cal) x (g °C/4.18 J) x (1/150 g) = **0.233°C decrease**

6.29 500 g x (4.18 J g^{-1} °C^{-1}) x 24.0°C = **5.02 x 10^4 J**
5.02 x 10^4 J x (1 cal/4.18 J) = **1.20 x 10^4 cal**

6.30 250 g x (4.18 J g^{-1} °C^{-1}) x (45.0°C - 30.0°C) = **1.57 x 10^4 J**

6.31 Heat lost by the penny = (Use T to indicate final temperature)
3.14 g x (0.387 J g^{-1} °C^{-1}) x (100°C - T) = ?
Heat gained by the water =
10.0 g x (4.18 J g^{-1} °C^{-1}) x (T - 25.0°C) = ?
The two amounts of heat must be equal; therefore, they can be set equal and the unknown "T" can be solved.

3.14 x 0.387 x (100 - T) = 10.0 x 4.18 x (T - 25.0)
121.5 - 1.215 T = 41.8 T - 1045
43.0 T = 1166.5
T = 27.1°C

6.32 Heat lost by the copper bar =
22.5 g x (0.387 J/g °C) x (100.0°C - 23.1°C) = 670 J
670 J equals the heat gained by the alcohol.
670 J = 20.0 g x (specific heat of alcohol) x (23.1°C - 10.0°C)
specific heat of methyl alcohol = **2.56 J g^{-1} °C^{-1}**
molar heat cap. of methyl alcohol = (2.56 J g^{-1} °C^{-1})(32.0 g mol^{-1})
= **81.9 J mol^{-1} °C^{-1}**

6.33 Heat lost by the metal specimen =
25.467 g x specific heat x (100.0°C - 31.2°C) = ?
Heat gained by the water =
15.0 g x 4.18 J g^{-1} °C^{-1} x (31.2°C - 24.3°C) = ?
Set the two equations equal:
25.467 x specific heat x 68.8 = 15.0 x 4.18 x 6.9
specific heat = **0.25 J g^{-1} °C^{-1}**

6.34 A calorimeter is an apparatus that is used to measure the heat of a reaction. A bomb calorimeter consists of a strong steel container in which the reaction is made to take place. The steel container is immersed in an insulated bath.

6.35 Heat capacity = 1347 J/(26.135°C - 25.000°C) = **1187 J/°C**

6.36 14.3×10^3 J = (1.78×10^4 J/°C) x (T-25.000°C) T = **25.803°C**

6.37 (a) -(97.1 kJ/°C)(27.282°C - 25.000°C) = -222 kJ or **-2.22×10^5 J**
(b) 222 kJ ÷ 1.00 mol = **-222 kJ mol^{-1}**

6.38 -162 kJ/0.500 mol = **-324 kJ mol^{-1}**

6.39 (a) Energy change = -(45.06 kJ/°C) x (26.413°C - 25.000°C) x (1000 J/kJ)
= -6.367×10^4 J or **6.367×10^4 J liberated**

(b) (6.367×10^4 J/1.500 g C_7H_8) x (92.15 g C_7H_8/mol C_7H_8)
= **3.911×10^6 J mol^{-1} liberated**

6.40 (a) The system is that part of the universe we happen to be studying.
(b) Everything outside the system is the surroundings.
(c) The state of the system is the description of such variables as temperature, pressure, number of moles and liquid, solid or gaseous state.
(d) An isothermal change is one that occurs at a constant temperature.
(e) An adiabatic change is one in which there has been no net flow of heat into or out of the system.

6.41 (a) If a system is isothermal, then not only does temperature not change, but average kinetic energy must also be constant since they are directly related.
(b) In an adiabatic system if the potential energy of the chemical substances decreases, the temperature of the system will increase. Therefore, in a reaction that loses potential energy, we will observe a warming (increased kinetic energy) if the total energy is kept constant (adiabatic).

6.42 A "coffee cup" calorimeter is an adiabatic system.

6.43 A state function is a quantity whose value depends only on the current state of a system and not on prior history. Energy change is always final energy content minus initial energy content and it will always be the same value provided that the states of the initial and final values do not change. Therefore, energy change is a state function. Otherwise one could create energy by different paths and break the law of conservation of energy.

6.44 Change in V = $\Delta V = V_{final} - V_{initial}$

6.45 (a) K.E. = 4.18 J g^{-1} °C^{-1} x 18.0 g mol^{-1} x (45°C - 25°C) = **1.5 x 10^3 J mol^{-1}**
(b) K.E. = 4.18 J g^{-1} °C^{-1} x 18.0 g mol^{-1} x (80°C - 25°C) = **4.1 x 10^3 J mol^{-1}**
(c) K.E. = 4.18 J g^{-1} °C^{-1} x 18.0 g mol^{-1} x (45°C - 80°C) = **-2.6 x 10^3 J mol^{-1}**
If one adds (b) and (c), one obtains (4.1 x 10^3 J mol^{-1} - 2.6 x 10^3 J mol^{-1}) or 1.5 x 10^3 J mol^{-1} - the same value as (a). Therefore, the net change of going from 25°C → 80°C → 45°C is the same as going directly from 25°C to 45°C.

6.46 Since we have no fixed motionless point (zero kinetic energy system) as a reference, we cannot determine exactly how much kinetic energy something has. We can only compare one system to another. For potential energy we need to know all the attractions and repulsions felt by all the molecules in a system.

6.47 A perpetual motion machine is one that would produce energy during a cyclic process. Such a machine is impossible because the change in energy is independent of path (a state function). Therefore, if no energy is lost to or gained from the surroundings, the net change in energy in going from state A to some other state then back to A must be zero. A perpetual motion machine would violate the law of conservation of energy.

6.48 Hess's law of heat summation is: The net value of ΔH for the overall process is merely the sum of all the enthalpy changes that take place along the way.

6.49 A substance is in its standard state when it is at a temperature of 25°C and a pressure of 1 atm.

6.50 A thermochemical equation is a balanced equation that also indicates the quantity of energy change. Fractional coefficients are allowed in thermochemical equations because in that application the coefficients indicate moles not molecules, etc., and fractional moles are sometimes very convenient.

6.51 $SO_2 + 1/2O_2 \rightarrow SO_3$ should not be labeled as a ΔH_f since it is not the formation of SO_3 from its elements.

6.52. (a) ΔH° = -77.4 kJ/2 = **-38.7 kJ**
(b) ΔH° = (-77.4 kJ) x (-1) x 3 = **232 kJ**

6.53

$1/2HCHO_2(\ell) + 1/2H_2O(\ell)$	$\rightarrow 1/2CH_3OH(\ell) + 1/2O_2(g)$	205.5 kJ
$1/2CO(g) + H_2(g)$	$\rightarrow 1/2CH_3OH(\ell)$	-64. kJ
$1/2HCHO_2$	$\rightarrow 1/2CO(g) + 1/2H_2O(\ell)$	-16.5 kJ
$HCHO_2(\ell) + H_2(g)$	$\rightarrow CH_3OH(\ell) + 1/2O_2(g)$	**ΔH° = 125 kJ**

6.54 Since equations (1) and (2) add to give $O_3 + O \rightarrow 2O_2$, the addition of the ΔH°'s for equations (1) and (2) will give ΔH° for $O_3 + O \rightarrow 2O_2$. ΔH° for this equation is -126 kJ + (-268 kJ) = -394 kJ or $\mathbf{O_3 + O \rightarrow 2O_2}$ **ΔH° = -394 kJ**

6.55
$1/2CaO + 1/2Cl_2 \rightarrow 1/2CaOCl_2$ ΔH° = -55.45 kJ
$1/2H_2O + 1/2CaOCl_2 + NaBr \rightarrow NaCl + 1/2Ca(OH)_2 + 1/2Br_2$ ΔH° = -30.1 kJ
$1/2Ca(OH)_2 \rightarrow 1/2CaO + 1/2H_2O$ ΔH° = +32.55 kJ

$1/2Cl_2 + NaBr \rightarrow NaCl + 1/2Br_2$ ΔH° = **-53.0 kJ**

6.56
$1/2Cu_2S(s) + O_2(g) \rightarrow CuO(s) + 1/2SO_2$ ΔH° = -263.8 kJ
$1/2SO_2(g) \rightarrow 1/2S(s) + 1/2O_2(g)$ ΔH° = +148.5 kJ
$Cu(s) + 1/2S(s) \rightarrow 1/2Cu_2S(s)$ ΔH° = -39.8 kJ

$Cu(s) + 1/2O_2(g) \rightarrow CuO(s)$ ΔH° = **-155.1 kJ**

6.57
$12/7NH_3(g) + 3O_2(g) \rightarrow 12/7NO_2(g) + 18/7H_2O(g)$ ΔH° = -485.1 kJ
$12/7NO_2(g) + 16/7NH_3(g) \rightarrow 2N_2(g) + 24/7H_2O(g)$ ΔH° = -782.9 kJ

$4NH_3(g) + 3O_2(g) \rightarrow 2N_2(g) + 6H_2O(g)$ ΔH° = **-1268 kJ**

6.58
$3Mg(s) + 2NH_3(g) \rightarrow Mg_3N_2(s) + 3H_2(g)$ ΔH° = -371 kJ
$N_2(g) + 3H_2(g) \rightarrow 2NH_3(g)$ ΔH° = -92 kJ

$3Mg(s) + N_2(g) \rightarrow Mg_3N_2(s)$ ΔH° = **-463 kJ**

6.59
$6Mg(s) + 3O_2(g) \rightarrow 6MgO(s)$ ΔH° = -3(1203 kJ)
$3Mg(s) + N_2(g) \rightarrow Mg_3N_2(s)$ ΔH° = -1(463 kJ)
$6MgO(s) + Mg_3N_2(s) \rightarrow 8Mg(s) + Mg(NO_3)_2(s)$ ΔH° = -1(-3884 kJ)

$Mg(s) + N_2(g) + 3O_2(g) \rightarrow Mg(NO_3)_2(s)$ ΔH° = **-188 kJ**

6.60 (1) $2NO(g) + O_2(g) \rightarrow 2NO_2(g)$ $\Delta H° = [2(+34)] - [2(+90.4) + 1(0)]$
$= $ **-113 kJ**

(2) $NO_2(g) \rightarrow NO(g) + O(g)$ $\Delta H° = 1(+90.4) + 1(1/2 \times 498)] - [1(+34)]$
$= $ **305 kJ**

(3) $O_2(g) + O(g) \rightarrow O_3(g)$ $\Delta H° = [1(+143)] -[1(0) + (1/2 \times 498)]$
$= $ **-106 kJ**

6.61

Reaction	ΔH°
(1) $2N_2(g) + 5O_2(g) \rightarrow 2N_2O_5(g)$	?
(2) $2N_2(g) + 6O_2(g) + 2H_2(g) \rightarrow 4HNO_3(\ell)$	4 (-174 kJ)
(3) $4HNO_3(\ell) \rightarrow 2N_2O_5(g) + 2H_2O(\ell)$	- 2 (-76.6 kJ)
(4) $2H_2O(\ell) \rightarrow 2H_2(g) + O_2(g)$	-1 (-571.5 kJ)

(2) + (3) + (4) = (1) $\Delta H° = 4(-174) + [-2(-76.6)] + [-1(-571.5)]$

ΔH° = 29 kJ

6.62 ΔH° (reaction) = (sum $\Delta H_f°$ products) - (sum $\Delta H_f°$ reactants)

ΔH° (per mol) = [1(-2020 kJ)]-[1(-1573 kJ)+(3/2)(-286 kJ)] = **-18 kJ**

6.63 $H_2O(\ell) \rightarrow H_2O(g)$ ΔH° (reaction) = ΔH° (evaporation)

$\Delta H_{vap} = (-242) - (-286) = +44 \text{ kJ mol}^{-1}$

(44 kJ mol^{-1}) (10.0 g H_2O) (1 mol /18.02 g) = **24 kJ**

6.64

Reaction	ΔH°
(1) $2C(s) + 3H_2(g) + 1/2O_2(g) \rightarrow C_2H_5OH(\ell)$	?
(2) $2CO_2(g) + 3H_2O(\ell) \rightarrow C_2H_5OH(\ell) + 3O_2(g)$	1.37×10^3 kJ
(3) $2C(s) + 2O_2(g) \rightarrow 2CO_2(g)$	2 (-394 kJ)
(4) $3H_2(g) + 3/2O_2(g) \rightarrow 3H_2O(\ell)$	3 (-286 kJ)

Reaction (2) + (3) + (4) = Reaction 1 = -276 kJ

$\Delta H_f°$ (C_2H_5OH) = -276 kJ /mol (The value in Table 6.1 is -278 kJ)

6.65 $\Delta H° = (\text{sum } \Delta H_f° \text{ products}) - (\text{sum } \Delta H_f° \text{ reactants})$

(a) $\Delta H° = [(-1676) + 2(0)] - [2(0) + (-822.2)] =$ **-854 kJ**

(b) $\Delta H° = [(-910.0 + 2(-242)] - [(+33) + 0] =$ **-1427 kJ**

(c) $\Delta H° = [-1433] - [(-635.5) + (-396)] =$ **-402 kJ**

(d) $\Delta H° = [(0) + (-242)] - [(-155) + 0] =$ **-87 kJ**

(e) $\Delta H° = [-84.5] - [(+51.9) + 0] =$ **-136.4 kJ**

6.66 $\Delta H° = (\text{sum } \Delta H_f° \text{ products}) - (\text{sum } \Delta H_f° \text{ reactants})$

(a) $\Delta H° = (+51.9) - [(+227) + (0)] =$ **-175 kJ**

(b) $\Delta H° = (-813.8) - [(-396) + (-286)] =$ **-132 kJ**

(c) $\Delta H° = (-1280) - [(-924.7) + (2 \times -92.5)] =$ **-170 kJ**

(d) $\Delta H° = [(-110) + (-242)] - [(-394) + (0)] =$ **42 kJ**

(e) $\Delta H° = [(10 \times 0) + (3 \times -394) + (4 \times -242)] - [(10 \times +81.5)$

$+ (1 \times -104)] =$ **-2861 kJ**

6.67 $\Delta H = 4.18 \text{ J g}^{-1} \text{ °C}^{-1} \times 350 \text{ mL} \times 1 \text{ g mL}^{-1} \times (30.00°\text{C} - 25.00°\text{C})$

$= 7315$ J or 7.315 kJ for the reaction of $0.150 \text{ L} \times 1.00 \text{ mol L}^{-1}$ HCl

(HCl is the limiting reactant)

$\Delta H = 7.315 \text{ kJ} / 0.150 \text{ mol} =$ **48.8 kJ /mol H^+**

6.68

$4/3CuS(s) + 2/3CuO(s) \rightarrow Cu_2S(s) + 1/3SO_2$	$\Delta H° = 1/3(-13.1 \text{ kJ})$
$1/3Cu(s) + 1/3S(s) \rightarrow 1/3CuS(s)$	$\Delta H° = 1/3(-53.1 \text{ kJ})$
$1/3SO_2(g) \rightarrow 1/3S(s) + 1/3O_2(g)$	$\Delta H° = -1/3(-297 \text{ kJ})$
$2/3Cu(s) + 1/3O_2(g) \rightarrow 2/3CuO(s)$	$\Delta H° = 1/3(-155 \text{ kJ})$
$CuS(s) + Cu(s) \rightarrow Cu_2S(s)$	$\Delta H° =$ **25.3 kJ**

6.69 Using the value for heat of combustion one can construct the following:

$$\frac{1\text{ mol } C_6H_{12}O_6}{2820\text{ kJ (total)}} \times \frac{100\text{ kJ (total)}}{60\text{ kJ (available as heat)}} \times \frac{5900\text{ kJ (as heat)}}{\text{hr}}$$

$$\times \frac{180.2\text{ g } C_6H_{12}O_6}{\text{mol } C_6H_{12}O_6} = 628.4\text{ g } C_6H_{12}O_6\text{ hr}^{-1}$$

or to two significant figures **630 g hr^{-1}**

6.70 $FeO(s) + 1/3CO_2(g) \rightarrow 1/3Fe_3O_4(s) + 1/3CO(g)$ $\Delta H = -1/3(+38\text{ kJ})$

$1/3Fe_3O_4(s) + 1/6CO_2(g) \rightarrow 1/2Fe_2O_3(s) + 1/6CO(g)$ $\Delta H = -\ 1/6(-59\text{kJ})$

$1/2Fe_2O_3(s) + 3/2CO(g) \rightarrow Fe(s) + 3/2CO_2(g)$ $\Delta H = 1/2(-28\text{ kJ})$

$FeO(s) + CO(g) \rightarrow Fe(s) + CO_2(g)$ **$\Delta H = -17$ kJ**

6.71 $\Delta H° = (\text{sum } \Delta H_f° \text{ products}) - (\text{sum } \Delta H_f° \text{ reactants})$

$FeO(s) + CO(g) \rightarrow Fe(s) + CO_2(g)$ $\Delta H = -17\text{kJ}$

$\Delta H°\text{ (reaction)} = [\Delta H_f°{}_{Fe} + \Delta H_f°{}_{CO_2}] - [\Delta H_f°{}_{FeO} + \Delta H_f°{}_{CO}]$

$-17\text{ kJ} = [0 + (-394)] - [\Delta H_f°{}_{FeO} + (-110)]$

$\Delta H_f°$ (FeO(s)) = -267 kJ /mol

6.72 $CaC_2 + 2H_2O \rightarrow Ca(OH)_2 + C_2H_2$ $\Delta H° = -126\text{ kJ}$

$CaO + 3C \rightarrow CaC_2 + CO$ $\Delta H° = +462.3\text{ kJ}$

$Ca(OH)_2 \rightarrow CaO + H_2O$ $\Delta H° = -1(-65.3\text{ kJ})$

$CO \rightarrow C + 1/2O_2$ $\Delta H° = -1/2(-220\text{ kJ})$

$H_2 + 1/2O_2 \rightarrow H_2O$ $\Delta H° = -1/2(+572\text{ kJ})$

$2C(s) + H_2(g) \rightarrow C_2H_2(g)$ **$\Delta H_f° = 226$ kJ mol^{-1}**

7 ELECTRONIC STRUCTURE AND THE PERIODIC TABLE

7.1 See Figure 7.1. Wavelength, λ, is the distance between consecutive peaks or troughs in a wave. Frequency, ν, is the number of peaks passing a given point per second. They are related to each other by the equation $\lambda \cdot \nu = c$ (equation 7.1) where c is the speed of light.

7.2 c = speed of light = 3.00×10^8 m s^{-1}

7.3 SI unit of frequency is the hertz, 1 Hz = 1 s^{-1}. Units for wavelengths are chosen so that the numbers are simple to comprehend. Thus, 320 nm is easier to comprehend than 3.20×10^{-7} m. The visible region of the spectrum runs from about 400 nm to 700 nm.

7.4 Infrared light has a longer wavelength, a lower frequency and is less energetic than is visible light. Ultraviolet light has a shorter wavelength, has a higher frequency and is more energetic than visible light.

7.5 (Shortest wavelengths to longest wavelengths) gamma rays, x rays, ultraviolet light, visible light, infrared light, microwaves, TV waves

7.6 AM radio broadcasts have frequencies between ~500 and ~1700 kHz, while **FM uses frequencies between ~85 and ~110 MHz.** (M= 10^6; k = 10^3)

7.7 Wavelength x frequency = speed of light
(a) wavelength x (8.0×10^{15} s^{-1}) = (3.0×10^8 m s^{-1})
wavelength = 3.75×10^{-8} m or 2 significant figures: 3.8×10^{-8} m or **38 nm**

(continued)

7.7 (continued)

(b) 200.0 nm (10^{-9} m/nm) x frequency = speed of light in m s^{-1}

(200.0×10^{-9} m) x ν = (3.00×10^{8} m s^{-1})

$\nu = 1.50 \times 10^{15}$ s^{-1} or **1.50×10^{15} Hz** (Only 3 significant figures because the speed of light is not an exact number.)

7.8 (a) $\lambda \cdot \nu = c$ $\lambda \times (9.40 \times 10^{9}\ s^{-1}) = (3.00 \times 10^{8}\ m\ s^{-1})$ λ = **0.0319 m**

(b) λ = **3.19 cm**

7.9 (a) **FM** 101.1 MHz = 101.1×10^{6} Hz = 101.1×10^{6} s^{-1}

$\lambda \times \nu = c$ $\lambda \times (101.1 \times 10^{6}\ s^{-1}) = (3.00 \times 10^{8}\ m\ s^{-1})$ λ = **2.97 m** (3 significant figures)

(b) **AM** 880 kHz = 880×10^{3} Hz = 880×10^{3} s^{-1}

$\lambda \times (880 \times 10^{3}\ s^{-1}) = 3.00 \times 10^{8}\ m\ s^{-1}$ λ = **341 m**

7.10 341 m x (100 cm/m) x (1 in./2.54 cm) x (1 ft / 12 in.) = 1120 ft (>3 football fields)

7.11 A line spectrum results when the light emitted does not contain radiation of all wavelengths as is needed for a continuous spectrum. A continuous spectrum can be obtained by passing sunlight through a prism. If light emitted by a gas discharge tube is passed through a prism, a line spectrum is obtained.

7.12 From the point of view of atomic structure, the lines in the emission line spectrum are light whose energies are equal to the energy given off by excited electrons in the atom going to less energetic states. Since only certain excited and ground states are found, this led Bohr to propose the existence of energy levels within an atom.

7.13 The line spectra is also known as the atomic emission spectra, emission spectra and atomic spectra.

7.14 546 nm = 546×10^{-9} m; $\lambda \cdot \nu = c$; (546×10^{-9} m) x ν = (3.00×10^{8} m s^{-1})

$\nu = 5.49 \times 10^{14}$ s^{-1} or **5.49×10^{14} Hz**

7.15 The visible lines of the atomic spectrum of hydrogen are at 410.3 nm, 432.4 nm, 486.3 nm, and 656.4 nm. The frequency of each of these is:

(a) (for 410.3 nm) $\lambda \times \nu = c$: (410.3×10^{-9} m) x ν = 3.00×10^{8} m s^{-1}

ν = **7.31×10^{14} Hz**

(b) (for 432.4 nm) ν = **6.94×10^{14} Hz**

(c) (for 486.3 nm) ν = **6.17×10^{14} Hz**

(d) (for 656.4 nm) ν = **4.57×10^{14} Hz**

7.16 Sodium vapor lamps emit intense light at a wavelength of **589 nm**. The frequency of the light is: $\lambda \times \nu = c$; $(589 \times 10^{-9}$ m$) \times \nu = (3.00 \times 10^{8}$ m s$^{-1})$;

ν = **5.09 x10^{14} Hz**

7.17 Each element has its own characteristic set of emission spectrum lines which can be used to identify the presence of an element in the presence of other elements.

7.18 Planck's constant is a proportionality constant equal to energy divided by frequency (E/ν) which has the units of energy x time.

7.19 A photon is a tiny packet, or quanta, of light energy.

7.20 The photoelectric effect is the ability of light to knock electrons off the surface of certain metals. The ejected electrons can be detected electronically. Since light of low frequency is incapable of causing the photoelectric effect, the photoelectric effect supports the concept that the energy of light depends upon the frequency of the light and that light is composed of particle-like photons with discrete energy values, independent of the intensity of the light.

7.21 (a) $E = h\nu = (6.63 \times 10^{-34}$ J s$)(3 \times 10^{15}$ s$^{-1}) =$ **2 x 10^{-18} J**

(b) $E = hc/\lambda = 2 \times 10^{-20}$ J

$(6.63 \times 10^{-34}$ J s$)$ $(3.00 \times 10^{8}$ m s$^{-1})/\lambda = 2 \times 10^{-20}$ J

Solving for λ gives: λ = **1 x 10^{-5} m**

7.22 (a) 589 nm = 589×10^{-9} m

$E = hc/\lambda = (6.63 \times 10^{-34}$ J s$)(3.00 \times 10^{8}$ m s$^{-1})/(589 \times 10^{-9}$ m$)$ = **3.38 x 10^{-19} J**

(b) $(3.38 \times 10^{-19}$ J/photon$) \times (6.02 \times 10^{23}$ photons/mol$) = 2.03 \times 10^{5}$ J /mol

= **203 kJ /mol**

(c) The specific heat of water is 4.184 J/g °C

(4.184 J/g °C) x ?g x ?°C = ? J

(4.184 J/g °C) x (10.0 kg x 1000 g/kg) x ? °C = $(2.03 \times 10^{5}$ J/mol$) \times 1$ mol

? °C = **4.85°C**

7.23 (a) $E = h\nu = (6.63 \times 10^{-34}$ J s/photon$)(2.6 \times 10^{14}$ s$^{-1}) = 1.7 \times 10^{-19}$ J/photon

$(1.7 \times 10^{-19}$ J/photon$) \times (6.02 \times 10^{23}$ photons/mol$)$ = **1.0 x 10^{5} J /mol**

= 100 kJ /mol

(b) $E = hc/\lambda = (6.63 \times 10^{-34}$ J s/photon$)(3.00 \times 10^{8}$ m s$^{-1})/(546 \times 10^{-9}$ m$)$

$= 3.64 \times 10^{-19}$ J/photon

$(3.64 \times 10^{-19}$ J/photon$) \times (6.02 \times 10^{23}$ photons/mol$)$ = **2.19 x 10^{5} J/mol**

= 219 kJ /mol

7.24 Planck's relationship (Equation 7.3) and the existence of energy levels in atoms can be used to derive an equation for the atomic emission spectra of hydrogen. Light is emitted by an atom only at certain frequencies and Planck's relationship shows, therefore, that an electron can have only certain discrete amounts of energy with none between. This suggests that the electron is restricted to specific energy levels in the atom.

7.25 The Bohr model imagined that the electrons travel around the nucleus in orbits of fixed size and energy. For this model an equation could be mathematically derived for the wavelengths of the light emitted by hydrogen when it produced its atomic spectrum. The model (theory) failed to correctly calculate energies for any atoms more complex than hydrogen.

7.26 In Bohr's theory, when energy is absorbed by an atom, an electron is raised in energy from one level to another and when the electron returns to a lower energy level, light is emitted whose energy is equal to the energy difference between the two levels.

7.27 (a) $\frac{1}{\lambda} = 109{,}678\ \text{cm}^{-1}\left[\frac{1}{2^2} - \frac{1}{4^2}\right]$

$\lambda = 4.86273 \times 10^{-5}\ \text{cm} =$

$\lambda = 4.86273 \times 10^{-5}\ \text{cm} \times 10^{-2}\ \text{m/cm} \times 1\ \text{nm}/10^{-9}\ \text{m} = \mathbf{486.273\ nm}$

(b) $\frac{1}{\lambda} = 109{,}678\ \text{cm}^{-1}\left[\frac{1}{3^2} - \frac{1}{6^2}\right]$ $\lambda = \mathbf{1.09411 \times 10^{-4}\ cm = 1094.11\ nm}$

7.28 $\Delta E = A\left(\frac{1}{n_2^2} - \frac{1}{n_1^2}\right) = 2.18 \times 10^{-18}\ \text{J}\left(\frac{1}{1^2} - \frac{1}{3^2}\right)$

$= 2.18 \times 10^{-18}\ \text{J}\ (0.8889) = \mathbf{1.94 \times 10^{-18}\ J}$

7.29 (a) $E = -A/n^2 = -2.18 \times 10^{-18}\ \text{J}/(3)^2 = \mathbf{-2.42 \times 10^{-19}\ J}$
(b) $E = -A/n^2 = -2.18 \times 10^{-18}\ \text{J}/(2)^2 = \mathbf{-5.45 \times 10^{-19}\ J}$
(c) $\Delta E = (-2.42 \times 10^{-19}\ \text{J}) - (-5.45 \times 10^{-19}\ \text{J}) = \mathbf{3.03 \times 10^{-19}\ J}$
(d) $\nu = E/h = 3.03 \times 10^{-19}\ \text{J}/6.63 \times 10^{-34}\ \text{J s} = \mathbf{4.57 \times 10^{14}\ Hz}$
(e) $\lambda = c/\nu = (3.00 \times 10^{8}\ \text{m s}^{-1}/4.57 \times 10^{14}\ \text{s}^{-1}) \times 1\ \text{nm}/10^{-9}\ \text{m} = \mathbf{656\ nm}$
(f) 656 nm is between the orange and red regions of the visible spectrum (Figure 7.2) Therefore, one would guess that the color of light having a wavelength of 656 nm would be **red-orange or red**.

7.30 The deBroglie relationship treats light as both particles and waves. Using Einstein's equation and Planck's equation, deBroglie derived for particles the equation $\lambda = h/m\nu$ which contains both wavelength and mass.

7.31 $\lambda = \frac{h}{m\nu}$ $\nu = c = 3.00 \times 10^8 \text{ m s}^{-1}$; $\lambda = 589 \text{ nm} = 589 \times 10^{-9} \text{ m}$

$$h = 6.63 \times 10^{-34} \text{ J s} = 6.63 \times 10^{-34} \text{ kg m}^2/\text{s}$$

$$m = \frac{6.63 \times 10^{-34} \text{ kg m}^2/\text{s}}{(589 \times 10^{-9} \text{ m})(3.00 \times 10^8 \text{ m s}^{-1})} = 3.75 \times 10^{-36} \text{ kg (per photon)}$$

$$m = \left(\frac{3.75 \times 10^{-36} \text{ kg}}{\text{photon}}\right) \times \left(\frac{6.02 \times 10^{23} \text{ photon}}{\text{mole}}\right) \times \left(\frac{1000 \text{ g}}{\text{kg}}\right) = \mathbf{2.26 \times 10^{-9} g}$$

7.32 K. E. = $(1/2)m\nu^2$, $\lambda = h/m\nu$ or $\nu = h/m\lambda$ Mass electron= 9.11×10^{-31} kg

$$\text{Therefore, K.E.} = \tfrac{1}{2}m\left(\frac{h^2}{m^2\lambda^2}\right) = \frac{h^2}{2m\lambda^2}$$

$$= \frac{(6.63 \times 10^{-34} \text{ J s})^2}{(2)(9.11 \times 10^{-31} \text{ kg})(0.10 \text{ nm})^2} \times \left(\frac{10^9 \text{ nm}}{1 \text{ m}}\right)^2 \times \left(\frac{1 \text{ kg m}^2 \text{ s}^{-2}}{1 \text{ J}}\right)$$

$$= \mathbf{2.4 \times 10^{-17} J}$$

7.33 Because the wavelengths are too short to be detected.

7.34 As stated in the marginal comment in section 7.3, "When the waves are in phase, they undergo **constructive interference** and their intensities add. When out of phase, the waves undergo **destructive interference** and their intensities cancel." The pattern of intensified areas and cancelled areas of converging light waves is called a diffraction pattern.

7.35 Electrons and other subatomic particles can be used to produce diffraction patterns. Diffraction patterns of electrons can only be explained as the result of the wave properties of the electrons.

7.36 A wave whose nodes are stationary is a standing wave. A node is a position where the amplitude of a wave is zero. For an orbital, the node is a place where the probability of finding the electron is zero.

7.37 The principle quantum number symbol is **n**. Allowed values of n are: **1, 2, 3, 4, etc.** The larger the value of n, the greater the average energy of the levels belonging to its associated shell and the larger the size of the wave function.

7.38 The azimuthal quantum number symbol is **ℓ**. Allowed values of ℓ are: **0, 1, 2, etc., to a maximum value of n-1.** The azimuthal quantum number determines the shape of an orbital and, to a certain degree, its energy.

7.39 The magnetic quantum number symbol is **m_ℓ**. Allowed values of m_ℓ are: integer values that range from **-ℓ to +ℓ including zero.** The magnetic quantum number serves to determine an orbital's orientation in space relative to the other orbitals.

7.40 An orbital in an atom has a characteristic energy and shape which can be viewed as a region around the nucleus where the electron can be expected to be found. The region can be described by a wave function.

7.41 **4**

7.42 **s, p, d, f, and g**

7.43 **f**

7.44 **25 orbitals:** 1 s, 3 p's, 5 d's, 7 f's, and 9 g's.

7.45 The ground state is the state of lowest energy.

7.46 (a) Max. electron population: **s 2, p 6, d 10, f 14, g 18, and h 22**
(b) First shell with h subshell has **n = 6**
(c) **-5, -4, -3, -2, -1, 0, 1, 2, 3, 4 and 5**

7.47 In order of increasing energy they are: s < p < d < f

7.48 Bohr's theory could work for hydrogen since hydrogen's electron can populate only one orbital at a given time. His theory failed with an element more complex than hydrogen because it did not account for the many possibilities that exist other than changes in n value for the excited states.

7.49 The spin of the electron causes the electron to act as a tiny electromagnet.

7.50 **$+\frac{1}{2}$ and $-\frac{1}{2}$**

7.51 The Pauli exclusion principle states that no two electrons in any one atom can have all four quantum numbers the same. This limits the number of electrons in any given orbital to two -- one with a positive spin, the other with a negative spin.

7.52

electron no.	n	ℓ	m_ℓ	m_s
1	2	0	0	+1/2
2	2	0	0	-1/2
3	2	1	-1	+1/2
4	2	1	-1	-1/2
5	2	1	0	+1/2
6	2	1	0	-1/2
7	2	1	+1	+1/2
8	2	1	+1	-1/2

7.53 **18**

7.54 Two electrons are said to be paired when they are in the same orbital and their spins are in opposite directions.

7.55 Unpaired electrons give atoms, molecules or ions that are paramagnetic. Paramagnetic species are weakly attracted to a magnetic field. Atoms, molecules or ions that have no unpaired electrons are slightly repelled by a magnetic field and are said to be diamagnetic.

7.56 Hund's rule states that: electrons entering a subshell containing more than one orbital will be spread out over the available equal-energy orbitals with their spins in the same direction.

7.57 Predicted electron configurations based on position in periodic table.

P; $1s^2\ 2s^2\ 2p^6\ 3s^2\ 3p^3$
Ni; $1s^2\ 2s^2\ 2p^6\ 3s^2\ 3p^6\ 3d^8\ 4s^2$
As; $1s^2\ 2s^2\ 2p^6\ 3s^2\ 3p^6\ 3d^{10}\ 4s^2\ 4p^3$
Ba; $1s^2\ 2s^2\ 2p^6\ 3s^2\ 3p^6\ 3d^{10}\ 4s^2\ 4p^6\ 4d^{10}\ 5s^2\ 5p^6\ 6s^2$
Rh; $1s^2\ 2s^2\ 2p^6\ 3s^2\ 3p^6\ 3d^{10}\ 4s^2\ 4p^6\ 4d^7\ 5s^2$ expected ($4d^8\ 5s^1$ actual)
Ho; $1s^2\ 2s^2\ 2p^6\ 3s^2\ 3p^6\ 3d^{10}\ 4s^2\ 4p^6\ 4d^{10}\ 4f^{11}\ 5s^2\ 5p^6\ 5d^0\ 6s^2$
Ge; $1s^2\ 2s^2\ 2p^6\ 3s^2\ 3p^6\ 3d^{10}\ 4s^2\ 4p^2$

7.58 Rb; $1s^2\ 2s^2\ 2p^6\ 3s^2\ 3p^6\ 3d^{10}\ 4s^2\ 4p^6\ 5s^1$
Sn; $1s^2\ 2s^2\ 2p^6\ 3s^2\ 3p^6\ 3d^{10}\ 4s^2\ 4p^6\ 4d^{10}\ 5s^2\ 5p^2$
Br; $1s^2\ 2s^2\ 2p^6\ 3s^2\ 3p^6\ 3d^{10}\ 4s^2\ 4p^5$
Cr; $1s^2\ 2s^2\ 2p^6\ 3s^2\ 3p^6\ 3d^5\ 4s^1$
Cu; $1s^2\ 2s^2\ 2p^6\ 3s^2\ 3p^6\ 3d^{10}\ 4s^1$

7.59 K $4s^1$; Al $3s^2 3p^1$; F $2s^2 2p^5$; S $3s^2 3p^4$; Tl $6s^2 6p^1$; Bi $6s^2 6p^3$

7.60 Si $3s^23p^2$; Se $4s^24p^4$; Sr $5s^2$; Cl $3s^23p^5$;
O $2s^22p^4$; S $3s^23p^4$; As $4s^24p^3$; Ga $4s^24p^1$

7.61 Since there are no d or f orbitals in periods 1 and 2 and the 3d subshell does not begin to fill until the 3p and 4s subshells have been filled, there are no orbitals to fill between the 3s orbital of Mg and the 3p orbital of Al.

7.62 (a) 15 ($2p^6\ 3p^6\ 4p^3$)
(b) 8 ($2p^6\ 3p^2$)
(c) 18 ($2p^6\ 3p^6\ 4p^6$)

7.63 (a) P $\underset{1s}{\underline{\uparrow\downarrow}}\ \underset{2s}{\underline{\uparrow\downarrow}}\ \underset{2p}{\underline{\uparrow\downarrow}\,\underline{\uparrow\downarrow}\,\underline{\uparrow\downarrow}}\ \underset{3s}{\underline{\uparrow\downarrow}}\ \underset{3p}{\underline{\uparrow}\,\underline{\uparrow}\,\underline{\uparrow}}$

(b) Ca $\underset{1s}{\underline{\uparrow\downarrow}}\ \underset{2s}{\underline{\uparrow\downarrow}}\ \underset{2p}{\underline{\uparrow\downarrow}\,\underline{\uparrow\downarrow}\,\underline{\uparrow\downarrow}}\ \underset{3s}{\underline{\uparrow\downarrow}}\ \underset{3p}{\underline{\uparrow\downarrow}\,\underline{\uparrow\downarrow}\,\underline{\uparrow\downarrow}}\ \underset{4s}{\underline{\uparrow\downarrow}}$

7.64 (a) Sn $\underset{5s}{\underline{\uparrow\downarrow}}\ \underset{5p}{\underline{\uparrow}\,\underline{\uparrow}\,\underline{\quad}}$ (b) Br $\underset{4s}{\underline{\uparrow\downarrow}}\ \underset{4p}{\underline{\uparrow\downarrow}\,\underline{\uparrow\downarrow}\,\underline{\uparrow}}$

(c) Ba $\underset{6s}{\underline{\uparrow\downarrow}}$

7.65 Cd, Sr and Kr are diamagnetic since they have no unpaired electrons.

7.66 Sc [Ar] $\underset{3d}{\underline{\uparrow}\,\underline{\quad}\,\underline{\quad}\,\underline{\quad}\,\underline{\quad}}\ \underset{4s}{\underline{\uparrow\downarrow}}$ 1 unpaired electron, paramagnetic

Ti [Ar] $\underset{3d}{\underline{\uparrow}\,\underline{\uparrow}\,\underline{\quad}\,\underline{\quad}\,\underline{\quad}}\ \underset{4s}{\underline{\uparrow\downarrow}}$ 2 unpaired electrons, paramagnetic

V [Ar] $\underset{3d}{\underline{\uparrow}\,\underline{\uparrow}\,\underline{\uparrow}\,\underline{\quad}\,\underline{\quad}}\ \underset{4s}{\underline{\uparrow\downarrow}}$ 3 unpaired electrons, paramagnetic

Cr [Ar] $\underset{3d}{\underline{\uparrow}\,\underline{\uparrow}\,\underline{\uparrow}\,\underline{\uparrow}\,\underline{\uparrow}}\ \underset{4s}{\underline{\uparrow}}$ 6 unpaired electrons, paramagnetic

Mn [Ar] $\underset{3d}{\underline{\uparrow}\,\underline{\uparrow}\,\underline{\uparrow}\,\underline{\uparrow}\,\underline{\uparrow}}\ \underset{4s}{\underline{\uparrow\downarrow}}$ 5 unpaired electrons, paramagnetic

Fe [Ar] $\underset{3d}{\underline{\uparrow\downarrow}\,\underline{\uparrow}\,\underline{\uparrow}\,\underline{\uparrow}\,\underline{\uparrow}}\ \underset{4s}{\underline{\uparrow\downarrow}}$ 4 unpaired electrons, paramagnetic

(continued)

7.66 (continued)

Co [Ar] $\frac{\uparrow\downarrow}{}\frac{\uparrow\downarrow}{}\frac{\uparrow}{3d}\frac{\uparrow}{}\frac{\uparrow}{}$ $\frac{\uparrow\downarrow}{4s}$ 3 unpaired electrons, paramagnetic

Ni [Ar] $\frac{\uparrow\downarrow}{}\frac{\uparrow\downarrow}{}\frac{\uparrow\downarrow}{3d}\frac{\uparrow}{}\frac{\uparrow}{}$ $\frac{\uparrow\downarrow}{4s}$ 2 unpaired electrons, paramagnetic

Cu [Ar] $\frac{\uparrow\downarrow}{}\frac{\uparrow\downarrow}{}\frac{\uparrow\downarrow}{3d}\frac{\uparrow\downarrow}{}\frac{\uparrow\downarrow}{}$ $\frac{\uparrow}{4s}$ 1 unpaired electron, paramagnetic

Zn [Ar] $\frac{\uparrow\downarrow}{}\frac{\uparrow\downarrow}{}\frac{\uparrow\downarrow}{3d}\frac{\uparrow\downarrow}{}\frac{\uparrow\downarrow}{}$ $\frac{\uparrow\downarrow}{4s}$ no unpaired electrons, diamagnetic

7.67 Yes, the value of ℓ does reflect the shape of an orbital. When $\ell = 0$, we are describing a spherical s orbital. When $\ell = 1$, the double lobed p orbital, etc. Greater values of n represent larger orbitals.

7.68 See Figure 7.17.

7.69 The size increases from 1s to 2s, etc., and more nodes occur. Their overall shape, however, is spherical.

7.70 The s orbital is spherical while the p orbital is dumbbell-shaped. (See Figures 7.13 and 7.15)

7.71 See Figure 7.18

7.72 Three electrons, which have the same charge, will tend to stay as far as possible from each other. This they can do by occupying separate p orbitals.

7.73 An atom or ion has no fixed outer limits. Atomic and ionic sizes are usually given in angstroms, nanometers, or picometers.

7.74 (a) 1 Å = 10^{-10} m; therefore, 4.06 Å x (10^{-10} m/1 Å) = **4.06 x 10^{-10} m**
(b) 4.06 Å = 4.06 x 10^{-10} m = 4.06 x 10^{-10} m x (10^9 nm/m) = **0.406 nm**
(c) 4.06 Å = 4.06 x 10^{-10} m = 4.06 x 10^{-10} m x (10^{12} pm/m)
= **4.06 x 10^2 pm = 406 pm**

7.75 Effective nuclear charge refers to the residual net charge felt by the outer valence electrons.

7.76 (a) Mg, 2+ (b) Al, 3+ (c) Si, 4+ (d) S, 6+ (e) Cl, 7+

7.77 (a) Be, 2+ (b) Mg, 2+ (c) Ca, 2+ (d) Sr, 2+ (e) Ba, 2+

7.78 As we move from left to right across a period, the atomic size decreases because the effective nuclear charge increases. Therefore, the electrons are attracted more by the increased effective nuclear charge and are held closer to the nucleus. Thus, the atoms become smaller as we move across a period of the periodic table.
Going down within a group on the periodic table the atoms become larger because the effective nuclear charge is about constant but the outermost orbitals are larger as the value of n increases.

7.79 (a) Ca^{2+} [Ar] (b) S^{2-} [Ar] (c) Cl^- [Ar] (d) K^+ [Ar]
These are all $1s^2\ 2s^2\ 2p^6\ 3s^2\ 3p^6$

7.80 (a) Cr^{3+} [Ar] $3d^3$ (b) Mn^{2+} [Ar] $3d^5$ (c) Mn^{3+} [Ar] $3d^4$
(d) Co^{2+} [Ar] $3d^7$ (e) Co^{3+} [Ar] $3d^6$ (f) Ni^{2+} [Ar] $3d^8$
(orbital diagrams should also be given)

7.81 Size trends within the periodic table would predict that Sn is largest.

7.82 (a) Se (b) C (c) Fe^{2+} (d) O^- (e) S^{2-}

7.83 The ions N^{3-}, O^{2-}, and F^- are isoelectronic (i.e., they have identical electron configurations). The effective nuclear charge is increasing N < O < F. This leads to greater attraction for the outer-shell electrons which are pulled closer to the nucleus decreasing the size of the ion. The electron-electron repulsions are the same so the variation of effective nuclear charge is primarily responsible for their size variation.

7.84 The lanthanide contraction is the gradual decrease in the sizes of the lanthanide elements that occurs upon the filling of the inner 4f subshell in the lanthanides. As a result, Hf is nearly the same size as Zr, but with a much larger nuclear charge pulling on the outermost electrons. Outer electrons are held more tightly.

7.85 When there are fewer electrons, the interelectron repulsions are less and the outer shell can contract in size under the influence of the nuclear charge.

7.86 Ionization energy is the amount of energy needed to remove an electron from an isolated gaseous atom or ion in its ground state.
Electron affinity is the amount of energy released or absorbed when an electron is added to a gaseous atom or ion in its ground state.

7.87 As we move from left to right across a period, the increased effective nuclear charge causes the shell to shrink in size and also makes it more difficult to remove an electron. Therefore, ionization energy increases from left to right across the periodic table with irregularities due to filled and half-filled subshells.

7.88 (a) Be (b) Be (c) N (d) N (e) Ne (f) S^+ (g) Na^+

7.89 (a) Cl (b) S (c) P (predicted from general trends; actually, this is an exception; the EA for As is more exothermic.) (d) S

7.90 The second electron that is added must be forced into an already negative ion. This requires work and, therefore, the second electron affinity for an element is an endothermic process.

7.91 $$\frac{1}{\lambda} = 109{,}678 \text{ cm}^{-1}\left(\frac{1}{1^2} - \frac{1}{\infty^2}\right) = 109{,}678 \text{ cm}^{-1}$$

$$E = h\nu = \frac{hc}{\lambda} = \left(6.63 \times 10^{-34} \text{ J s}\right)\left(3.00 \times 10^{8} \text{ m s}^{-1}\right)\left(109{,}678 \text{ cm}^{-1}\right)\left(\frac{100 \text{ cm}}{1 \text{ m}}\right)$$

$$= 2.18 \times 10^{-18} \text{ J}$$

$$\left(2.18 \times 10^{-18} \frac{\text{J}}{\text{atom}}\right)\left(6.02 \times 10^{23} \frac{\text{atom}}{\text{mol}}\right)\left(\frac{1 \text{ kJ}}{1000 \text{ J}}\right) = 1310 \text{ kJ/mol}$$

vs 1312 kJ/mol in Table 7.4

7.92 $$\Delta E = 2.18 \text{ X } 10^{-18} \text{ J}\left[\frac{1}{1^2} - \frac{1}{4^2}\right] = 2.18 \text{ X } 10^{-18} \text{ J}\left[\frac{1}{1} - \frac{1}{16}\right]$$

$$\Delta E = \mathbf{2.04 \text{ X } 10^{-18} \text{ J}}$$

$$\nu = \frac{E}{h} = \frac{2.04 \text{ X } 10^{-18} \text{ J}}{6.63 \times 10^{-34} \text{ J s}} = \mathbf{3.08 \times 10^{15} \text{ Hz}}$$

$$\lambda = \frac{c}{\nu} = \frac{3.00 \text{ X } 10^{8} \text{ m s}^{-1}}{3.08 \times 10^{15} \text{ s}^{-1}} \text{ x } \frac{1 \text{ nm}}{10^{-9} \text{ m}} = \mathbf{97.4 \text{ nm}}$$

7.93 No. There is no way that all electrons can be paired if there is an odd number of electrons. Therefore, if there is an odd number of electrons, the atom cannot be diamagnetic.

7.94 Gd^{3+} would have an electron configuration of [Xe] $4f^7$ and would have **7 unpaired electrons.**

7.95 Elements at the center of period 6 are so dense due to the **lanthanide contraction.**

7.96 **[Rn] $5f^{14}$ $6d^4$ $7s^2$**

7.97 Z = 114 would belong to group **IVA**. Its configuration would be:

[Rn] $5f^{14}$ $6d^{10}$ $7s^2$ $7p^2$

7.98 **[Ar] $4s^1$**

7.99 Each graph will show a large jump in ionization energy as one goes beyond the removal of valence electrons. Until all valence electrons are removed, the increase in ionization energy is a gradually increasing phenomenon.

7.100 Ionization energy of H = 1,312 kJ /mol

heat cap. x mass x change in temp. = energy

[4.184 J/(1g(H_2O) x 1°C)] x [mass (H_2O)] x (25°C)
= (1,312 kJ/mol H) x (1 mol H) x (10^3 J/kJ)

mass(H_2O) = **1.3 x 10^4 g H_2O**

8 CHEMICAL BONDING: GENERAL CONCEPTS

8.1 An <u>ionic bond</u> results from the attraction between oppositely charged ions.

8.2 (a) Ba^{2+} $1s^2 2s^2 2p^6 3s^2 3p^6 3d^{10} 4s^2 4p^6 4d^{10} 5s^2 5p^6$ or [Xe]
(b) Se^{2-} $1s^2 2s^2 2p^6 3s^2 3p^6 3d^{10} 4s^2 4p^6$ or [Kr]
(c) Al^{3+} $1s^2 2s^2 2p^6$ or [Ne]
(d) Na^+ $1s^2 2s^2 2p^6$ or [Ne]
(e) Br^- $1s^2 2s^2 2p^6 3s^2 3p^6 3d^{10} 4s^2 4p^6$ or [Kr]

8.3 (a) $Ba^{2+} \Leftrightarrow$ [Xe] (b) $Se^{2-} \Leftrightarrow$ [Kr] (c) $Al^{3+} \Leftrightarrow$ [Ne]
(d) $Na^+ \Leftrightarrow$ [Ne] (e) $Br^- \Leftrightarrow$ [Kr]

8.4 In (b) Na_2O and (d) Mg_3N_2 the cation and anion have exactly the same electron configuration. ([Ne])

8.5 $AlCl_4$ does not form because aluminum does not form the Al^{4+} ion under normal circumstances. Likewise, Na_3O does not form because oxygen does not form a three minus ion.

8.6 Lattice energy is the potential energy released as isolated ions are allowed to approach each other and to form crystals. For ionic compounds the lattice energy more than compensates for the energy required for the creation of ions.

8.7 (a)

$Na(g) \rightarrow Na^+(g) + e^-$	495.8 kJ
$Cl(g) + e^- \rightarrow Cl^-(g)$	-348 kJ
$Na(g) + Cl(g) \rightarrow Na^+(g) + Cl^-(g)$	**148 kJ**

(continued)

8.7 (continued)

(b) $Na(g) \rightarrow Na^{+}(g) + e^{-}$ 496 kJ

$Na^{+}(g) \rightarrow Na^{2+}(g) + e^{-}$ 4565 kJ

$2Cl(g) + 2e^{-} \rightarrow 2Cl^{-}(g)$ 2(-348 kJ)

$Na(g) + 2Cl(g) \rightarrow Na^{2+}(g) + 2Cl^{-}(g)$ **4,365 kJ**

The lattice energy of $NaCl_2$ would have to be more than 29.5 times the lattice energy of NaCl before $NaCl_2$ would become more stable than NaCl.

8.8 The tendency for ions of many of the representative elements to achieve an outer-shell configuration with a total of eight electrons (the stable electronic structure of noble gases) forms the basis of the octet rule. The octet rule states that when metals and nonmetals of the A-groups react, they tend to gain or lose electrons until there are eight electrons in the outer shell.

8.9 In Li^{+} $[:\ddot{\underset{..}{F}}:]^{-}$ the **fluoride obeys** the octet rule; **lithium does not.**

8.10 Many transition metals form ions with neither a noble gas configuration nor a pseudonoble gas configuration. Depending on the particular circumstances, such a metal atom loses electrons until the extra energy needed to take off one more electron is greater than can be made up by the lattice energy. Therefore, such metal atoms can form different ions depending on the amount of lattice energy released from the formation of crystals with different anions.

8.11 KF(s) is more stable than K(s) and $F_2(g)$ because of the large lattice energy value.

8.12 Pb^{2+} [Xe] $4f^{14}\,5d^{10}\,6s^2$
Pb^{4+} [Xe] $4f^{14}\,5d^{10}$
Mn^{2+} [Ar] $3d^5$
Mn^{3+} [Ar] $3d^4$
Sb^{3+} [Kr] $4d^{10}\,5s^2$
Sc^{3+} [Ar]
Ti^{2+} [Ar] $3d^2$

8.13 (a) Zn^{2+} [Ar] $3d^{10}$ (b) Sn^{2+} [Kr] $4d^{10}\,5s^2$ (c) Bi^{3+} [Xe] $4f^{14}\,5d^{10}\,6s^2$
(d) Cr^{2+} [Ar] $3d^4$ (e) Fe^{3+} [Ar] $3d^5$ (f) Ag^{+} [Kr] $4d^{10}$

8.14 $ns^2\,np^6\,nd^{10}$, Zn^{2+}, Cd^{2+}, Hg^{2+}, and Ag^{+}

8.15 (a) both are [Ar]
(d) both are [Kr]
(e) both are [Ne]

8.16 Looking at outer-shell configurations

K	$4s^1 \rightarrow 4s^0$	O	$2s^2\,2p^4 \rightarrow 2s^2\,2p^6$
Mg	$3s^2 \rightarrow 3s^0$	N	$2s^2\,2p^3 \rightarrow 2s^2\,2p^6$
Na	$3s^1 \rightarrow 3s^0$	S	$3s^2\,3p^4 \rightarrow 3s^2\,3p^6$
Ba	$6s^2 \rightarrow 6s^0$	Br	$4s^2\,4p^5 \rightarrow 4s^2\,4p^6$

8.17 For many of them, their outer shell has two electrons. Loss of the two electrons produces a pseudonoble gas configuration.

8.18 Lewis symbols help us to keep tabs on the outer shell electrons (valence shell electrons) which are the most important in bond formation.

8.19 (a) :Se· (b) :Br· (c) ·Al· (d) ·Ba· (e) ·Ge· (f) ·P·

8.20 Elements in a given group all have the same number of valence electrons. Therefore, elements in a given group should all have generalized Lewis symbols that are the same. In a group on the periodic table, the atomic symbols change, but the number of dots or valence electrons does not.

8.21 Mg: + Mg: + ·C· $\rightarrow$ $2Mg^{2+}$, $[:C:]^{4-}$ or Mg_2C

8.22 (a) $Ba^{2+}[:O:]^{2-}$ (b) $2Na^+, [:O:]^{2-}$ (c) $K^+[:F:]^-$

(d) $Ca^{2+}[:S:]^{2-}$ (e) $3Li^+, [:N:]^{3-}$

8.23 (a) **Two** (b) ·C· **Four** (c) Lewis symbols are written in a way that reflects the number of unpaired electrons that are typically involved in bonding and not necessarily the number of unpaired electrons on the atom in its ground state.

8.24 The ionic bond results from attraction between oppositely charged ions that can be traced to the transfer of electrons between neutral atoms or groups. A covalent bond results from the sharing of a pair of electrons between atoms.

8.25 Neither Cl nor F can form a stable cation. For both Cl and F the first ionization energy is much greater than the electron affinity. Both strive to achieve an octet by gaining an electron, not by losing electrons. Therefore, no energetically favored combination of cation and anion can be formed from Cl and F.

8.26 (a) **4** (b) **2** (c) **1** (d) **4** (e) **3**

8.27 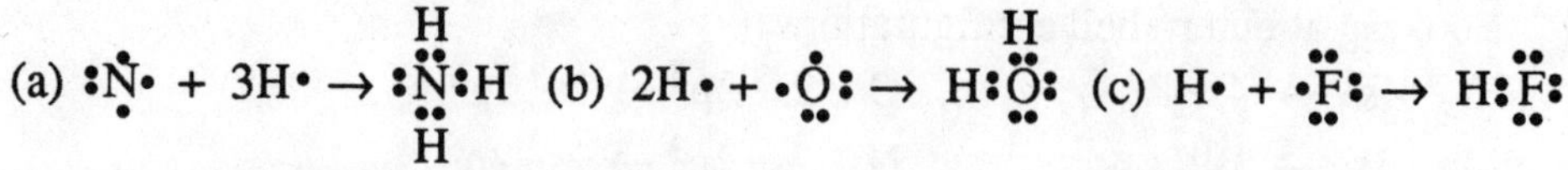

8.28 A <u>double bond</u> is the sharing of two pairs of electrons between two atoms. A <u>triple bond</u> is the sharing of three pairs of electrons.

8.29 There are no unpaired electrons in any of these compounds. NH_3, H_2O, and HCl each have unshared but not unpaired electrons. (See Question 8.27).

8.30 <u>Bond energy</u> is the depth of the energy minimum obtained as two nuclei (atoms) approach each other (See Figure 8.2). It is also the energy needed to separate two covalently bonded atoms. <u>Bond length</u> is the distance between the nuclei when the energy is a minimum.

8.31 (a) 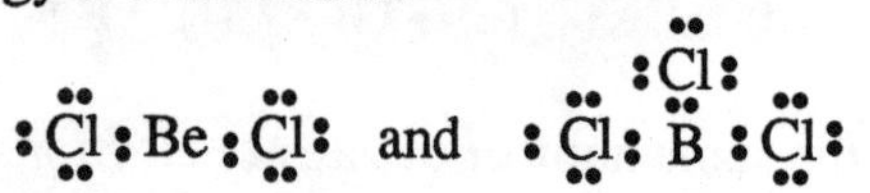 (b) **4** in $BeCl_2$ and **6** in BCl_3

8.32 The second shell can contain a maximum of 8 electrons because it contains only an s and three p orbitals to work with; it does not contain unused d orbitals like all following shells.

8.33 (a) **4** (b) **4** (c) **9** (d) **9**

8.34 (a) **4** (b) Yes (c) Because in some compounds the d orbitals of phosphorus are utilized. Nitrogen does not possess d orbitals in its valence shell.

8.35 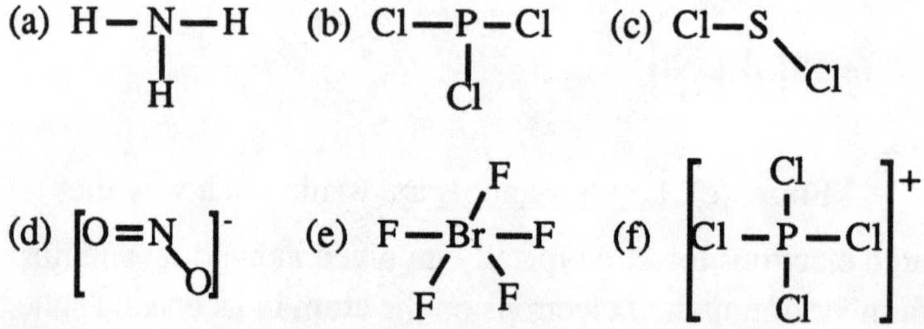

8.36

Elements	Number of valence electrons
(a) As	5
(b) I	7
(c) Si	4
(d) Sn	4
(e) S	6

8.37 (a) **8** (b) **26** (c) **20** (d) **18** (e) **42** (f) **32**

8.38 Hydrogen generally forms only a single bond. The central atom must be able to bond to more than one other atom which hydrogen is not capable of doing.

8.39 The maximum number of covalent bonds formed by a hydrogen atom is **one.**

8.40 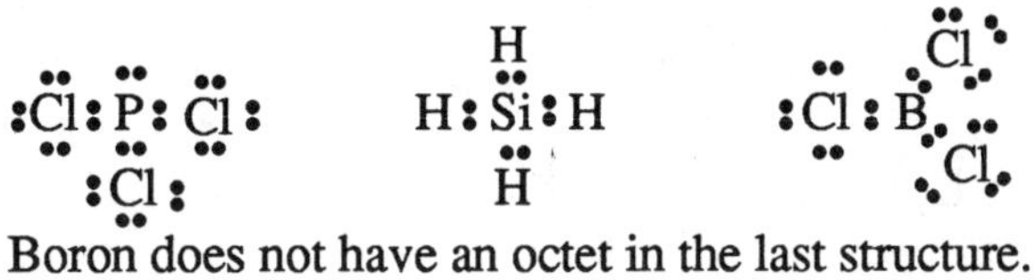

Boron does not have an octet in the last structure. To form an octet one of the chlorines would need to provide two more electrons for covalent bonding. Then that chlorine would have a double bond which chlorine does not normally form.

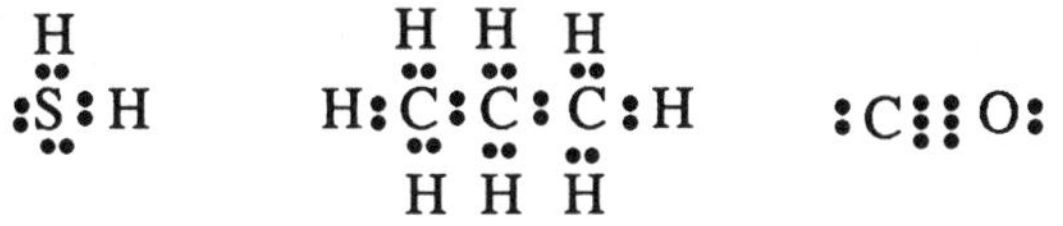

8.41 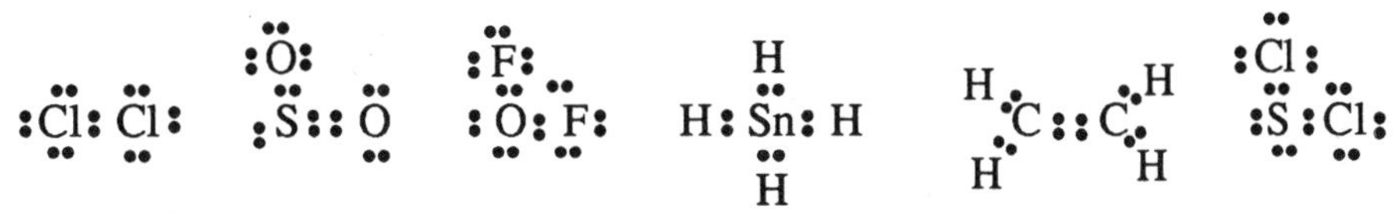

8.42 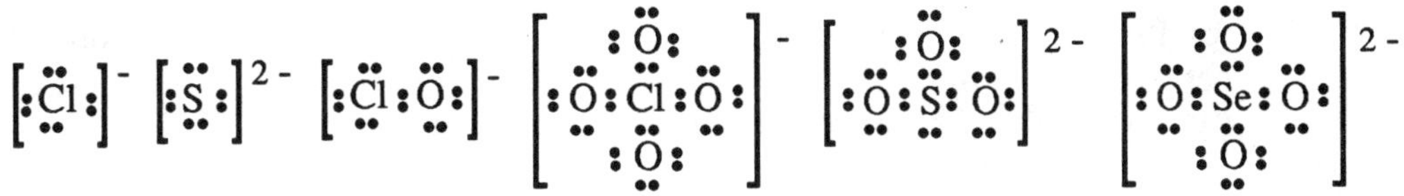

8.43 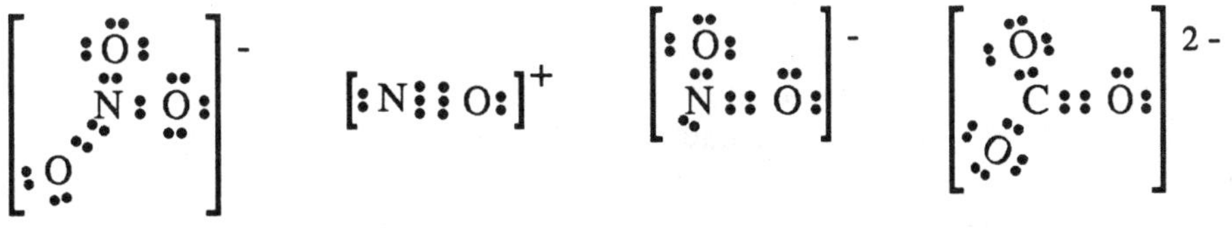

8.44 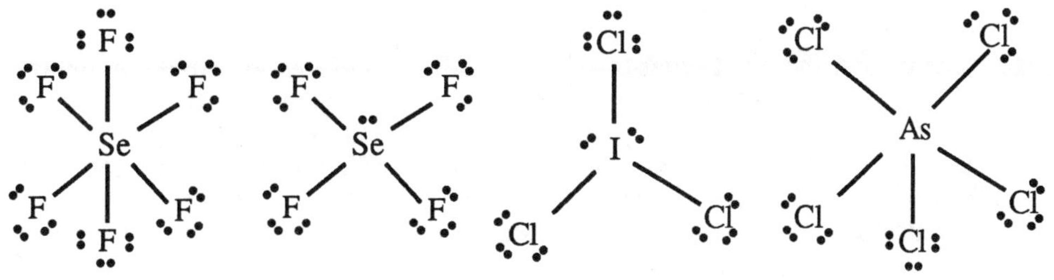

(continued)

8.44 (continued)

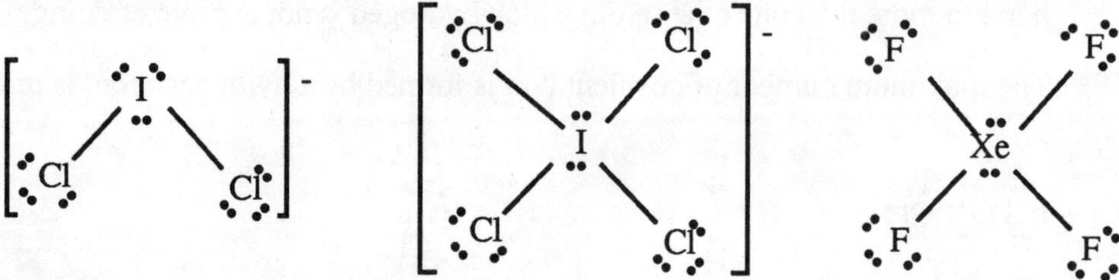

8.45 ClF_3, SF_4, IF_7, NO_2, BCl_3

8.46 Bond order is the number of covalent bonds that exist between a pair of atoms.

8.47 (a) Each C-Cl bond has a bond order of **1**.

(b) In HCN the H-C bond order is **1** and the C$\equiv$N bond order is **3.**

(c) In CO_2 each C=O has a bond order of **2**.

(d) In NO^+ the bond order is **3**.

(e) In CH_3NCO the bond order between each hydrogen and the carbon is **1**, between one of the carbons and nitrogen it is **1**, between nitrogen and the other carbon it is **2** and between carbon and the oxygen it is **2**.

8.48 Bond energy **increases as bond order increases** because the additional electron density between the nuclei increases the attraction of the nuclei for that electron density.

8.49 Bond length **decreases as bond order increases** because the additional electron density between the nuclei causes the nuclei to be pulled together.

8.50 Bending motions are those that change the bond angles while vibrational motions change bond lengths. Strength of bonds will influence the frequency with which a molecule will vibrate.

8.51 The greater the bond order the shorter and stronger the bond and the greater the vibrational frequency. Strong bonds will vibrate faster with shorter amplitude just as will a strong spring compared to a weak spring. Infrared absorption spectra are used to measure bond vibrational frequencies. When infrared radiation interacts with a substance, the infrared frequencies that are the same as the vibrational frequencies of the bonds in the substance are absorbed.

8.52 154pm, 146pm, 140pm and 137pm. Bond energy also increases in the same order.

8.53 A resonance hybrid is the true structure of a compound. It is a composite of the contributing structures that one can draw. We use resonance because it is impossible to draw a single electron-dot formula that obeys the octet rule and is consistent with experimental facts at the same time.

8.54

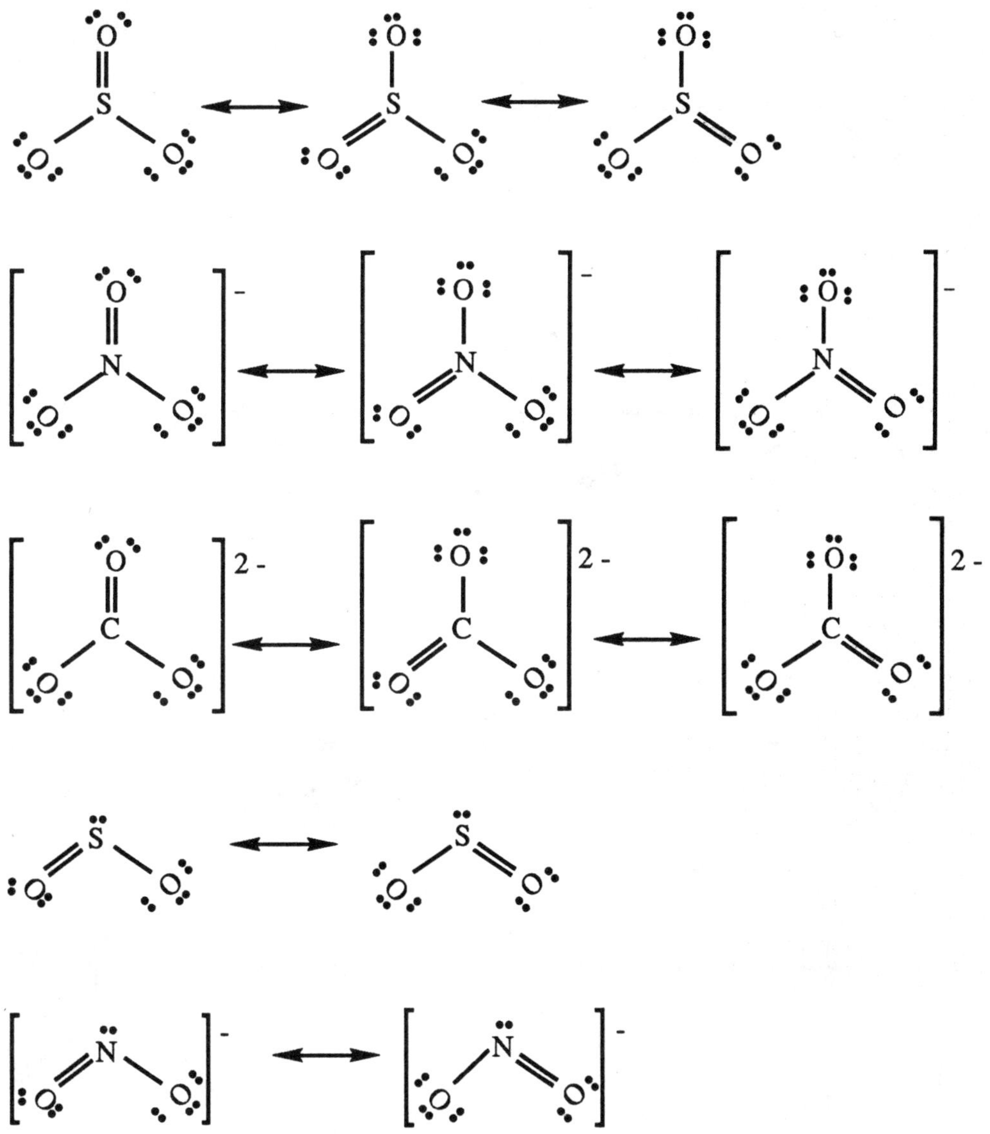

8.55 H–O–N(=O)–O ⟷ H–O–N(–O)=O

O=Se–O ⟷ O–Se=O

O=Se(–O)–O ⟷ O–Se(=O)–O ⟷ O–Se(–O)=O

O=N(–O)–N(–O)=O ⟷ O–N(=O)–N(–O)=O ⟷ O=N(–O)–N(=O)–O ⟷

O–N(=O)–N(=O)–O

8.56 [Lewis structures: resonance forms of $C_2O_4^{2-}$, CH_3COO^-, and N_3^-]

$$[\,\ddot{N}=N=\ddot{N}\,]^- \longleftrightarrow [\,:N\equiv N-\ddot{N}:\,]^- \longleftrightarrow [\,:\ddot{N}-N\equiv N:\,]^-$$

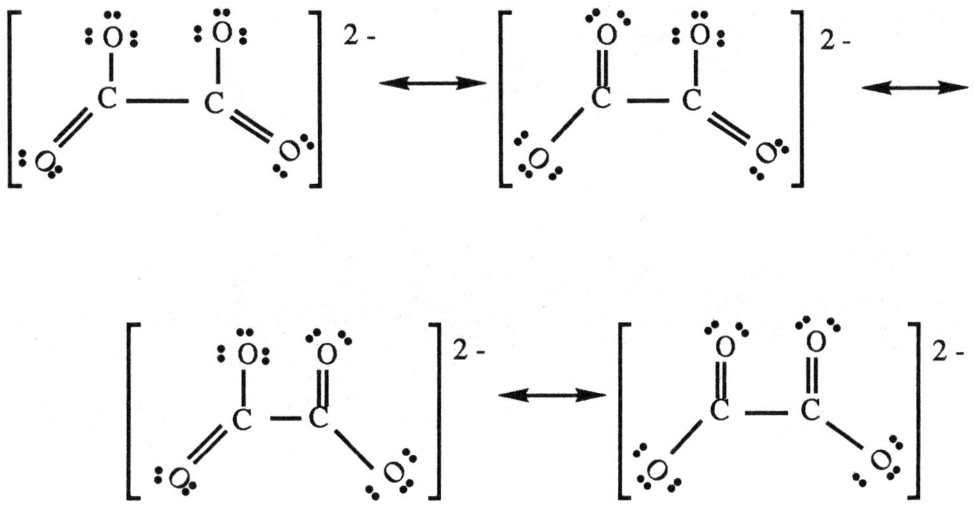

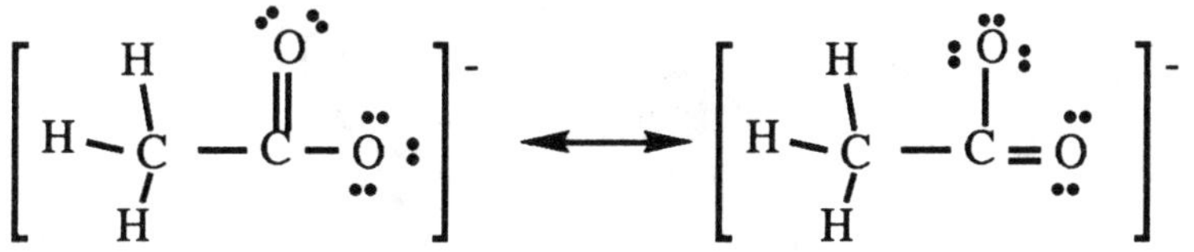

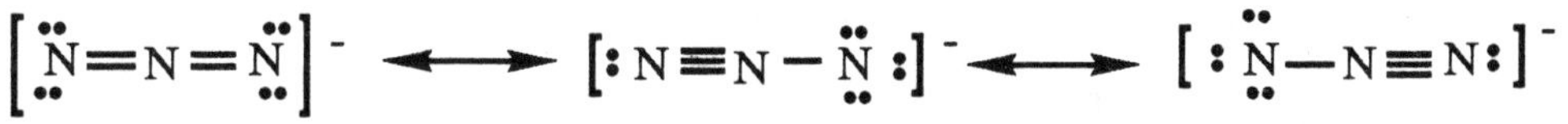

8.57 The N-O bond order in NO_2^- is an average of 1.5 and in NO_3^- it is an average of 1.3. Therefore, the **bond length** should be a little **shorter** and the **bond energy** a little **higher** in NO_2^-.

8.58

Molecule	Average Bond Order	Bond Length	Bond Energy	Vib.Frequency
CO	3			
CO_2	2	↓	↓	↓
CH_3COO^-	1.5	increases	decreases	decreases
CO_3^{2-}	1.3	↓	↓	↓
CH_3CH_2OH	1			

8.59

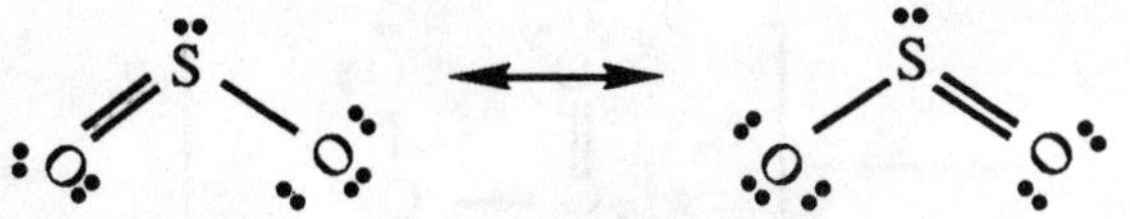

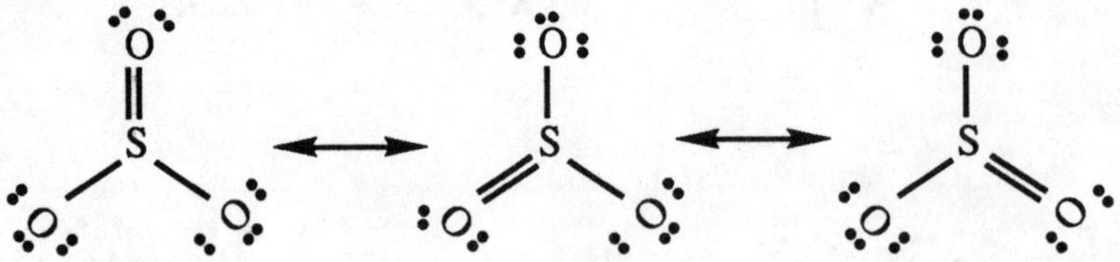

Molecule	Average Bond Order	Bond Length	Bond Energy	Vib.Frequency
SO_2	1.5	shorter	more	higher
SO_3	1.3	longer	less	lower

8.60 or

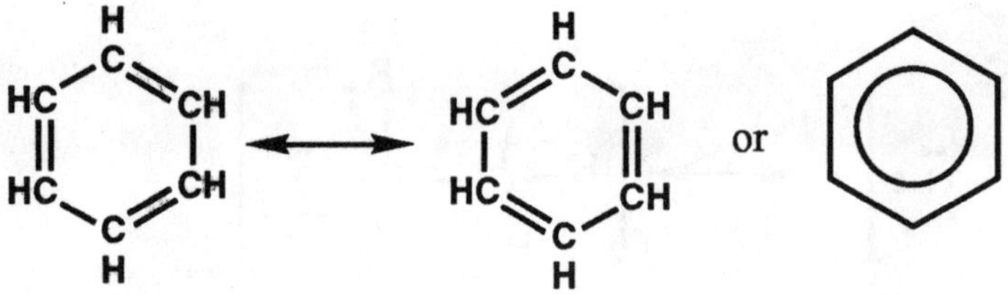

8.61 (See question 8.59)

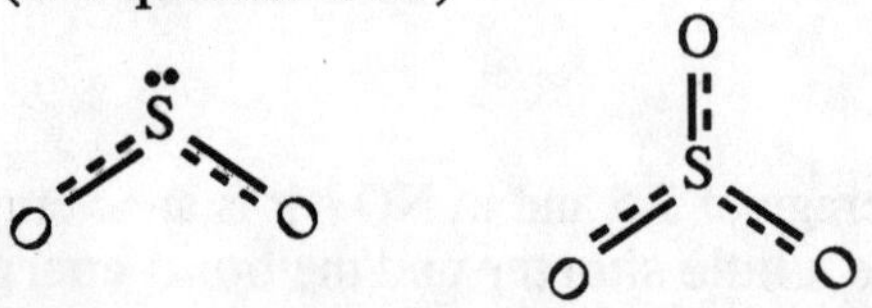

8.62 For a molecule that exists as a hybrid of two or more resonance structures the energy of the actual molecule is less than that of any of the individual resonance structures.

8.63 The lowering of the energy of the molecule compared to the energy of any one of its contributing resonance structures is called resonance energy.

8.64 Formal charge = # of e^-'s in valence-shell of isolated atom - # of bonds - # of unshared e^-'s.

8.65

```
                :O:
         ..      ||          2-
      [ :O ——— S = S: ]
         ..      |    ..
                :O:
                 ..
```

	Formal Charge
center S	6 - 6 = 0
top O	6 - 2 - 4 = 0
right S	6 - 2 - 4 = 0
left and bottom O's	2(6 - 1 -6 =-1)
Total	**- 2**

8.66 (Student answers to this question and the following question should include Lewis dot structures.)

(a) Cl = 7 - 3 -2 = **+2**
each O = 6 - 1 - 6 = **-1**

(b) Br = 7 - 2 - 4 = **+1**
each O = 6 - 1 - 6 = **-1**

(c) P = 5 - 4 = **+1**
each O = 6 - 1 - 6 = **-1**

(d) Se = 6 -3 - 2 = **+1**
each O = 6 - 1 - 6 = **-1**

(e) Cl = 7 - 4 = **+3**
each O = 6 - 1 - 6 = **-1**

8.67 (a) H = 1 - 1 = **0**
Cl = 7 - 4 = **+3**
O (between Cl and H) = 6 - 2 - 4 = **0**
O (other 3 oxygens) = 6 - 1 - 6 = **-1**

(b) H = 1 - 1 = **0**
Cl = 7 - 2- 4 = **+1**
O (between Cl and H) = 6 - 2 - 4 = **0**
O (other oxygen) = 6 - 1 - 6 = **-1**

(c) H = 1 - 1 = **0**
S = 6 - 3 - 2 = **+1**
O (2 between H's and the S) = 6 - 2 - 4 = **0**
O (not bonded to H) = 6 - 1 - 6 = **-1**

(d) H = 1 - 1 = **0**
P = 5 - 4 = **+1**
O (3 between H's and the P) = 6 - 2 - 4 = **0**
O (not bonded to H) = 6 - 1 - 6 = **-1**

8.68 In each of the ions in Question 8.66 a better Lewis structure can be obtained if the octet rule is not obeyed.

(a) 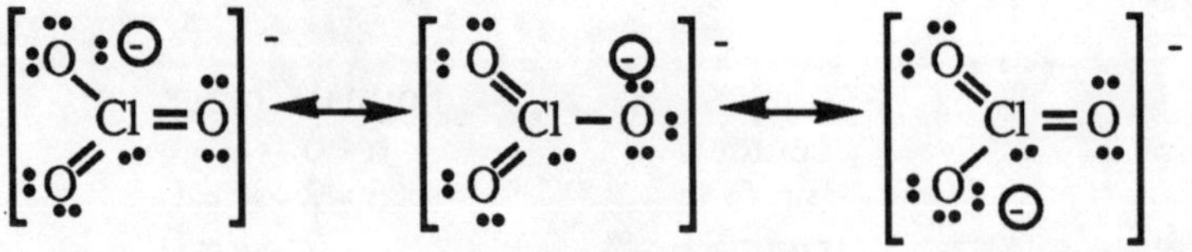 [Lewis resonance structures of ClO_3^-]

This has each atom, except the indicated oxygen, with a formal charge of zero. Remember: "Those (structures) with the smallest formal charges are the most stable and are preferred."

(b) 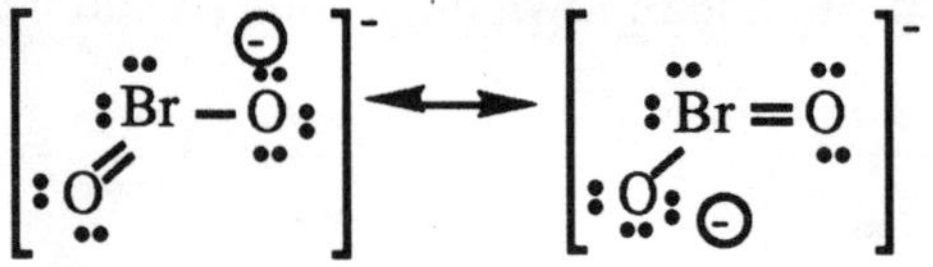 [Lewis resonance structures of BrO_2^-]

(c) 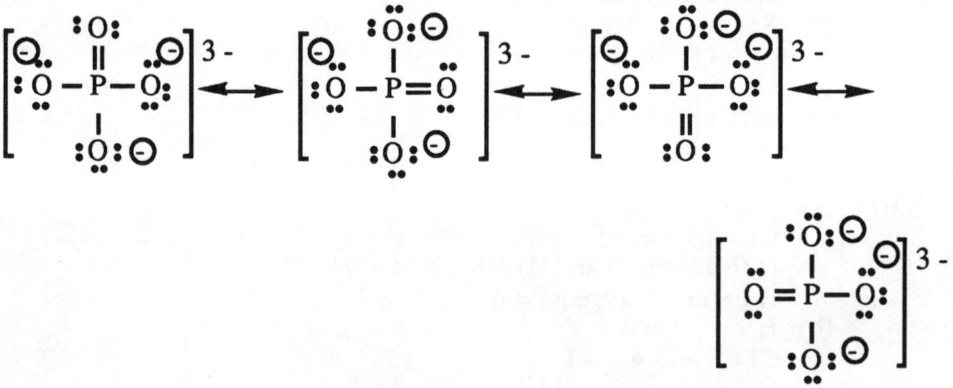 [Lewis resonance structures of PO_4^{3-}]

(d) 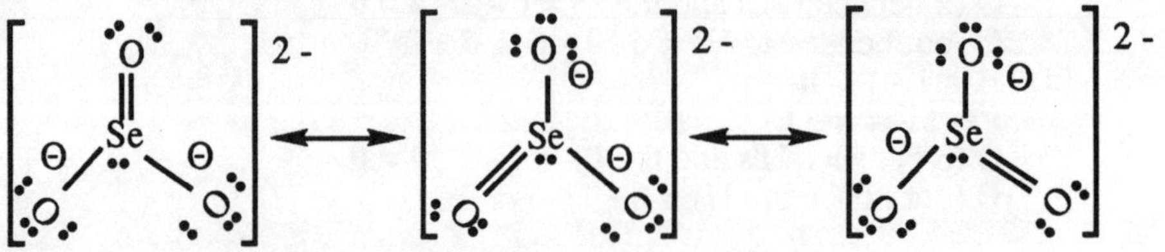 [Lewis resonance structures of SeO_3^{2-}]

(continued)

8.68 (continued)

(e)

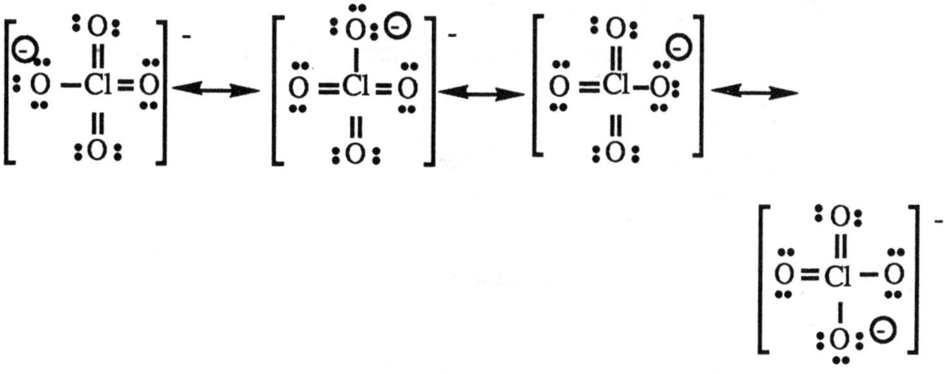

8.69 In each of the molecules in Question 8.67 a better Lewis structure can be drawn if the octet rule is not obeyed.

(a)

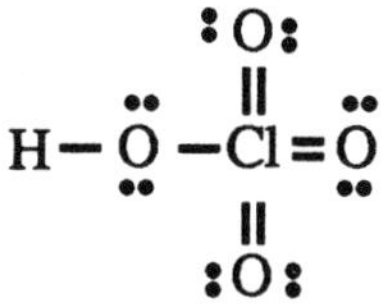

no formal charges other than zero
no resonance structures

(b)

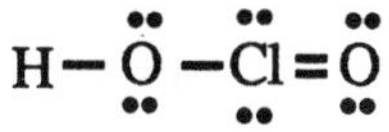

no formal charges other than zero
no resonance structures

(c)

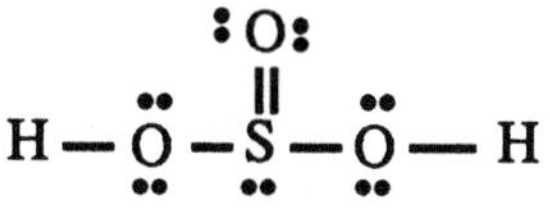

no formal charges other than zero
no resonance structures

(d)

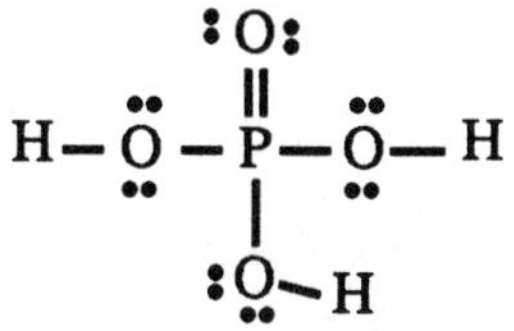

no formal charges other than zero
no resonance structures

8.70 Nitrogen and other elements in the second period cannot share more than eight electrons. The reason they cannot go beyond the octet is that they do not have d orbitals (vacant or partially filled) in their valence shell.

8.71 $\left[:N\equiv \overset{\oplus}{N}-\ddot{\underset{\cdot\cdot}{N}}:\right]^{-}_{(-2)} \longleftrightarrow \left[:\ddot{\underset{\ominus}{N}}=\underset{\oplus}{N}=\ddot{\underset{\ominus}{N}}:\right]^{-} \longleftrightarrow \left[:\ddot{\underset{\cdot\cdot}{N}}-\overset{\oplus}{N}\equiv N:\right]^{-}_{(-2)}$

The most stable structure would be the second structure.

8.72 $:\ddot{O}=C=\ddot{O}:$ has all atoms having formal charges of zero while $^{\oplus}:O\equiv C-\ddot{\underset{\cdot\cdot}{O}}:^{\ominus}$ does not. Therefore, the first structure is the one usually used to describe carbon dioxide.

8.73 The structure given in this question is rejected due to its having larger formal charges than that structure given in Question 8.70. (Also, since oxygen is one of the most electronegative elements, it does not assume a positive formal charge except when bonded to F. The structure in Question 8.73 does have a positive formal charge on the oxygen between the H and the N).

8.74 $H-\ddot{\underset{\cdot\cdot}{O}}-\overset{\oplus}{\ddot{\underset{\cdot\cdot}{F}}}-\overset{\ominus}{\ddot{\underset{\cdot\cdot}{O}}}:$ Since F is more electronegative than O, any compound, such as this one, that would force the F to have a positive formal charge would be predicted to be a high energy compound.

8.75 (a) Average bond order = 5/3 = 1.67

(b) Average bond order = 3/2 =1.5

(c) Average bond order = 5/4 = 1.25

(d) Average bond order = 4/3 = 1.33

(a) Average bond order = 7/4 = 1.75

8.76 See the structure in the answer to Question 8.69 (d); the average bond is 5/4.

8.77 SO_2Cl_2: O=S(=O) with Cl–S–Cl (Lewis structure) All formal charges = zero

8.78 A coordinate covalent bond is one where a pair of electrons from one atom is shared by the two atoms in a bond. It differs from normal covalent bonds in that electrons being shared between two atoms are both coming from one atom instead of one electron from each. It is really only a bookkeeping device; once formed, the coordinate covalent bond is the same as any other covalent bond.

8.79 $AlCl_3 + [:Cl:]^- \longrightarrow [Cl_3Al \leftarrow Cl]^-$

8.80 Since it has neither an unbonded pair of electrons to donate nor any way to accept a pair of electrons, the molecule CH_4 would not be expected to participate in the formation of a coordinate covalent bond.

8.81 $BF_3 + [:F:]^- \longrightarrow [F_3B \leftarrow F]^-$

8.82 Electronegativity is the attraction an atom has for electrons in a chemical bond. Electron affinity, which is an energy term referring to an isolated atom, is the energy released or absorbed when an electron is added to a neutral gaseous atom.

8.83 A polar molecule is one that has its positive and negative charges separated by a distance. The resulting molecule is said to be a dipole. The dipole moment is the product of the charge on either end of the dipole times the distance between the charges.

8.84 Fluorine, upper right-hand corner, is the element highest in electronegativity. The values generally decrease down a group and right to left across a period. Elements with low ionization energies generally have low electronegativities and those with high ionization energies also have high electronegativities.

8.85 (a) P—F (b) Al—Cl (c) Se—Cl

8.86 MgO, Al_2O_3 and CsF

8.87 NH_3, BCl_3, $MgCl_2$, BeI_2 and NaH

8.88 $F_2 < H_2Se < H_2S < OF_2 < SO_2 < ClF_3 < SF_2$

8.89 Rb

8.90 Being a nonsymmetrical molecule (bent) with polar bonds, it will be a polar molecule. Since the oxygen is the more electronegative element, the oxygen atoms will have a negative charge and the sulfur end or side will carry the positive charge.

8.91 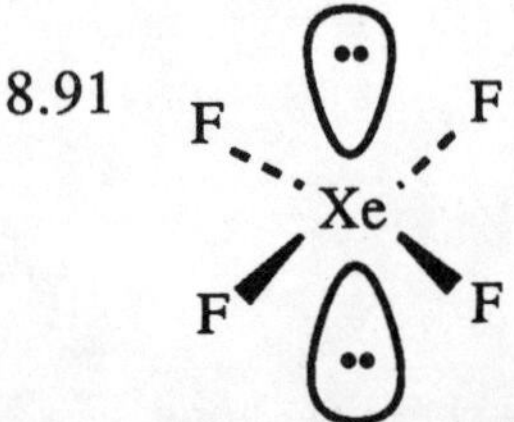

The bonds are polar but, since it is symmetrical, it is a nonpolar molecule.

8.92

Reaction	Energy
$K(s) \rightarrow K(g)$	+90.0 kJ
$1/2Cl_2(g) \rightarrow Cl(g)$	+119 kJ
$K(g) \rightarrow K^+(g) + e^-$	+419 kJ
$e^- + Cl(g) \rightarrow Cl^-(g)$	-348 kJ
$K^+(g) + Cl^- \rightarrow KCl(s)$	-704 kJ
$K(s) + 1/2Cl_2(g) \rightarrow KCl(s)$	**-424 kJ**

The lattice energy of KCl(s) and the electron affinity of Cl both contribute to the fact that the formation of KCl(s) is exothermic.

8.93

Reaction	Energy
1. $Ca(s) \rightarrow Ca(g)$	192 kJ
2. $Ca(g) \rightarrow Ca^+(s) + e^-$	589.5 kJ
3. $Ca^+(g) \rightarrow Ca^{2+}(g) + e^-$	1146 kJ
4. $Cl_2(g) \rightarrow 2Cl(g)$	238 kJ
5. $2Cl(g) + 2e^- \rightarrow 2Cl^-(g)$	2(-348) kJ
6. $Ca^{2+}(g) + 2Cl^-(g) \rightarrow CaCl_2(s)$	lattice energy
7. $Ca(s) + Cl_2(g) \rightarrow CaCl_2(s)$	-795 kJ

The sum of reactions 1-6 equals reaction 7. Therefore, 192 kJ + 589.5 kJ + 1146 kJ + 238 kJ + 2(-348) kJ + lattice energy must equal -795 kJ.
Lattice energy = **-2,264 kJ** per mol $CaCl_2$.

8.94

Reaction	Energy
1. $Na(s) \rightarrow Na(g)$	109 kJ
2. $1/2Br_2(\ell) \rightarrow 1/2Br_2(g)$	1/2(31) kJ
3. $Na(g) \rightarrow Na^+(g) + e^-$	495.8 kJ
4. $1/2Br(g) \rightarrow Br(g)$	1/2(192) kJ
5. $Br(g) + e^- \rightarrow Br^-(g)$	electron affinity
6. $Na^+(g) + Br^-(g) \rightarrow NaBr(s)$	-734.3 kJ
7. $Na(s) + 1/2Br_2(\ell) \rightarrow NaBr(s)$	-360 kJ

The sum of reactions 1-6 equals reaction 7. Therefore, 109 kJ +15.5 kJ + 495.8 kJ + 96 kJ + electron affinity + (-734.3) kJ must equal -360 kJ.
Electron affinity = **-342 kJ** per mol Br(g).

8.95 The absolute difference between a calculated bond energy and an experimental bond energy is proportional to the electronegativity difference between the bonded atoms.

Compounds	HF	HCl	HBr	HI
Difference between calc. and exp. bond energies (kJ/mol)	270	94	50	10

Since the electronegativity of hydrogen is constant and less than the electronegativity of the element to which it is bonded, the differences must be due to the presence of the other elements. The change in bond energy indicates a decrease in electronegativity from F to I.

8.96 (See answer to Question 8.95)

Compounds	LiH	NaH	KH	RbH
Difference between calc. and exp. bond energies (kJ/mol)	34	56	64	82

Since the bond energy difference is increasing, the electronegativity difference must be increasing. Since the electronegativity of hydrogen is constant, and larger than the electronegativity of any alkali metal, the electronegativities of the alkali metals must decrease from lithium to rubidium.

9 COVALENT BONDING AND MOLECULAR STRUCTURE

9.1 See the figures in Section 9.1 for drawing of the linear, planar triangular, tetrahedral, trigonal bipyramidal and octahedral molecular shapes.

9.2 **180°**

9.3 planar triangular, **120°**; tetrahedral, **109.5°**; octahedral, **90°**.

9.4 Within triangular plane, **120°**; between triangular plane and apex, **90°**.

9.5 (a) **4** (b) **6** (c) **5**

9.6 The valence shell electron-pair repulsion theory is based on the basic postulate that the geometric arrangements of atoms or groups of atoms about some central atom is determined solely by the mutual repulsion between the electron pairs (both bonding and lone pairs) present in the valence shell of the central atom.

9.7 (a) MX_3E; trigonal pyramidal (b) MX_4; tetrahedral (c) MX_3; planar triangular

(d) MX_2E; nonlinear or V-shaped (e) MX_4E; unsymmetrical tetrahedral

(f) MX_2E_3; linear (g) MX_5E; square pyramidal (h) MX_4; tetrahedral

(i) MX_6; octahedral (j) MX_5; trigonal bipyramidal

(k) MX_2E; nonlinear, bent, angular or V-shaped

9.8 See examples in Figures 9.2 through 9.5 for guidance in making sketches of the molecules and ions in Questions 9.7

9.9

	(1)	(2)
(a)	planar triangular	planar triangular
(b)	tetrahedral	trigonal pyramidal
(c)	octahedral	octahedral
(d)	tetrahedral	trigonal pyramidal
(e)	tetrahedral	trigonal pyramidal
(f)	tetrahedral	tetrahedral
(g)	planar triangular	planar triangular
(h)	tetrahedral	tetrahedral
(i)	tetrahedral	nonlinear (bent)
(j)	octahedral	square planar

9.10

	(1)	(2)
(a)	tetrahedral	bent
(b)	tetrahedral	bent
(c)	octahedral	octahedral
(d)	tetrahedral	tetrahedral
(e)	octahedral	square planar
(f)	tetrahedral	tetrahedral
(g)	planar triangular	planar triangular
(h)	trigonal bipyramidal	T-shaped
(i)	planar triangular	planar triangular
(j)	tetrahedral	trigonal pyramidal
(k)	tetrahedral	tetrahedral

9.11 (a) planar triangular to tetrahedral
(b) trigonal bipyramidal to octahedral
(c) T-shaped to square planar
(d) nonlinear (V-shaped) to unsymmetrical tetrahedral
(e) linear to planar "double" bent $H_2C{=}CH_2$

(Each half is planar triangular with the total molecule being coplanar.)
(f) planar triangular to linear

9.12 Figure 9.2, the X - M - X angle of the bottom compound will be less than 120° (e.g., SO_2).

In the two compounds at the bottom of Figure 9.3, the X - M - X angles will be compressed to less than the 109.5° found in a perfect tetrahedron (e.g., H_2O and NH_3).

(continued)

9.12 (continued)
Figure 9.4, for the unsymmetrical tetrahedral the distortion is:

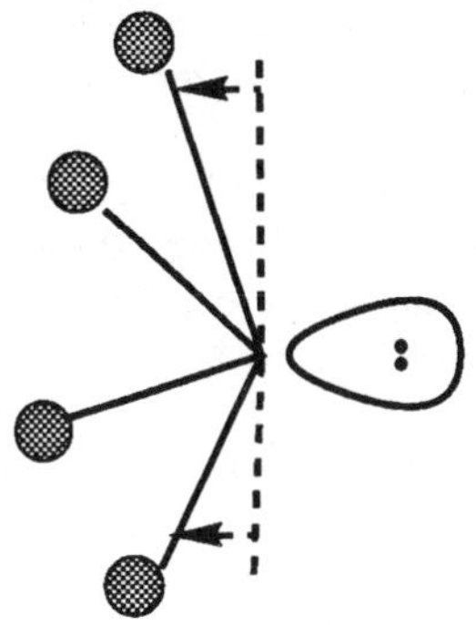

For the T-shape the distortion is:

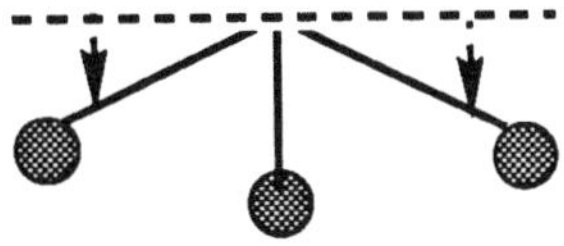

The distortion in Figure 9.5, for the square pyramidal, is :

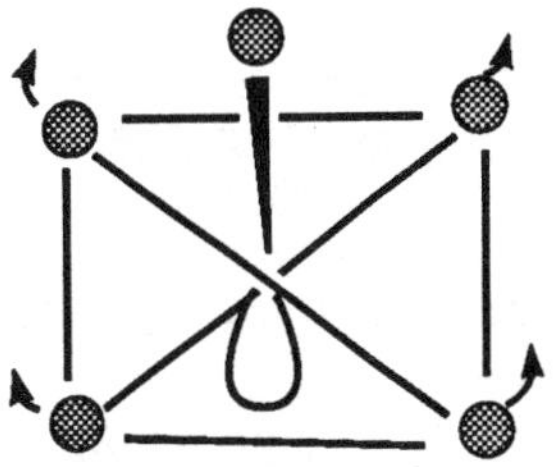

9.13 When two atoms joined by a covalent bond differ in electronegativity, the bond will be polar due to the resulting uneven distribution of electrons. A dipole exists when opposite ends of the molecule carry opposite electrical charges.

9.14 Molecules with polar bonds may be nonpolar molecules if there is complete cancellation of the effects of the bond dipoles due to a symmetrical arrangement of the bonds.

9.15 Because SO_2 is V-shaped (bent), its bond dipoles do not cancel each other. The result is a polar molecule. The dipoles in the planar triangular arrangement in SO_3 do cancel and the molecule is nonpolar.

9.16 (a), (c), (d), (g), (h) and (j)

9.17 Planar triangular

9.18 (1) When two atoms come together to form a covalent bond, an atomic orbital of one atom overlaps with an atomic orbital of the other and a pair of electrons is shared between the two atoms in the region of overlap. (2) The electrons can be shared between the two overlapping orbitals. (3) The strength of the covalent bond is proportional to the amount of overlap of the atomic orbitals.

9.19 Orbital overlap means that the two orbitals share some common region in space. The pair of electrons associated with a covalent bond is shared between the two atoms in this region and the strength of the covalent bond is proportional to the amount of overlap.

9.20 The p orbital with one electron on one Cl atom overlaps with the p orbital with one electron on the other Cl atom similar to the overlap of p_z orbitals. For the drawing imagine two atoms like the fluorine in Figure 9.9 arranged so that the half-filled p orbitals overlap. The spins of the electrons must be paired if bonding is to occur.

9.21 The s orbital on hydrogen with one electron overlaps with the p orbital with one electron on the Cl atom. See Figure 9.9. Chlorine and fluorine are interchangeable in this type of consideration.

9.22 See Figure 8.2.

9.23 Hybrid orbitals are produced when two or more atomic orbitals are mixed, producing a new set of orbitals. In hybrid orbitals one lobe is much larger than the other. Some atoms prefer to use hybrid orbitals because hybrid orbitals tend to form stronger covalent bonds (overlap more effectively) than those from ordinary atomic orbitals and hybridization allows the formation of equivalent bonds.

9.24 (a) linear (b) planar triangular (c) tetrahedral

(d) trigonal bipyramidal (e) octahedral

9.25 (a) 109.5° (b) 120° (c) 180° (d) 90°

9.26 B $\underset{2s}{\underline{\uparrow\downarrow}}$ $\underset{2p}{\underline{\uparrow}\ \underline{\quad}\ \underline{\quad}}$ (unhybridized) B $\underset{sp^2}{\underline{\uparrow}\ \underline{\uparrow}\ \underline{\uparrow}}$ (hybridized) $\underset{2p}{\underline{\quad}}$ (unhybridized)

See Figure 9.15 for the shape of sp^2 hybridized orbitals. The overlap of the three sp^2 orbitals with p orbitals of chlorine will yield a planar triangular structure for BCl_3. With this knowledge, a sketch of BCl_3 that shows the bonding can be made.

9.27 We must employ hybrid orbitals in order to account for the H - C - H bond angle of 109.5° in CH_4 and the equivalence of the four C - H bonds.

9.28 One explanation of the 104.5° bond angle in H_2O considers the overlap of s orbitals of the hydrogens with the half-filled p orbitals of oxygen followed by repulsion of the electron clouds of the hydrogens to make the angle greater than 90°. The other explanation involves the use of hybridized sp^3 orbitals on the oxygen and the angle is less than the 109.5 angle of a tetrahedral set due to the space requirements of the non-bonded lone pairs.

9.29 Bond angle measurements indicate that the lone pair of electrons project out from the central atom rather than being spread symmetrically as in the s orbital. The reaction of the alkaline properties of NH_3, studied in a later chapter, will also lend experimental evidence to the sp^3 hybridization of nitrogen in NH_3.

9.30 The overlap of the unpaired electrons in the p orbitals of As with the unpaired electron in the s orbital of the hydrogens will yield bonding with a bond angle of close to 90°.

9.31 p orbitals

9.32 The bond angle of 104° is much closer to the 109.5° of sp^3 orbital bonding than it is to the 90° of p orbital bonding.

(unhybridized) P $\underset{3s}{\uparrow\downarrow}$ $\underset{3p}{\uparrow\ \uparrow\ \uparrow}$ (hybridized) P $\underset{sp^3}{\uparrow\downarrow\ \uparrow\ \uparrow\ \uparrow}$

F $\underset{2s}{\uparrow\downarrow}$ $\underset{2p}{\uparrow\downarrow\ \uparrow\downarrow\ \uparrow}$ F $\underset{2s}{\uparrow\downarrow}$ $\underset{2p}{\uparrow\downarrow\ \uparrow\downarrow\ \uparrow}$ F $\underset{2s}{\uparrow\downarrow}$ $\underset{2p}{\uparrow\downarrow\ \uparrow\downarrow\ \uparrow}$

The hybridized sp^3 orbitals of P have 3 unpaired electrons. Each of the sp^3 orbitals of P that contains only a single electron can bond with an unpaired electron in a 2p orbital of one of the F's. The result will be three bonds each formed by the overlap of an sp^3 and a p orbital.

9.33 No. Our explanation is no better than the quality of the experimental evidence about this compound and very similar compounds. There is considerable evidence that indicates that fluorine uses hybrid orbitals.

9.34 Sn has the valence shell configuration, $5s^2\ 5p^2$. If the Cl atoms' orbital overlap with the p orbitals of Sn, a nonlinear (bent) molecule should result with a bond angle of about 90° (actual Cl - Sn - Cl angle in $SnCl_2$ is 95°).

9.35 (a) planar triangular, sp^2 bonding
(b) tetrahedral, sp^3 bonding
(c) trigonal bipyramidal, sp^3d bonding
(d) octahedral, sp^3d^2 bonding
(e) linear, sp bonding
(f) octahedral, sp^3d^2 bonding
(g) trigonal pyramidal, sp^3 bonding
(h) unsymmetrical tetrahedral, sp^3d bonding
(i) tetrahedral, sp^3 bonding

9.36 (a) sp^3 (b) sp^3 (c) sp^2 (d) sp^2 (e) sp^3d (f) sp^3d (g) sp^3d^2 (h) sp^3 (i) sp^3d^2 (j) sp^3d (k) sp^2

9.37 (a) sp^2 (b) sp^3 (c) sp^3d^2 (d) sp^3 (e) sp^3 (f) sp^3 (g) sp^2 (h) sp^3 (i) sp^3 (j) sp^3d^2

9.38 Sb $\underset{5s}{\underline{\uparrow\downarrow}}$ $\underset{5p}{\underline{\uparrow}\ \underline{\uparrow}\ \underline{\uparrow}}$ $\underset{5d}{\underline{\ \ }\ \underline{\ \ }\ \underline{\ \ }\ \underline{\ \ }\ \underline{\ \ }}$ $SbCl_5$ $\underset{sp^3d}{\underline{\uparrow x}\ \underline{\uparrow x}\ \underline{\uparrow x}\ \underline{\uparrow x}\ \underline{\uparrow x}}$ $\underline{\ \ }\ \underline{\ \ }\ \underline{\ \ }\ \underline{\ \ }$

(unhybridized 5d orbitals)

x = Cl electrons

In $SbCl_5$, sp^3d hybrid orbitals are involved in bonding

9.39 Si has a relatively low-energy d subshell in its valence shell, whereas C does not have a d subshell in its valence shell.

9.40 Boron in BCl_3 is sp^2 hybridized. Nitrogen in NH_3 is sp^3 hybridized. In order for boron to accept two electrons from the nitrogen to form Cl_3BNH_3, it must first become sp^3 hybridized. The geometry of B changes from planar triangular to tetrahedral. The N does not change geometry nor hybridization.

9.41 A coordinate covalent bond "exists" in NH_4^+, $AlCl_6^{3-}$, $SbCl_6^-$ and ClO_4^-. After formation, the coordinate covalent bond is identical to the "normal" covalent bond.

9.42 Sn is sp^3 hybridized. The Sn - Cl bond would be formed by sp^3-p overlap.

9.43 In NH_3, the N - H bonds are polar with the negative ends of the bond dipoles at the nitrogen. The lone pair, if it were in a nonbonded sp^3 orbital, would also produce a contribution to the dipole moment of the molecule, with its negative end pointing <u>away</u> from the nitrogen.

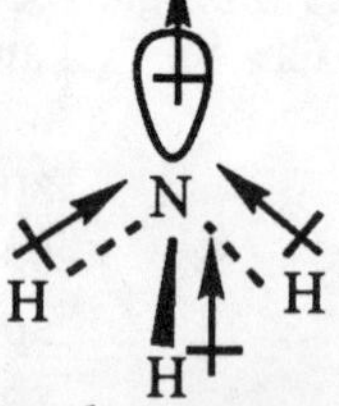

N H H H

All dipoles are additive and produce a large net molecular dipole. In NF_3, the fluorines are more electronegative than nitrogen, producing bond dipoles with their positive ends at nitrogen. These, therefore, tend to offset the contribution of the lone pair in the sp^3 orbital on nitrogen, thereby giving a very small net dipole moment for the NF_3 molecule.

(continued)

9.43 (continued)

(The bond dipoles tend to cancel the effects of the lone pair dipole. This would not be the case if an s orbital were used for the lone pair and p orbitals were used for the sigma bonds.)

9.44 The σ-bond results from a head-on overlap of atomic orbitals that concentrates electron density along the imaginary line joining the bonded nuclei. A π-bond is produced by the sideways overlap of atomic p orbitals providing electron density above and below the line connecting the bound nuclei. A double bond consists of one σ-bond and one π-bond. A triple bond is made by overlap of orbitals to give one σ and two π-bonds.

9.45 H—C≡C—C(H)=C(H)—C(=O)—O—C(H)(H)—H   (sp, sp^2, sp^2, sp^3)

H—C≡C—C(H)=C(H)—C(=O)—O—C(H)(H)—H   (σ, σ + 2π, σ + π, σ + π, σ, σ)

9.46 Valence bond theory would show that the N_2 triple bond consists of one sp-sp σ and two p-p π-bonds.

9.47 Valence bond theory shows that the CN^- triple bond consists of one sp-sp σ bond and two p-p π-bonds.

9.48 $H_2C=O$: (σ + π, σ) $H_2C=O$: (s - sp^2 σ; p - p π; sp^2 - sp^2 σ)

9.49 [:O::N:O:]$^-$ ⟷ [:O:N::O:]$^-$

sp^2 hybrid orbitals are used for σ bonds

p orbitals are used for π bond

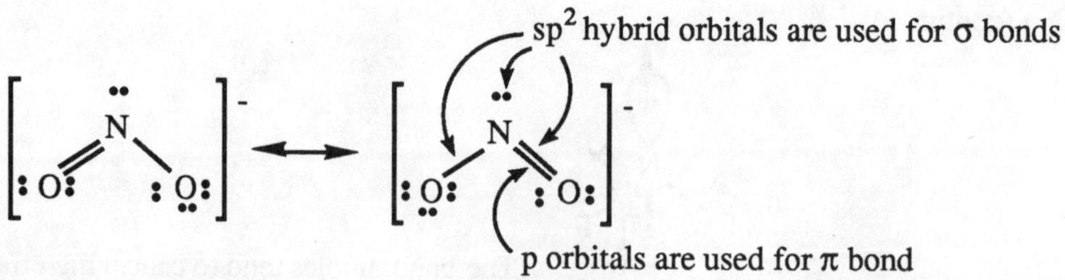

9.50 Period-2 nonmetals are smaller than period-3 nonmetals and are able to approach each other more closely. As a result, more effective sideways overlap of the p orbitals can occur. Therefore, period-2 nonmetals form much stronger pi bonds. Since period-3 nonmetals do not form very strong pi bonds, they tend to form sigma bonds with several atoms rather than multiple bonds to a single atom.

9.51 Because they can exist without involving a pi bond, as they need to form only one bond (a σ bond).

9.52 Allotropes are different forms of the same element as the result of differences in molecular structure as with O_2 and O_3 or as the result of differences in packing of molecules in the solid state.

9.53 Diatomic oxygen (O_2) and ozone (O_3).

9.54 The most stable form of elemental sulfur contains 8 sulfur atoms, each sigma bonded to two neighbors, to form an 8-member, puckered crown-like ring. (See Figure 9.27)

9.55 White phosphorus consists of P_4 molecules in which each phosphorus atom lies at a corner of a tetrahedron. (See Figure 9.28)

9.56 White phosphorus is very reactive because the very small P-P-P bond angle (~60°) does not allow effective overlap of orbitals and results in a bond that is easily broken or, stated another way, is very reactive.

9.57 Compare the structures shown in Figures 9.28 and 9.30.

9.58 In diamond each carbon atom is sigma bonded to 4 other carbon atoms. This produces a gigantic three-dimensional network made up of sp^3-sp^3 σ bonds. In graphite each carbon atom is sigma bonded to 3 other carbon atoms. The atoms are arranged in the form of hexagonal rings connected together in large planar sheets. The carbons are sp^2 hybridized in graphite with one p-electron on each carbon being involved in π bonding. (See Figs. 9.23, 9.24, 9.25 and 9.26)

9.59 Because Si has no apparent tendency to form multiple (π) bonds.

10 CHEMICAL REACTIONS AND THE PERIODIC TABLE

10.1 Metals have positive oxidation states in nearly all of their compounds.

10.2 Metals are elements that have low ionization energies and small electron affinities. They, therefore, tend to react by the loss of electrons to form positive ions, i.e., they become oxidized and serve as reducing agents.

10.3 Since the only effective oxidizing agent in aqueous solutions of HCl is H^+, HCl is called a non-oxidizing acid.

10.4 $Mn(s) + 2H^+(aq) \rightarrow H_2(g) + Mn^{2+}(aq)$

10.5 $2Al(s) + 6HBr(aq) \rightarrow 2AlBr_3(aq) + 3H_2(g)$

10.6 Since a solution of HNO_3 contains not only H^+ but also the nitrate ion, which is a stronger oxidizing agent than H^+, HNO_3 is called an oxidizing acid.

10.7 (a) $Ag(s) + NO_3^-(aq) + 2H^+(aq) \rightarrow Ag^+(aq) + NO_2(g) + H_2O$

(b) $3Ag(s) + NO_3^-(aq) + 4H^+(aq) \rightarrow 3Ag^+(aq) + NO(g) + 2H_2O$

10.8 If silver metal is placed in a solution of hydrochloric acid, nothing happens. A stronger oxidizing agent than H^+, such as NO_3^-, is needed to oxidize silver.

10.9 $4Zn(s) + NO_3^-(aq) + 10H^+(aq) \rightarrow 4Zn^{2+}(aq) + NH_4^+(aq) + 3H_2O$

10.10 $4Zn(s) + H_2SO_4(aq) + 8H^+(aq) \xrightarrow{hot} 4Zn^{2+}(aq) + H_2S(g) + 4H_2O$

10.11 (a) $Mg(s) + 2HCl(aq) \rightarrow MgCl_2(aq) + H_2(g)$
(b) $2Al(s) + 6HCl(aq) \rightarrow 2AlCl_3(aq) + 3H_2(g)$

10.12 (a) $2Cr(s) + 6HCl(aq) \rightarrow 2CrCl_3(aq) + 3H_2(g)$
(b) $Ni(s) + H_2SO_4(aq) \rightarrow NiSO_4(aq) + H_2(g)$

10.13 For 10.11: (a) $Mg(s) + 2H^+(aq) \rightarrow Mg^{2+}(aq) + H_2(g)$
(b) $2Al(s) + 6H^+(aq) \rightarrow 2Al^{3+}(aq) + 3H_2(g)$

For 10.12: (a) $2Cr(s) + 6H^+(aq) \rightarrow 2Cr^{3+}(aq) + 3H_2(g)$
(b) $Ni(s) + 2H^+(aq) \rightarrow Ni^{2+}(aq) + H_2(g)$

(The oxidizing agent is H^+ in each of these reactions.)

10.14 A single displacement reaction is a reaction in which one element displaces another element from a compound.

10.15 In order of increasing ease of oxidation: Ag < Cu < Sn < Cd < Mg

10.16 (a) $2Al(s) + 3Zn^{2+}(aq) \rightarrow 2Al^{3+}(aq) + 3Zn(s)$
(b) $Sn(s) + Cu^{2+}(aq) \rightarrow Sn^{2+}(aq) + Cu(s)$
(c) $Ag(s) + Co^{2+}(aq) \rightarrow$ no reaction
(d) $Mn(s) + Pb^{2+}(aq) \rightarrow Mn^{2+}(aq) + Pb(s)$
(e) $Cu(s) + Mg^{2+}(aq) \rightarrow$ no reaction
(f) $Hg(\ell) + H^+(aq) \rightarrow$ no reaction
(g) $Ni(s) + 2H^+(aq) \rightarrow Ni^{2+}(aq) + H_2(g)$
(h) $Cd(s) + H_2O \rightarrow Cd(OH)_2(s) + H_2(g)$
(i) $Ba(s) + 2H_2O \rightarrow Ba^{2+}(aq) + 2OH^-(aq) + H_2(g)$
(j) $H_2(g) + Pt^{2+}(aq) \rightarrow Pt(s) + 2H^+(aq)$

10.17 The easily oxidized metals are located on the left side of the periodic table including IA metals and IIA metals except beryllium. The least easily oxidized metals are located in Period 6 just to the right of center in the block of transition metals.

10.18 (a) Rb (b) Rb (c) Na (d) Ca

10.19 They all react with water to liberate hydrogen.

10.20 Because they are very resistant to oxidizing agents that attack other metals.

10.21 Ca should react more rapidly because it has lower ionization energies than Mg and, therefore, should be more reactive.

10.22 1 part concentrated HNO_3 and 3 parts concentrated HCl, by volume, dissolves gold and is known as aqua regia.

10.23 (a) $2Na(s) + 2H_2O \rightarrow H_2(g) + 2Na^+(aq) + 2OH^-(aq)$

(b) $2Rb(s) + 2H_2O \rightarrow H_2(g) + 2Rb^+(aq) + 2OH^-(aq)$

(c) $Sr(s) + 2H_2O \rightarrow H_2(g) + Sr^{2+}(aq) + 2OH^-(aq)$

10.24 (a) The oxidizing ability of nonmetals increases to the right across a period.
(b) The oxidizing ability of nonmetals decreases down a group.

10.25 (a) $C < N < O < F$ (b) $I < Br < Cl < F$

10.26 (a) $F_2 + 2Cl^- \rightarrow 2F^- + Cl_2$ (b) $Br_2 + Cl^- \rightarrow$ N.R.

(c) $I_2 + Cl^- \rightarrow$ N.R. (d) $Br_2 + 2I^- \rightarrow 2Br^- + I_2$

10.27 $2Al(s) + 3Br_2(\ell) \rightarrow 2AlBr_3(s)$ $Zn(s) + Br_2(\ell) \rightarrow ZnBr_2(s)$

10.28 Combustion is a rapid oxidation of a substance with oxygen that evolves a large amount of heat and light.

10.29 Rust is the product of a direct reaction of iron with oxygen in the presence of moisture to form an iron oxide whose crystals contain water molecules in variable amounts.

$2Fe(s) + 3/2O_2(g) + xH_2O(\ell) \rightarrow Fe_2O_3{\bullet}xH_2O(s)$

10.30 The product of the corrosion of aluminum is aluminum oxide, Al_2O_3. This thin oxide layer covers and adheres tightly to the aluminum surface and effectively protects it from further oxidation.

10.31 $2Mg(s) + O_2(g) \rightarrow 2MgO(s)$

10.32 (a) $4Fe(s) + 3O_2(g) \rightarrow 2Fe_2O_3(s)$ (b) $4Li(s) + O_2(g) \rightarrow 2Li_2O(s)$

(c) $2Ca(s) + O_2(g) \rightarrow 2CaO(s)$ (d) $2Mg(s) + O_2(g) \rightarrow 2MgO(s)$

(e) $4Al(s) + 3O_2(g) \rightarrow 2Al_2O_3(s)$

10.33 (a) $C(s) + O_2(g) \rightarrow CO_2(g)$ (b) $S(s) + O_2(g) \rightarrow SO_2(g)$

(c) $P_4(s) + 5O_2(g) \rightarrow P_4O_{10}(s)$

10.34 $2C(s) + O_2(g) \rightarrow 2CO(g)$

10.35 (a) $C_9H_{20} + 14O_2 \rightarrow 9CO_2 + 10H_2O$ (b) $2C_2H_4(OH)_2 + 5O_2 \rightarrow 4CO_2 + 6H_2O$
(c) $2(CH_3)_2S + 9O_2 \rightarrow 4CO_2 + 6H_2O + 2SO_2$

10.36 $2CH_4 + 3O_2 \rightarrow 2CO + 4H_2O$ (somewhat limited supply of O_2)
$CH_4 + O_2 \rightarrow C + 2H_2O$ (extremely limited supply of O_2)

10.37 $2C_{20}H_{42} + 21O_2 \rightarrow 40C + 42H_2O$
$2C_{20}H_{42} + 41O_2 \rightarrow 40CO + 42H_2O$
$2C_{20}H_{42} + 61O_2 \rightarrow 40CO_2 + 42H_2O$

10.38 $(C_6H_{10}O_5)_x (s) + 6x\ O_2(g) \rightarrow 6x\ CO_2(g) + 5x\ H_2O(g)$

10.39 Brønsted-Lowry definitions of acids and bases
Acid: substance that donates a proton (a hydrogen ion H^+) to some other substance.
Base: substance that accepts a proton from an acid.

10.40 (a) NH_2^- (b) NH_3 (c) $C_2H_3O_2^-$
(d) $H_2PO_4^-$ (e) NO_3^-

10.41 (a) H_2SO_4 (b) HSO_4^- (c) H_3O^+
(d) HCl (e) $HCHO_2$ (or HCOOH)

10.42 The Brønsted-Lowry definition is less restrictive than the Arrhenius concept because it recognizes acid-base phenomena in other than just aqueous solutions.

10.43 Acid-base conjugate pairs (acid written first in each pair)
(a) $HC_2H_3O_2$, $C_2H_3O_2^-$ and H_2O, OH^-
(b) HF, F^- and NH_4^+, NH_3
(c) HSO_4^-, SO_4^{2-} and $H_2PO_4^-$, HPO_4^{2-}
(d) $Al(H_2O)_6^{3+}$, $Al(H_2O)_5OH^{2+}$ and H_2O, OH^-
(e) $N_2H_5^+$, N_2H_4 and H_2O, OH^-
(f) NH_3OH^+, NH_2OH and HCl, Cl^-
(g) OH^-, O^{2-} and H_2O, OH^-
(h) H_2, H^- and H_2O, OH^-
(i) NH_3, NH_2^- and N_2H_4, $N_2H_3^-$
(j) HNO_3, NO_3^- and $H_3SO_4^+$, H_2SO_4

10.44 Acid-base conjugate pairs (acids written first in each pair)
(a) $HClO_4$, ClO_4^- and $N_2H_5^+$, N_2H_4
(b) H_3PO_3, $H_2PO_3^-$ and H_2SO_3, HSO_3^-
(c) $C_5H_5NH^+$, C_5H_5N and $(CH_3)_3NH^+$, $(CH_3)_3N$
(d) H_2O, OH^- and HCO_3^-, CO_3^{2-}
(e) $HCHO_2$, CHO_2^- and $C_7H_5O_2H$, $C_7H_5O_2^-$
(f) $H_2C_2O_4$, $HC_2O_4^-$ and $CH_3NH_3^+$, CH_3NH_2
(g) H_2CO_3, HCO_3^- and H_3O^+, H_2O
(h) C_2H_5OH, $C_2H_5O^-$ and NH_3, NH_2^-
(i) $N_2H_5^+$, N_2H_4 and HNO_2, NO_2^-
(j) H_2CN^+, HCN and H_2SO_4, HSO_4^-

10.45 (a) $2H_2O \rightleftharpoons H_3O^+ + OH^-$

(b) $2NH_3 \rightleftharpoons NH_4^+ + NH_2^-$

(c) $2HCN \rightleftharpoons H_2CN^+ + CN^-$

10.46 acid, $(CH_3)_2NH_2^+$; base, $(CH_3)_2N^-$

10.47 $HCO_3^- + H_2O \rightleftharpoons CO_3^{2-} + H_3O^+$ (as an acid)

$HCO_3^- + H_2O \rightleftharpoons H_2CO_3 + OH^-$ (as a base)

10.48 The highly charged Cr^{3+} ion polarizes the O - H bonds of the water molecules attached to it, thereby making it easier for the H^+ to be transferred to a neighboring H_2O molecule.

$Cr(H_2O)_6^{3+} + H_2O \rightleftharpoons Cr(H_2O)_5(OH)^{2+} + H_3O^+$

10.49 (a) $HClO_3$ (b) HNO_3 (c) H_3PO_4 (d) H_2SO_4
(e) $HClO_3$ (f) $HBrO_3$ (g) H_2Se (h) HBr
(i) PH_3

10.50 (a) The one on $CHCl_2CO_2H$
(b) The one on CH_2FCO_2H

10.51 CH_3SH This compound and CH_3OH act like weak binary acids. The CH_3SH has a much weaker S-H bond than is the O-H bond in CH_3OH.

10.52 The additional oxygen in HNO_3 draws more electron density from the nitrogen which gives it a greater partial positive charge. The nitrogen in turn draws electron density from the N-OH bond. This in turn draws the electron density from the O-H bond making HNO_3 a stronger acid than HNO_2.

10.53 Cl^- is a larger ion than F^- so the HCl bond is much weaker than the HF bond, thus making it more acidic.

10.54 HCl is much more polar than H_2S.

10.55 By drawing Lewis structures and then indicating electron shift due to varying electronegativity, one can show an increased negative charge of the central atoms as one goes from $H_2PO_3^-$ to HSO_4^- to ClO_4^-. The shift of electrons away from the oxygen is responsible for $HClO_4$ being a stronger acid than H_2SO_4 which in turn is stronger than H_3PO_4. The question asks for Lewis structures. You should draw them. You may also wish to look at resonance structures of the anions.

10.56 $HClO_4$ is a stronger acid than $HClO_3$ because the presence of another electronegative oxygen in $HClO_4$ makes the O - H bond more polar. The anion from $HClO_4$ will also have the negative charge spread over a greater number of atoms. (See answer to Question 10.55).

10.57 Same arguments as those presented in the answer to 10.56. The anion of the stronger acid will be the weaker base. Therefore, HSO_3^- is a stronger base than HSO_4^-.

10.58 <u>Lewis base</u>: a substance that can donate a pair of electrons to the formation of a covalent bond.
<u>Lewis acid</u>: a substance that can accept a pair of electrons to form a covalent bond.

10.59 Ammonia acts as a Lewis base by donating a pair of electrons to form a covalent bond with H^+ which acts as a Lewis acid.

10.60 BCl_3 (acid) + $:O(C_2H_5)_2$ (base) $\longrightarrow$ $Cl_3B \leftarrow O(C_2H_5)_2$

acid base

10.61 $H_2O + CO_2 \longrightarrow H-O(H)-C(=O)-O \longrightarrow H-O-C(=O)-O-H$

Water acts as a Lewis base by donating a pair of electrons to form a covalent bond with CO_2.

10.62 $H_3N\!: \rightarrow Ag^+ \leftarrow :NH_3 \longrightarrow [H_3N\text{-}Ag\text{-}NH_3]^+$
NH3 is a Lewis base which forms coordinate covalent bonds to the Lewis acid, Ag^+.

10.63

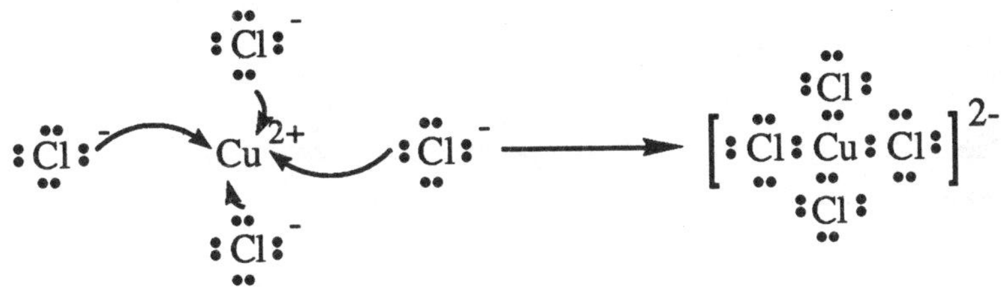

$$
\begin{array}{ccccccc}
 & & :Cl:^- & & & & :Cl: \\
:Cl:^- & \rightarrow & Cu^{2+} & \leftarrow & :Cl:^- & \longrightarrow & \left[:Cl:Cu:Cl:\right]^{2-} \\
 & & :Cl:^- & & & & :Cl:
\end{array}
$$

The chloride ions are the Lewis base in the above reaction. Copper ion is the Lewis acid.

10.64 (a) Ligands are Lewis bases that attach themselves to metal ions.
(b) Complexes are ionic or neutral species in which a metal ion has anions or neutral molecules attached via coordinate covalent bonds.
(c) A donor atom is the atom of a ligand that donates the electron pair to the metal to form a coordinate covalent bond in a complex.
(d) The metal ion in complexes is the acceptor ion.
(e) A polydentate ligand is one that has two or more donor atoms that are able to simultaneously bind to a metal ion.
(f) A bidentate ligand is one that has two such donor atoms.
(g) A monodentate ligand has only one donor atom.

10.65 Complexes are also called coordination compounds because they involve coordinate covalent bonds.

10.66 Ligands are anions or neutral molecules that contain lone pairs of electrons that they can share. They are Lewis bases.

10.67 (a) The CN^- displaces the H_2O of $Zn(H_2O)_4{}^{2+}$ to form $Zn(CN)_4{}^{2-}$. The CN^- is a stronger ligand (Lewis base) than is water. The Lewis acid is Zn^{2+}.
(b) The Cl^- displaces an NH_3 to form $Pt(NH_3)_3Cl^+$ and free NH_3. The Lewis acid is the Pt^{2+}.

10.68 See the drawing in Section 10.7.

10.69 See the drawing in Section 10.7. EDTA has 6 possible donor atoms. Identify them. EDTA is used in a limited way as a food preservative, in shampoos as a water softener, to prevent clotting of blood samples, and as an antidote for poisoning by heavy metals. It is relatively non-toxic.

10.70 Solutions of copper salts often contain the $Cu(H_2O)_4^{2+}$ species which is pale blue. This ion is often formed in solution via the reaction between Cu^{2+} and water.

10.71 (a) $Cu(NH_3)_4^{2+}$ is a deep blue color
(b) $Cu(H_2O)_4^{2+}$ is a pale blue color
(c) $Co(H_2O)_6^{2+}$ is a reddish-pink color
(d) $Ni(H_2O)_6^{2+}$ is a green color

10.72 The water molecules are probably held via coordinate covalent bonds to the manganese ion in the compound $MnCl_2{\bullet}6H_2O$ as $Mn(H_2O)_6^{2+}$ ions.

10.73 (a) AgI_2^- (b) $Ag(NH_3)_2^+$ (c) $Co(C_2O_4)_3^{3-}$
(d) $Co(NO_2)_6^{3-}$ (e) $Cr(EDTA)^-$ (f) $Ni(CN)_4^{2-}$
(g) $Fe(SCN)_6^{3-}$

10.74 (a) 2+ (b) 3+ (c) 2+ (d) 3+ (e) 2+

11 PROPERTIES OF GASES

11.1 Pressure is force per unit area. Pressure is an intensive property. As long as there is a space above the mercury column, the pressures acting along the reference level will be the same regardless of the size and length of the tube. If the diameter of the tube is doubled, there will be four times the weight of Hg acting over four times the area. The ratio of F/A remains unchanged.

11.2 The SI unit of pressure is the pascal; 1 atm = 101,325 Pa

11.3 (a) $1.50 \text{ atm} \times \frac{760 \text{ torr}}{\text{atm}} = \mathbf{1{,}140 \text{ torr}}$ or $\mathbf{1.14 \times 10^3 \text{ torr}}$

(b) $785 \text{ torr} \times \frac{1 \text{ atm}}{760 \text{ torr}} = \mathbf{1.03 \text{ atm}}$

(c) $3.45 \text{ atm} \times \frac{101{,}325 \text{ Pa}}{\text{atm}} = 350{,}000 \text{ Pa} = \mathbf{3.50 \times 10^5 \text{ Pa}}$

(d) $3.45 \text{ atm} \times \frac{101.325 \text{ kPa}}{\text{atm}} = \mathbf{350 \text{ kPa}}$

(e) $165 \text{ torr} \times \frac{1 \text{ atm}}{760 \text{ torr}} \times \frac{101{,}325 \text{ Pa}}{\text{atm}} = \mathbf{2.20 \times 10^4 \text{ Pa}}$

(f) $342 \text{ kPa} \times \frac{1 \text{ atm}}{101.3 \text{ kPa}} = \mathbf{3.38 \text{ atm}}$

(g) $11.5 \text{ kPa} \times \frac{1 \text{ atm}}{101.3 \text{ kPa}} \times \frac{760 \text{ torr}}{\text{atm}} = \mathbf{86.3 \text{ torr}}$

11.4 $1 \text{ atm} \times \frac{76.0 \text{ cm Hg}}{\text{atm}} \times \frac{13.6 \text{ g}}{\text{mL}} \times \frac{1 \text{ mL}}{\text{cm}^3} \times (2.54 \text{ cm/in.})^2 \times \frac{1 \text{ lb}}{454 \text{ g}}$

$= \mathbf{14.7 \text{ lb/in.}^2}$

11.5 The sketch would be similar to that shown in Figure 11.4(c). The difference in the height of mercury in the two arms would be 25 mm.

11.6 A closed-end manometer is often used because of its compact size and, when one uses the closed-end manometer, it is not necessary to measure the atmospheric pressure.

11.7 Mercury is often used in barometers and manometers because it has a very high density and, therefore, pressure differences result in convenient height differences. Other reasons for its use are that it is a liquid over a large temperature range and it has a very low vapor pressure.

11.8 $15.8 \text{ cm} \times \frac{1 \text{ torr}}{\text{mm}} \times \frac{10 \text{ mm}}{\text{cm}} = \mathbf{158 \text{ torr}}$

11.9 (65 mm x 1 torr/mm) + 733 torr = **798 torr**

11.10 (774 torr - 535 torr) x 1 mm/torr = **239 mm**

11.11 (755 mm Hg + 17 mm Hg) x 1 torr/mm Hg = **772 torr**

11.12 $836 \text{ torr} + (74.0 \text{ cm} \times \frac{0.847 \text{ g oil}}{\text{mL}} \times \frac{1 \text{ mL}}{13.6 \text{ g Hg}} \times \frac{10 \text{ torr}}{\text{cm}}) = \mathbf{882 \text{ torr}}$

11.13 $1 \text{ atm} \times \frac{76.0 \text{ cm Hg}}{\text{atm}} \times \frac{13.6 \text{ g Hg}}{\text{mL Hg}} \times \frac{1 \text{ mL } H_2O}{1 \text{ g } H_2O} \times \frac{1 \text{ in.}}{2.54 \text{ cm}} \times \frac{1 \text{ ft}}{12 \text{ in.}} = 33.9 \text{ ft}$
(maximum height of a column of water in a sealed column at 1 atm of external pressure)

No, the person will not be able to draw water a height of 35 ft.

11.14 The combined gas law is: $P_iV_i/T_i = P_fV_f/T_f$

11.15 Charles's law is: V/T = constant. Boyle's law is: PV = constant. Gay-Lussac's law is: P/T = constant. PV/T = constant is the combined gas law. If T is constant, then PV = constant, i.e.; Boyle's law. If P is constant, then V/T = constant, i.e.; Charles's law. If V is constant, then P/T = constant, i.e.; Gay-Lussac's law.

11.16 (Charles's law states that at a constant pressure, the volume of a given quantity of a gas is directly proportional to its absolute temperature.) A temperature of -273.15°C represents that temperature below which gases would have a negative volume. That is impossible.

11.17 Heating the can causes the pressure of the gas inside to increase. This may cause the can to explode.

11.18 During the trip the tires become warm and the air pressure in them increases (Gay-Lussac's law).

11.19 Cooling the gas causes it to contract in volume. This means that a given volume will contain more oxygen.

11.20 The hot air in the balloon has a lower density than the cooler air surrounding the balloon.

11.21 Boyle's law states that at a constant temperature, the volume occupied by a fixed quantity of gas is inversely proportional to the applied pressure (or PV = constant). All gases do not always exactly obey Boyle's law. A gas that does would be called an ideal gas.

11.22 $P_iV_i = P_fV_f$ $\qquad$ (740 torr) x (350 mL) = (900 torr) x (? mL)

? mL = **288 mL**

11.23 $P_iV_i = P_fV_f$ $\qquad$ (2.75 atm) x (1.45 L) = [(800/760)atm] x (? L)

? L = **3.79 L**

11.24 $P_iV_i = P_fV_f$ $\qquad$ (475 torr) x (540 mL) = (? torr) x (320 mL)

? torr = **802 torr**

11.25 $\frac{P_i}{T_i} = \frac{P_f}{T_f}$ $\qquad \frac{350 \text{ torr}}{(273 + 20) \text{ K}} = \frac{? \text{ torr}}{(273 + 40) \text{ K}}$ $\qquad$? torr = **374 torr**

11.26 $\frac{P_i}{T_i} = \frac{P_f}{T_f}$ $\qquad \frac{655 \text{ torr}}{(273 + 25) \text{ K}} = \frac{825 \text{ torr}}{(273 + ?^\circ\text{C}) \text{ K}}$ $\qquad$? °C = **102°C**

11.27 29 lb /in.^{2} gauge pressure is a total pressure of (14.7 + 29) lb /in.^{2}

65°F = ? °C $\qquad$? °C = (5/9)(65 - 32) = 18°C

130°F = ? °C $\qquad$? °C = (5/9)(130 - 32) = 54°C

$\frac{P_i}{T_i} = \frac{P_f}{T_f}$ $\qquad \frac{14.7 + 29}{18 + 273^\circ\text{C}} = \frac{14.7 + ?\ P}{54 + 273^\circ\text{C}}$ $\qquad$? P = **34 lb/in.^{2}**

11.28 $\frac{V_i}{T_i} = \frac{V_f}{T_f}$ $\qquad \frac{1.50 \text{ L}}{(273 + 25) \text{ K}} = \frac{? \text{ L}}{(273 + 100) \text{ K}}$ $\qquad$? L = **1.88 L**

11.29 $\frac{V_i}{T_i} = \frac{V_f}{T_f}$ $\qquad \frac{2.0\ L}{(273 + 25)\ K} = \frac{?\ L}{(273 - 28.9)\ K}$ $\qquad ?\ L = \mathbf{1.6\ L}$

11.30 $\frac{V_i}{T_i} = \frac{V_f}{T_f}$ $\qquad \frac{2.00\ L}{(273 + 26)\ K} = \frac{?\ L}{(273 + 100)\ K}$ $\qquad ?\ L = \mathbf{2.49\ L}$

11.31 $\frac{V_i}{T_i} = \frac{V_f}{T_f}$ $\qquad \frac{285\ mL}{(273 + 25)\ K} = \frac{350\ mL}{(273 + ?°C)\ K}$ $\qquad ?\ °C = \mathbf{93°C}$

11.32 $\frac{V_i}{T_i} = \frac{V_f}{T_f}$ $\qquad \frac{400\ mL}{(273 + 32)\ K} = \frac{850\ mL}{(273 + ?°C)\ K}$ $\qquad ?\ °C = \mathbf{375°C}$

11.33 $P_iV_i = P_fV_f$ $\qquad (200\ kPa) \times (350\ cm^3) = (?\ kPa) \times (400\ cm^3)$

$(?\ kPa) = \mathbf{175\ kPa}$

11.34 $P_iV_i = P_fV_f$ $\qquad (1\ atm) \times (75.0\ cm \times area) = (5.50\ atm) \times (?\ length \times area)$

? length = 13.6 cm

length of downstroke = 75.0 cm - 13.6 cm = **61.4 cm**

11.35 STP stands for standard temperature and pressure, 0°C (273 K) and one standard atmosphere (760 torr). It is a reference set of conditions.

11.36 (a) $P_iV_i = P_fV_f$ $\qquad (645\ torr)(50.0\ mL) = (?\ torr)(65.0\ mL)$ $\qquad ?\ torr = \mathbf{496\ torr}$

(b) $\frac{P_iV_i}{T_i} = \frac{P_fV_f}{T_f}$ $\qquad \frac{(645\ torr)(50.0\ mL)}{(273 + 25)\ K} = \frac{(?\ torr)(65.0\ mL)}{(273 + 35)\ K}$

$?\ torr = \mathbf{513\ torr}$

11.37 $\frac{(450\ torr)(300\ mL)}{(273 + 27)\ K} = \frac{(?\ torr)(200\ mL)}{(273 + 20)\ K}$ $\qquad ?\ torr = \mathbf{659\ torr}$

11.38 $\frac{(700\ torr)(2.00\ L)}{(273 + 25)\ K} = \frac{(585\ torr)(5.00\ L)}{(273 + ?°C)\ K}$ $\qquad ?\ °C = \mathbf{350°C}$

11.39 $\frac{(1\ atm)(1\ L)}{(273 + 0)\ K} = \frac{(650/760)\ atm(?\ L)}{(273 + 25)\ K}$ $\qquad ?\ L = \mathbf{1.28\ L}$

Density = 1.96 g/1.28 L = **1.53 g/L**

11.40 $\frac{(450 \text{ torr})(50.0 \text{ mL})}{(273 + 35) \text{ K}} = \frac{(760 \text{ torr})(?\text{mL})}{(273 + 0) \text{ K}}$? mL = **26.2 mL**

11.41 atm liter^2 K^2/mol

11.42 The units used for pressure and volume determine which R you should use.

11.43 $\frac{0.08206 \text{ L atm}}{\text{mol K}} \times \frac{10^3 \text{cm}^3}{\text{L}} \times \frac{\text{m}^3}{(100 \text{ cm})^3} \times \frac{1.013 \times 10^5 \text{ Pa}}{\text{atm}}$

$= \mathbf{8.31 \text{ Pa m}^3 \text{ mol}^{-1}\text{K}^{-1}}$

11.44 (a) $0.200 \text{ mol} \times \frac{22.4 \text{ L (STP)}}{\text{mol}} = \mathbf{4.48 \text{ L (STP)}}$

or from Table 11.1 $0.200 \text{ mol } O_2 \times \frac{22.397 \text{ L (STP)}}{\text{mol}} = \mathbf{4.48 \text{ L (STP)}}$

(b) $12.4 \text{ g } Cl_2 \times \frac{1 \text{ mol } Cl_2}{70.9 \text{ g } Cl_2} \times \frac{22.4 \text{ L (STP)}}{\text{mol}} = \mathbf{3.92 \text{ L (STP)}}$

(c) $0.150 \text{ mol (total)} \times \frac{22.4 \text{ L}}{\text{mol}} = \mathbf{3.36 \text{ L (total at STP)}}$

11.45 $245 \text{ mL (STP) } SO_2 \times \frac{1 \text{ mol}}{22.4 \text{ L (STP)}} \times \frac{1 \text{ L}}{1{,}000 \text{ mL}} \times \frac{64.1 \text{ g } SO_2}{\text{mol}} = \mathbf{0.701 \text{ g } SO_2}$

11.46 For C_4H_{10} $\frac{58.1 \text{ g}}{\text{mol}} \times \frac{1 \text{ mol}}{22.4 \text{ L (STP)}} = \mathbf{2.59 \text{ g/L (STP)}}$

11.47 $\frac{1.96 \text{ g}}{\text{L (STP)}} \times \frac{22.4 \text{ L (STP)}}{\text{mol}} = \mathbf{43.9 \text{ g/mol}}$

11.48 $n = \frac{PV}{RT} = \frac{760 \text{ torr}}{760 \text{ torr/atm}} \times \frac{250 \text{ mL}}{1000 \text{ mL/L}} \times \frac{\text{mol K}}{0.0821 \text{ L atm}} \times \frac{1}{(273 + 25)\text{K}} \Leftrightarrow 0.0102 \text{ mol}$

$\frac{0.164 \text{ g}}{0.0102 \text{ mol}} = \mathbf{16.1 \text{ g/mol}}$

11.49 $P = \frac{nRT}{V} = 25.0 \text{ kg} \times \frac{1000 \text{ g}}{\text{kg}} \times \frac{1 \text{ mol } H_2O}{18.0 \text{ g}} \times 0.0821 \text{ L atm mol}^{-1} \text{K}^{-1}$

$$\times (273 + 200) \text{ K} \times \frac{1}{1000 \text{ L}} = \mathbf{53.9 \text{ atm}} \text{ or } \mathbf{41{,}000 \text{ torr}}$$

11.50 $n = \frac{PV}{RT} = \frac{760 \text{ torr}}{760 \text{ torr/atm}} \times 1 \text{ L} \times \frac{1}{0.0821 \text{ L atm mol}^{-1} \text{K}^{-1}}$

$$\times \frac{1}{(273 + 30)\text{K}} = 0.0402 \text{ mol} \qquad \frac{1.81 \text{ g}}{0.0402 \text{ mol}} = \mathbf{45.0 \text{ g/mol}}$$

11.51 $V = \frac{nRT}{P} = \frac{0.234 \text{ g}}{17.03 \text{ g/mol}} \times 0.08206 \text{ L atm mol}^{-1} \text{K}^{-1}$

$$\times (273 + 30) \text{ K} \times \frac{1}{0.847 \text{ atm}} = 0.403 \text{ L} = \mathbf{403 \text{ mL}}$$

11.52 (a) For C: $\frac{80.0 \text{ g}}{12.01 \text{ g/mol}} = 6.66 \text{ mol}$

For H: $\frac{20.0 \text{ g}}{1.01 \text{ g/mol}} = 19.80 \text{ mol}$

Empirical Formula $C_{6.66/6.66}H_{19.80/6.66} = \mathbf{CH_3}$

(b) $n = \frac{PV}{RT} = \frac{1 \text{ atm} \times 0.500 \text{ L}}{0.0821 \text{ L atm mol}^{-1} \text{K}^{-1} \times 273 \text{ K}} = 0.0223 \text{ mol}$

$\text{M.M.} = \frac{0.6695 \text{ g}}{0.0223 \text{ mol}} = \mathbf{30.0 \text{ g/mol}}$

(c) $CH_3 \Leftrightarrow$ 15 g/formula   30 g/mol $\Leftrightarrow$ $\mathbf{C_2H_6}$

11.53 (a) For C: $0.482 \text{ g } CO_2 \times \frac{1 \text{ mol } CO_2}{44.01 \text{ g}} \times \frac{1 \text{ mol C}}{\text{mol } CO_2} \times \frac{12.01 \text{ g C}}{\text{mol}} = 0.132 \text{ g C}$

$$\% \text{ C} = \frac{0.132 \text{ g C}}{0.200 \text{ g sample}} \times 100\% = \mathbf{66.0\% \text{ C}}$$

For H: $0.271 \text{ g } H_2O \times \frac{1 \text{ mol } H_2O}{18.02} \times \frac{2 \text{ mol H}}{\text{mol } H_2O} \times \frac{1.01 \text{ g H}}{\text{mol}} = 0.0304 \text{ g H}$

$$\% \text{ H} = \frac{0.0304 \text{ g H}}{0.200 \text{ g sample}} \times 100\% = \mathbf{15.2\% \text{ H}}$$

For N: $n = \frac{PV}{RT} = \frac{(755/760)(0.0423)}{(0.0821)(273 + 26.5)} = 0.00171 \text{ mol } N_2$

$$0.00171 \text{ mol } N_2 \times \frac{2 \text{ mol N}}{\text{mol } N_2} \times \frac{14.01 \text{ g N}}{\text{mol N}} = 0.0479 \text{ g N}$$

$$\% \text{ N} = \frac{0.0479 \text{ g N}}{0.2500 \text{ g sample}} \times 100\% = \mathbf{19.2\% \text{ N}}$$

(or 100% - 66% - 15.2 % = 18.8%)

(b) mol C in 100 g of sample $= \frac{65.8 \text{ g C}}{12.01 \text{ g/mol}} = 5.48 \text{ mol C}$

mol H in 100 g of sample $= \frac{15.2 \text{ g H}}{1.01 \text{ g/mol}} = 15.0 \text{ mol H}$

mol N in 100 g of sample $= \frac{19.2 \text{ g N}}{14.01 \text{ g/mol}} = 1.37 \text{ mol N}$

Empirical Formula: $C_{5.48/1.37}H_{15.0/1.37}N_{1.37/1.37} = \mathbf{C_4H_{11}N}$

11.54 From the partial pressure of the O_2, the moles of the O_2 and the temperature, the volume of the flask can be calculated. Once the volume of the flask is known, the mass of N_2 can be calculated.

For O_2: $V = \frac{nRT}{P} = \frac{0.100\ g}{32.0\ g/mol} \times \frac{0.0821\ L\ atm}{mol\ K}$

$$\times\ (25 + 273)\ K \times \frac{1}{[(760\text{-}525)/760]\ atm} = 0.247\ L$$

For N_2: $n = \frac{PV}{RT} = (\ 525/760)\ atm \times 0.247\ L$

$$\times \frac{1}{0.0821\ L\ atm\ mol^{-1}K^{-1}} \times \frac{1}{298\ K} = 0.00697\ mol\ N_2$$

$$0.00697\ mol\ N_2 \times \frac{28.02\ g\ N_2}{mol} = \mathbf{0.195\ g\ N_2}$$

11.55 $1\ L \times 1\ atm \times \frac{101{,}325\ Pa}{atm} \times \frac{1\ N\ m^{-2}}{Pa} \times \frac{1\ J}{1\ N\ m} \times \frac{0.100000\ m^3}{L} = \mathbf{101.325\ J}$

$R = 0.0821\ L\ atm\ mol^{-1}\ K^{-1} \times 101.325\ J\ L^{-1}\ atm^{-1} = \mathbf{8.32\ J\ mol^{-1}K^{-1}}$

$R = 8.32\ J\ mol^{-1}K^{-1} \times \frac{1\ cal}{4.184\ J} = \mathbf{1.99\ cal\ mol^{-1}\ K^{-1}}$

11.56 $400\ mL\ NH_3 \times \frac{1\ L}{1{,}000\ mL} \times \frac{1.00\ mol\ NH_3}{22.4\ L\ NH_3} \times \frac{1\ mol\ N_2}{2\ mol\ NH_3}$

$$\times \frac{22.4\ L\ N_2}{1\ mol\ N_2} \times \frac{1{,}000\ mL\ N_2}{L} \Leftrightarrow \mathbf{200\ mL\ N_2}$$

$400\ mL\ NH_3 \times \frac{3\ mL\ H_2}{2\ mL\ NH_3} \Leftrightarrow \mathbf{600\ mL\ H_2}$

11.57 (a) $0.00140\ mol\ NO \times \frac{1\ mol\ N_2}{2\ mol\ NO} \times \frac{22.4\ L}{mol} \times \frac{1{,}000\ mL}{L} \Leftrightarrow \mathbf{15.7\ mL\ N_2}$ (STP)

(b) $1.3 \times 10^{-3}\ g\ H_2 \times \frac{1\ mol\ H_2}{2.02\ g\ H_2} \times \frac{1\ mol\ N_2}{2\ mol\ H_2} \times \frac{22.4\ mL}{mol} \times \frac{1{,}000\ mL}{L}$

$$\Leftrightarrow \mathbf{7.2\ mL\ N_2}\ \text{(STP)}$$

11.58 $2KClO_3 \rightarrow 2KCl + 3O_2$

Vapor pressure of water at 30°C = 31.8 torr. (Table 11.2)

(a) $n = \frac{PV}{RT} = \frac{(600 - 32)\text{ atm}}{760} \text{ x } \frac{0.150\text{ L}}{1} \text{ x } \frac{1\text{ mol K}}{0.0821\text{ L atm}} \text{ x } \frac{1}{303\text{ K}}$

$\Leftrightarrow 0.004507\text{ mol }O_2$

$0.004507\text{ mol }O_2 \text{ x } 32.0\text{ g/mol} = \mathbf{0.144\text{ g }O_2}$

(b) $0.004507\text{ mol }O_2 \text{ x } \frac{2\text{ mol }KClO_3}{3\text{ mol }O_2} \text{ x } \frac{122.5\text{ g }KClO_3}{\text{mol}} \Leftrightarrow \mathbf{0.368\text{ g }KClO_3}$

11.59 $10.0\text{ g }HNO_3 \text{ x } \frac{1\text{ mol }HNO_3}{63.02\text{ g}} \text{ x } \frac{3\text{ mol }NO_2}{2\text{ mol }HNO_3} \text{ x } \frac{22.4\text{ L (STP)}}{\text{mol}} \text{ x } \frac{760\text{ torr}}{770\text{ torr}}$

$\text{x } \frac{298\text{ K}}{273\text{ K}} \text{ x } \frac{1{,}000\text{ mL}}{\text{L}} \Leftrightarrow \mathbf{5{,}740\text{ mL }NO_2}$

11.60 For the initial amount of NH_3

$n = \frac{PV}{RT} = \frac{750\text{ atm}}{760} \text{ x } \frac{0.120\text{ L}}{1} \text{ x } \frac{1\text{ mol K}}{0.0821\text{ L atm}} \text{ x } \frac{1}{298\text{ K}} = 0.00484\text{ mol }NH_3$

For the initial amount of O_2

$n = \frac{PV}{RT} = \frac{635\text{ atm}}{760} \text{ x } \frac{0.165\text{ L}}{1} \text{ x } \frac{1\text{ mol K}}{0.0821\text{ L atm}} \text{ x } \frac{1}{323\text{ K}} = 0.00520\text{ mol }O_2$

Moles of NH_3 that would react with 0.00520 mol O_2 =

$0.00520\text{ mol }O_2 \text{ x } \frac{4\text{ mol }NH_3}{5\text{ mol }O_2} \Leftrightarrow 0.00416\text{ mol }NH_3$

O_2 is the limiting reactant.

0.00484 mol NH_3 present - 0.00416 mol NH_3 reacted = 0.00068 mol of excess NH_3

Moles of gaseous product =

$0.00520\text{ mol }O_2 \text{ x } \frac{10\text{ mol products}}{5\text{ mol }O_2} \Leftrightarrow 0.01040\text{ mol product}$

Total number of moles at the end of the reaction = moles of gaseous products

+ moles of excess NH_3 = 0.01040 mol + 0.00068 = 0.01108 mol.

$P = \frac{nRT}{V} = 0.01108\text{ mol x } \frac{0.0821\text{ L atm}}{\text{mol K}} \text{ x } \frac{423\text{ K}}{0.300\text{ L}} = 1.283\text{ atm}$ or **975 torr**

11.61 $2CO + O_2 \rightarrow 2CO_2$

Moles of CO before reaction =

$$\frac{PV}{RT} = \frac{760 \text{ atm}}{760} \times \frac{0.500 \text{ L}}{1} \times \frac{1 \text{ mol K}}{0.0821 \text{ L atm}} \times \frac{1}{288 \text{ K}} = 0.02115 \text{ mol CO}$$

Moles of O_2 before reaction =

$$\frac{PV}{RT} = \frac{770 \text{ atm}}{760} \times .500\text{L} \times \frac{1 \text{ mol K}}{0.0821 \text{ L atm}} \times \frac{1}{273 \text{ K}} = 0.02260 \text{ mol } O_2$$

The limiting reactant is CO.

$$\text{Moles } CO_2 = 0.02115 \text{ mol CO} \times \frac{2 \text{ mol } CO_2}{2 \text{ mol CO}} \Leftrightarrow 0.02115 \text{ mol } CO_2$$

$$\text{Volume } CO_2 = \frac{nRT}{P} = 0.02115 \text{ mol} \times \frac{0.0821 \text{ L atm}}{\text{mol K}} \times 301 \text{ K}$$

$$\times \frac{760 \text{ atm}^{-1}}{750} = 0.530 \text{ L or } \mathbf{530\ mL\ CO_2}$$

11.62 (a) $$0.420 \text{ mL } Na_2S_2O_3 \times \frac{0.0100 \text{ mol } Na_2S_2O_3}{\text{L } Na_2S_2O_3} \times \frac{1 \text{ L}}{1{,}000 \text{ mL}}$$

$$\times \frac{1 \text{ mol } I_2}{2 \text{ mol } Na_2S_2O_3} \Leftrightarrow \mathbf{2.10 \times 10^{-6}\ mol\ I_2}$$

(b) $\mathbf{2.10 \times 10^{-6}\ mol\ I_2}$

(c) $$2.10 \times 10^{-6} \text{ mol } I_2 \times \frac{1 \text{ mol } O_3}{\text{mol } I_2} \Leftrightarrow \mathbf{2.10 \times 10^{-6}\ mol\ O_3}$$

(d) $$2.10 \times 10^{-6} \text{ mol} \times \frac{22.4 \text{ L}}{\text{mol}} \times \frac{1{,}000 \text{ mL}}{\text{L}} = \mathbf{4.70 \times 10^{-2}\ mL}$$

(e) $$\frac{4.70 \times 10^{-2} \text{ mL}}{200{,}000 \text{ L}} \times \frac{1 \text{ L}}{1{,}000 \text{ mL}} \times 10^6 = \mathbf{0.000235\ ppm}$$

11.63 The sequence of calculations in this problem are: calculate the moles of NO, convert to moles of O_2, and then calculate the volume of the O_2.

$$\frac{PV}{RT} = \frac{750 \text{ atm}}{760} \times 100 \text{ L} \times \frac{\text{mol K}}{0.0821 \text{ L atm}} \times \frac{1}{773 \text{ K}} = 1.55 \text{ moles NO}$$

$$\text{moles } O_2 = 1.55 \text{ mol NO} \times \frac{5 \text{ mol } O_2}{4 \text{ mol NO}} \Leftrightarrow 1.94 \text{ mol } O_2$$

$$\text{vol. } O_2 = \frac{nRT}{P} = 1.94 \text{ mol} \times \frac{0.0821 \text{ L atm}}{\text{mol K}} \times 298 \text{ K} \times \frac{1}{0.895 \text{ atm}} = \mathbf{53.0\ L\ of\ O_2}$$

11.64 (a) Vapor pressure of water at 25°C = 23.8 torr.

Partial pressure of oxygen = 745 torr - 23.8 torr = 721 torr.

Moles of oxygen = n = PV/RT.

$$n = \frac{721 \text{ atm}}{760} \times 0.0350 \text{ L} \times \frac{\text{mol K}}{0.0821 \text{ L atm}} \times \frac{1}{298 \text{ K}} = \mathbf{1.36 \times 10^{-3} \text{ mol } O_2}$$

(b) $1.36 \times 10^{-3} \text{ mol } O_2 \times \frac{2 \text{ mol } KClO_3}{3 \text{ mol } O_2} \times \frac{122.6 \text{ g } KClO_3}{\text{mol } KClO_3} \Leftrightarrow \mathbf{0.111 \text{ g } KClO_3}$

(c) $\frac{0.111 \text{ g } KClO_3}{0.2500 \text{ g sample}} \times 100\% = \mathbf{44.4\%}$

11.65 Dalton's law states that the total pressure exerted by a mixture of gases is equal to the sum of the partial pressures of each gas in the mixture.

11.66 200 torr + 500 torr + 150 torr = **850 torr**

11.67 $P_{N_2} = 300 \text{ torr} \times \frac{2.00 \text{ L}}{1.00 \text{ L}} = 600 \text{ torr}$

$P_{H_2} = 80 \text{ torr} \times \frac{2.00 \text{ L}}{1.00 \text{ L}} = 160 \text{ torr}$

$P_T = P_{N_2} + P_{H_2} = 600 + 160 = 760$ torr or **1.00 atm**

11.68 $P_{N_2} = 740 \text{ torr} \times \frac{20.0 \text{ mL}}{50.0 \text{ mL}} = 296 \text{ torr}$

$P_{O_2} = 640 \text{ torr} \times \frac{30.0 \text{ mL}}{50.0 \text{ mL}} = 384 \text{ torr}$

$P_T = P_{N_2} + P_{O_2} = 296 + 384 =$ **680 torr**

11.69 Oxygen Data

	i	f
P	400 torr	? torr
V	50.0 mL	100 mL
T	60°C	50°C

$$\frac{(400\text{ torr})(50.0\text{ mL})}{(273+60)\text{ K}} = \frac{(?\text{ torr})(100\text{ mL})}{(273+50)\text{ K}}$$

P_{Oxygen} = ? torr = 194 torr

$P_{Nitrogen} = P_{Total} - P_{Oxygen}$ = 800 torr - 194 torr = 606 torr

Nitrogen Data

	i	f
P	400 torr	606 torr
V	X mL	100 mL
T	40°C	50°C

$$\frac{(400\text{ torr})(X\text{ mL})}{(273+40)\text{ K}} = \frac{(606)(100\text{ mL})}{(273+50)\text{ K}}$$

X mL of Nitrogen = **147 mL**

11.70 $P_{Total} = P_{dry\ gas} + P_{H_2O}$

700 torr = $P_{dry\ gas}$ + 23.8 torr

$P_{dry\ gas}$ = 676 torr

Dry Gas Data

	i	f
P	676 torr	760
V	100 mL	? mL
T	298 K	273 K

$$\frac{(676\text{ torr})(100\text{ mL})}{298\text{ K}} = \frac{(760\text{ torr})(?\text{ mL})}{273\text{ K}}$$

? mL = **81.5 mL**

11.71 (a) First find the partial pressure of nitrogen PV = nRT

$$\text{Pressure of } N_2 = \frac{(0.0020\text{ mol})(0.0821\text{ L atm mol}^{-1}\text{ K}^{-1})(308\text{ K})}{0.200\text{ L}}$$

= 0.253 atm or 192 torr

Use $P_A = X_A P_T$ or $X_A = P_A/P_T$ $X_{Nitrogen}$ = 192/720 = **0.27**

(b) **190 torr** (only 2 significant figures)

(c) 720 - 190 = 530 torr (only 2 significant figures)

(d) $$n = \frac{PV}{RT} = \frac{(530/760\text{ atm}^{-1})(0.200\text{ L})}{(0.0821\text{ L atm mol}^{-1}\text{K}^{-1})(308\text{ K})} = \mathbf{0.0055\ mol\ of\ O_2}$$

1.72 Use $P_A = X_AP_T$
P_T = 569 torr + 116 torr + 28 torr + 47 torr = 760 torr
mol fraction N_2 = 569 torr/760 torr = **0.749**
mol fraction O_2 = 116 torr/760 torr = **0.153**
mol fraction CO_2 = 28 torr/760 torr = **0.037**
mol fraction H_2O = 47 torr/760 torr = **0.062**

11.73 $P_T = P_{N_2} + P_{CO_2}$ 900 = 800 + ? pressure CO_2
Partial pressure CO_2 = 100 torr at 20°C and 500 mL
CO_2 Data

	i	f
P	700 torr	100 torr
V	? V	500 mL
T	303 K	293 K

$$\frac{(700\ \text{torr})(?\ \text{V})}{(303\ \text{K})} = \frac{(100\ \text{torr})(500\ \text{mL})}{(293\ \text{K})}$$

? V = **73.9 mL**

11.74 Partial pressure of O_2 = 740 torr - 21.1 torr = 719 torr

$$V = \frac{nRT}{P} = \frac{0.0244\ \text{g}}{32.0\ \text{g/mol}} \times (0.0821\ \text{L atm mol}^{-1}\text{K}^{-1}) \times 296\ \text{K}$$

$$\times \frac{1}{(719/760)\ \text{atm}} = 0.0196\ \text{L} = \mathbf{19.6\ mL}$$

11.75 (a) $n_T = \frac{P_TV}{RT} = \frac{(800/760)(10.0)}{(0.0821)(303)} = \mathbf{0.423\ mol}$

(b) First calculate the moles of N_2 present.

$$0.423\ \text{moles total} = \frac{8.00\ \text{g}\ CO_2}{44.0\ \text{g}\ CO_2/\text{mol}} + \frac{6.00\ \text{g}\ O_2}{32.0\ \text{g}\ O_2/\text{mol}} + \text{moles}\ N_2$$

moles N_2 = 0.423 - 0.182 - 0.188 = 0.053

$$\text{mol fraction}\ N_2 = \frac{0.053\ \text{mol}\ N_2}{0.423\ \text{mol total}} = \mathbf{0.13}$$

$$\text{mol fraction}\ CO_2 = \frac{0.182\ \text{mol}\ CO_2}{0.423\ \text{mol total}} = \mathbf{0.430}$$

$$\text{mol fraction}\ O_2 = \frac{0.188\ \text{mol}\ O_2}{0.423\ \text{mol total}} = \mathbf{0.444}$$

(c) partial pressure = mol fraction x total pressure
partial pressure N_2 = 0.13 x 800 torr = **100 torr**(only 2 significant figures)
partial pressure CO_2 = 0.430 x 800 torr = **344 torr**
partial pressure O_2 = 0.444 x 800 torr = **355 torr**

(d) $0.053\ \text{mol} \times \frac{28\ \text{g}}{\text{mol}} = \mathbf{1.5\ g\ N_2}$

11.76 Data of Dry Gas

	i	f
P	800 - 42	? torr
V	500 mL	250 mL
T	273 + 35	273 + 35

$$\frac{(800 - 42)(500)}{308} = \frac{(?\ \text{torr})(250)}{308}$$

? torr = 1,520 torr

1,520 torr + 42 torr = total pressure at final conditions
= 1,562 torr or **1,560 torr**

11.77 Vap. press. of H_2O at 31°C = 33.7 torr

$$n = \frac{PV}{RT} = \frac{(33.7/760)(1)}{(0.0821)(273 + 31)} = 0.00178 \text{ mol } H_2O$$

$$0.00178 \text{ mol } H_2O \times \frac{18.02 \text{ g } H_2O}{\text{mol } H_2O} = \mathbf{0.0320\ g\ H_2O}$$

11.78 V.P. (H_2O) = 17.5 torr at 20°C
Total P on gases = 763 torr - pressure of the column of H_2O

$$\text{Pressure of the column of water} = 28.4 \text{ mm } H_2O \times \frac{1 \text{ mm Hg}}{13.6 \text{ mm } H_2O}$$

= 2.09 mm Hg = 2.1 torr   Total P = 763 - 2.1 = 760.9

Partial pressure of gas = 760.9 torr - 17.5 torr = 743 torr
Data

	i	f
P	743	760
V	280 mL	? mL
T	293 K	273 K

$$\frac{(743)(280)}{293} = \frac{(760)(?\ \text{mL})}{273}$$

? mL = **255 mL**

11.79 **Effusion** is the escape of a gas, under pressure, through a very small opening, while **diffusion** is the spontaneous mixing of two gases placed in the same container.

11.80 Graham's law states that under identical conditions of temperature and pressure the rate of effusion of gases is inversely proportional to the square root of their densities.

11.81 (a) $\frac{V_{He}}{V_{Ne}} = \sqrt{\frac{M_{Ne}}{M_{He}}}$

(b) **He**

(c) $\frac{\text{Rate of effusion of He}}{\text{Rate of effusion of Ne}} = \sqrt{\frac{20.2}{4}}$ = **2.25 times faster**

11.82 $\frac{V_a}{V_b} = \sqrt{\frac{M_b}{M_a}}$ Let a = CO_2 and b = CH_4

$\frac{\text{speed of } CO_2}{1{,}000 \text{ miles/hr}} = \sqrt{\frac{16.05}{44.01}} = 0.6039$

Speed of CO_2 = 0.6039 x 1,000 = **603.9 miles/hr**

(1,000 miles/hr is assumed to have four significant figures)

11.83 $\frac{\text{rate of effusion of unknown}}{\text{rate of effusion of } NH_3} = 2.92 = \sqrt{\frac{\text{M.M. } (NH_3)}{\text{M. M. (unknown)}}}$

$(2.92)^2 = \frac{17.0}{\text{M.M. (unknown)}}$

$8.526 = \frac{17.0}{\text{M.M. (unknown)}}$

M.M. of the unknown = **1.99 g mol^{-1}**

11.84 A gas is composed of a large number of infinitesimally small particles that are in rapid random motion, and the average kinetic energy of these particles is directly proportional to the absolute temperature.

11.85 Pressure arises from the impacts of the molecules of the gas with the walls of the container.

11.86 At the same temperature, two gases will have the same average kinetic energies. Since K.E. = $1/2mv^2$, if the mass of the molecules of one is less than that of the other, the average velocity must be larger, so the product, $1/2mv^2$, can be the same for both.

11.87 Raising the temperature increases the average kinetic energy and the average velocity of the molecules of a given gas sample. If the molecules have a greater velocity, they will diffuse more rapidly.

11.88 More molecules are forced into a given volume when the volume is decreased. Decreased volume results in each portion of the wall surface having more molecules beside it, so there are more molecule-wall collisions per second and the pressure goes up.

11.89 Raising the temperature increases the pressure because the molecules hit the walls with more force and more often at the higher temperature.

11.90 The transfer of heat energy is from the warm object to the cool one. As the warm object loses heat its temperature decreases as well as the average K.E. of its molecules. The cooler object receiving heat energy begins to warm, increasing its temperature and the average kinetic energy of its molecules. This process will continue until the average kinetic energies of the molecules in both objects become equal. During the heat transfer process, high energy particles of the warm object collide with low energy particles of the cool object and transfer energy (heat) until the K.E. of the molecules of each object are equal.

11.91 See Figure 11.13. The curves are not symmetrical because molecules have a lower limit of speed (zero) but virtually no upper limit.

11.92 Gases that do not obey the combined gas law (or ideal gas law) are said to be non-ideal. This is most evident at high pressures and low temperatures.

11.93 As gases expand, the average distance of separation of the molecules increases. Since real molecules in a gas attract each other somewhat, moving the molecules further apart requires an increase in potential energy at the expense of kinetic energy. Thus, the average kinetic energy of the molecules decreases, which leads to a decrease in the temperature of the gas.

11.94 The a is the constant that corrects for the intermolecular attractive forces and b is the constant that corrects for the excluded volume of the molecules.

11.95 Van der Waals subtracted a correction from the value of the measured volume to exclude the volume occupied by the molecules. He added a correction to the measured pressure to correct for the pressure drop in real molecule systems caused by attractions between molecules.

11.96 $(P + \frac{n^2 a}{V^2})(V - nb) = nRT$

$$\left(P + \frac{1.000^2 \times 0.034}{22.400^2}\right)(22.400 - 1.000 \times 0.0237) = (1.000)(0.082057)(273.15)$$

$(P + 6.776 \times 10^{-5})(22.3763) = 22.414$

$P + 6.776 \times 10^{-5} = 22.414/22.3763 = 1.0017$

$P = 1.0017 - 6.776 \times 10^{-5} =$ **1.002 atm or 1.0 atm**

(This shows that He acts very nearly like an ideal gas.)

11.97 $(P + \frac{n^2 a}{V^2})(V - nb) = nRT$

$$(P + \frac{1.000^2 \times 5.489}{22.400^2})(22.400 - 1.000 \times 0.06380) = 1.000 \times 0.082057 \times 273.15$$

$(P + 0.0109)(22.336) = 22.414$

$P + 0.0109 = 22.414/22.336 = 1.003$

$P = 1.003 - 0.0109 =$ **0.992 atm** (ideal gas would have $P = 1$ atm)

11.98 $(P + \frac{n^2 a}{V^2})(V - nb) = nRT$

For 1 mol this equation becomes

$(P + \frac{a}{V^2})(V - b) = RT$ or

$V - b = \frac{RT}{P + a/V^2}$ or $V = \frac{RT}{P + a/V^2} + b$

The V^2 is part of the pressure term and its effect is small once it has been divided into a. Therefore, one can substitute 22.400 L in for V in the V^2 term and solve for V. This will give an answer slightly different than 22.400. A more precise answer can be obtained by substituting the value obtained for V into the V^2 term and re-solving for V. This should be repeated until the value of V does not change.

$$V_1 = \frac{(0.082057)(273.15)}{1.0000 + (1.36/22.400^2)} + 0.0318 = 22.385$$

$$V_2 = \frac{(0.082057)(273.15)}{1.0000 + (1.36/22.385^2)} + 0.0318 = 22.385$$

(The value did not change).

Answer: **22.385 L/mol at STP** The value in Table 11.1 is 22.397.

11.99 Use the equation $M + HCl \rightarrow 1/2H_2 + MCl$. The equivalent weight of the metal is that mass that will react with 1 mol of H^+ (actually it's that mass that will furnish 1 mol of electrons, but the 2 values are the same). First calculate moles of HCl; then calculate the grams of metal that react with 1 mol of HCl.

Mol $H_2 \Leftrightarrow PV/RT = (680 \text{ torr}/760 \text{ torr atm}^{-1})(0.348 \text{ L})/$
$(0.0821 \text{ L atm mol}^{-1} \text{ K}^{-1})(273 + 24) \text{ K} = 1.28 \times 10^{-2}$ mol H_2

Mol $H^+ = 1.28 \times 10^{-2} \times 2 = 2.56 \times 10^{-2}$

Eq. Wt. = $(0.230 \text{ g}/2.56 \times 10^{-2} \text{ mol}) \times (1 \text{ mol/eq}) =$ **8.98 g/eq**

The metal is aluminum. $Al + 3HCl \rightarrow 3/2\ H_2 + AlCl_3$

11.100 For C: $0.671\ g\ CO_2 \times \frac{1\ mol\ CO_2}{44.01\ g} \times \frac{1\ mol\ C}{mol\ CO_2} \times \frac{12.01\ g\ C}{mol} \Leftrightarrow 0.183\ g\ C$

$$\%\ C = \frac{0.183\ g\ C}{0.4500\ g\ sample} \times 100\% = \mathbf{40.7\%\ C}$$

For H: $0.345\ g\ H_2O \times \frac{1\ mol\ H_2O}{18.02\ g\ of\ H_2O} \times \frac{2\ mol\ H}{mol\ H_2O} \times \frac{1.01\ g\ H}{mol} \Leftrightarrow 0.0387\ g\ H$

$$\%\ H = \frac{0.0387\ g\ H}{0.4500\ g\ sample} \times 100\% = \mathbf{8.59\%\ H}$$

For N: $n = \frac{PV}{RT} = \frac{(740/760)(0.153)}{(0.0821)(273 + 25)} = 0.00609\ mol\ N_2$

$$0.00609\ mol\ N_2 \times \frac{2\ mol\ N}{mol\ N_2} \times \frac{14.01\ g\ N}{mol\ N} = 0.171\ g\ N$$

$$\%\ N = \frac{0.171\ g\ N}{0.3600\ g\ sample} \times 100\% = \mathbf{47.5\%\ N}$$

For O: % O = 100.0% - % N - % C - % H = 100.0 - 47.5 - 40.7 - 8.59 = 3.2% O

$$\text{mol C in 100 g of sample} = \frac{40.7\ g\ C}{12.01\ g/mol} = 3.39\ mol\ C$$

$$\text{mol H in 100 g of sample} = \frac{8.59\ g\ H}{1.01\ g/mol} = 8.50\ mol\ H$$

$$\text{mol N in 100 g of sample} = \frac{47.5\ g\ N}{14.01\ g/mol} = 3.39\ mol\ N$$

$$\text{mol O in 100 g of sample} = \frac{3.2\ g\ O}{16.0\ g/mol} = 0.20\ mol\ O$$

Empirical Formula: $C_{3.39/0.20}H_{8.50/0.20}N_{3.39/0.20}O_{0.20/0.20}$

$= C_{17}H_{42.5}N_{17}O_1 = \mathbf{C_{34}H_{85}N_{34}O_2}$

11.101 $mol\ O_2 = (P_{O_2})\frac{V}{RT} = ((740 - 32)/760)(0.250)/(0.0821)(273 + 30)$

$= 9.36 \times 10^{-3}\ mol\ O_2$

$mol\ N_2 = (P_{N_2})\frac{V}{RT} = ((780 - 24)/760)(0.300)/(0.0821)(273 + 25)$

$= 12.2 \times 10^{-3}\ mol\ N_2$

$mol_{(O_2+N_2)} = 9.36 \times 10^{-3} + 12.2 \times 10^{-3} = 21.6 \times 10^{-3}\ mol$

$P_{(N_2+O_2)} = (n_{(O_2+N_2)})\frac{RT}{V} = (21.6 \times 10^{-3})(0.0821)(273 + 35)/(0.500\ L)$

$= 1.09\ atm$

Total $P_{(N_2+O_2)} = (1.09\ atm \times 760\ torr/atm) + 42.2\ torr =$ **871 torr**

12 STATES OF MATTER AND INTERMOLECULAR FORCES

12.1 Intermolecular attractions are strong in liquids and solids but not in gases because in liquids and solids the particles are very close while in gases the particles are very far apart.

12.2 In gases differences in intermolecular attractive forces are so small that all gases behave pretty much alike, while in solids and liquids the variances in intermolecular attractions result in such differences among the physical properties that no general "liquid laws" comparable to the gas laws can be formulated.

12.3 Attractions between the positive end of one dipole and the negative end of another are called dipole-dipole attractions. They are weaker in a gas because the molecules are further apart.

12.4 Hydrogen bonds are particularly strong dipole-dipole attractions. They are most important for N-H, O-H, and F-H bonds because they contain a small hydrogen atom on a very small electronegative element.

12.5 Instantaneous dipole-induced dipole attractions create what we know as London forces.

12.6

	dipole-dipole	hydrogen bonding	London forces
(a) HCl	x		x
(b) Ar			x
(c) CH_4			x
(d) HF	x	x	x
(e) NO	x		x
(f) CO_2			x
(g) H_2S	x		x
(h) SO_2	x		x

12.7 Hydrogen bonding forces water molecules into a tetrahedral arrangement about each other which leads to a more "open", less dense structure for ice than for the liquid.

12.8 Surface tension, vapor pressure, ΔH_{vap}, boiling point, freezing point, and ΔH_{fus}.

12.9 London forces should increase from helium to neon to argon as the atoms become larger.

12.10 Polarizability is the ease of distortion of the electron cloud of an atom, molecule or ion. The greater the polarizability, the stronger the London forces.

12.11 See Figure 12.21. The boiling point comparisons to other compounds are evidence of hydrogen bonding in H_2O, HF and NH_3.

12.12

	density	rate of diffusion	compressibility	ability to flow
Solid	high	very low	very small	poor
Liquid	medium	medium	very small	good
Gas	low	high	high	good

12.13 Density, rate of diffusion, and compressibility are determined primarily by how tightly the molecules are packed.

12.14 The distance traveled by a molecule between collisions in a liquid is very short compared to a gas. It, therefore, takes a given molecule more time to move a given distance in the liquid.

12.15 At room temperature molecules within solids are very tightly packed and held quite rigidly in place. Therefore, diffusion in solids is virtually nonexistent. At high temperatures the molecules are not as tightly packed and some diffusion can take place.

12.16 Surface tension is proportional to the energy needed to increase the surface area of a liquid. Liquids tend to minimize their surface area and, therefore, lower their energy, by forming spherical droplets.

12.17 Surface tension tends to keep the surface area of the liquid from expanding. When a glass is filled to slightly above its rim, surface tension will keep the liquid from overflowing.

12.18 Soap bubbles strive to form a shape that gives the least surface area to the soap film. That shape is spherical. The lowest surface area will give the lowest possible potential energy state.

12.19 Draw a diagram that contains the two curves shown in Figure 12.13. One curve is at T_1 and the other is at T_2. It is clear from this diagram that more molecules possess the minimum K.E. for the evaporation at the higher temperature. Therefore, the liquid at that higher temperature will evaporate faster.

12.20 The higher energy molecules pass into the gas phase leaving the lower energy molecules in the liquid. Therefore, the average kinetic energy of molecules in the liquid is less and the temperature is lower.

12.21 On a dry day water evaporates rapidly because there is little water vapor in the air. The rate of return of H_2O to the clothes is slow compared to the rate of evaporation. When the wind is blowing, the air immediately surrounding the clothes does not have a chance to saturate. Thus, evaporation can continue at a rapid rate. Also the wind can replace some of the heat needed for evaporation.

12.22 They sublime at the low pressures found at high altitudes.

12.23 A greater fraction of molecules in the warm water have enough energy to escape the surface.

12.24 Sublimation

12.25 Increasing the surface area increases the overall rate of evaporation.

12.26 Methyl alcohol has the weaker intermolecular attractions at that temperature.

12.27 Surface area, temperature, and strengths of the intermolecular attractions determine the rate of evaporation of a liquid.

12.28 The molar heat of vaporization is the amount of energy needed to cause the evaporation of one mole of the liquid at a constant pressure. Because we are really only interested in the difference in energy between the liquid and solid, it is not important that we know the total energy of either state.

12.29 Substance X should have the higher boiling point. Substance Y would be less likely to hydrogen bond.

12.30 ΔH_{vap} and boiling points increase from CH_4 to $C_{10}H_{22}$ because of increased London forces in the same direction. Long chain-like molecules are attracted to one another in more places than shorter molecules since they are more complex than the smaller ones.

12.31 ΔH_{vap} should increase $PH_3 < AsH_3 < SbH_3$ due to the increased size.

12.32 ΔH_{vap} should increase $H_2S < H_2Se < H_2Te$ due to the increased size.

12.33 There are more hydrogen bonds in water than in HF because water has 2 hydrogen atoms whereas HF has only one.

12.34 The gasoline evaporates very rapidly, thereby removing heat faster than the evaporation of water.

12.35 The source of energy in a thunderstorm is the heat of vaporization that is liberated when H_2O condenses.

12.36 When steam contacts skin, it condenses to H_2O at 100°C. Heat equal to the heat of vaporization is given off to the skin. Then the skin is further burned by the hot water produced during condensation. Therefore, steam at 100°C can give up much more heat and cause much more severe burns than can water at 100°C.

12.37 $\frac{\Delta H vap}{B.P.} = \text{const.}$ For CH_4: $\frac{9.20 \text{ kJ/mol}}{(273 - 161)\text{K}} \times 10^3 \text{ J/kJ} = \mathbf{82.1\ J\ mol^{-1}\ K^{-1}}$

For C_2H_6: **76 J mol^{-1} K^{-1}** For C_3H_8: **74.5 J mol^{-1} K^{-1}**

For C_4H_{10}: **81.7 J mol^{-1} K^{-1}** For C_6H_{14}: **83.9 J mol^{-1} K^{-1}**

For C_8H_{18}: **85.2 J mol^{-1} K^{-1}** For $C_{10}H_{22}$: **82.7 J mol^{-1} K^{-1}**

(ΔH_{vap} is directly proportional to B.P. because they both depend on intermolecular attractions.)

12.38 $55.0 \text{ g} \times \frac{1 \text{ mol}}{46.08 \text{ g}} \times \frac{38.6 \text{ kJ}}{\text{mol}} \Leftrightarrow \mathbf{46.1\ kJ}$

12.39 $35.0 \text{ g} \times \frac{1 \text{ mol}}{78.12 \text{ g}} \times \frac{9.92 \text{ kJ}}{\text{mol}} \Leftrightarrow \mathbf{4.44\ kJ}$

12.40 $4.29 \text{ kJ} \times \frac{1}{14.5 \text{ g Hg}} \times \frac{200.6 \text{ g Hg}}{\text{mol Hg}} = \mathbf{59.3\ kJ/mol}$

$\frac{59.3 \text{ kJ}}{\text{mol}} \times \frac{1 \text{ kcal}}{4.184 \text{ kJ}} = \mathbf{14.2\ kcal/mol}$

12.41 K.E. = $1/2mv^2$ = 1/2(68.2 kg)(10.0 mi/hr)^2 (1 hr/60 min)^2 (1 min/60 s)^2

(5280 ft/mi)^2 (12 in./ft)^2 (2.54 cm/in.)^2 (1 m/100 cm)^2 (1 J s^2/kg m^2)

(continued)

12.41 (continued)
$= 6.81 \times 10^2$ J $\quad\quad$ $\Delta H_{fus.}(H_2O) = 5.98$ kJ/mol (Table 12.3)

$$6.81 \times 10^2 \text{J} \times \frac{1 \text{ mol}}{5.98 \text{ kJ}} \times \frac{1 \text{ kJ}}{1{,}000 \text{ J}} \times \frac{18.02 \text{ g}}{\text{mol}} = \mathbf{2.05\ g}$$

12.42 Energy change = heat released via condensation plus heat released via cooling

ΔH_{vap} (H_2O) = 40.6 kJ/mol (Table 12.3) $\quad\quad$ $\Delta H_{cond.}$ (H_2O) = -40.6 kJ/mol

Specific heat (H_2O) = 4.184 J/g °C

Energy change = ($\Delta H_{cond.}$ x mol_{H_2O}) + (sp. heat x g x change in temp.)

Energy change = (-40.6 kJ/mol)(1.00 g/18.02 g mol^{-1})(1000 J/kJ) +

(4.184 J/g °C)(1.00 g)(40°C - 100°C) = (-2253 J) + (-251 J) = -2504 J

= **2.50 kJ of energy absorbed**

12.43 Energy absorbed by the benzene = $\Delta H_{fus.}$ x $mol_{benzene}$

= 9.92 kJ/mol x 10.0 g /(78.06 g/mol) = 1.27 kJ

1.27 kJ = energy also released by the H_2O = sp. heat x g x change in temp.

1.27 kJ = (4.184 J/g °C) x (50.0 g) x (30.0 - ?)°C x (1.00 kJ/1000 J)

6.07 = (30.0 - ?) $\quad\quad$? = **23.9°C**

12.44 Heat gained by less energetic body ⇔ (-1) x heat lost by the more energetic body

($\Delta H_{fus.}$ x mol ice) + (sp. $heat_{(\ell)}$ x mass ice x change temp.) =

(-1)[($\Delta H_{cond.}$ x mol steam) + (sp. $heat_{(\ell)}$ x mass steam x change temp.)]

(5.98×10^3 J/mol x 50.0 g/18.02 g mol^{-1}) + [4.184 J/g °C x 50.0 g x (T - 0.00)°C]

= (-1){(-40.6×10^3 J/mol x 10.0 g/18.02 g mol^{-1}) + [4.184 J/g °C x 10.0 g x

(T - 100.0°C)]} or $1.659 \times 10^4 + (2.092 \times 10^2\ T) =$
$-1(-2.253 \times 10^4 + 41.84\ T - 4.184 \times 10^3)$

$2.51 \times 10^2\ T = 1.01 \times 10^4$ $\quad\quad$ T = **40.2°C**

12.45 The "equilibrium vapor pressure" is the pressure exerted by the gaseous molecules of a substance above the liquid in a closed system at equilibrium.

12.46 The nature of the attractive forces in the liquid phase and the temperature are the two principal factors that determine the magnitude of the vapor pressure.

12.47 Ethyl alcohol

12.48 The vapor pressure of a liquid increases with increasing temperature because at the higher temperature a larger fraction of molecules in the liquid state have sufficient kinetic energy to leave the liquid state. The greater the tendency to evaporate, the greater the vapor pressure.

12.49 The equilibrium vapor pressure depends on the rate of evaporation of the liquid, which is determined by the fraction of molecules having enough K.E. to escape. If the attractive forces are large, this fraction is small.

12.50 Decreasing the volume of a container will simply force more vapor molecules to condense into the liquid leaving the V.P. constant. See Figure 12.19.

12.51 Vapor pressure is the equilibrium pressure of the gas in the equilibrium, liquid $\rightleftharpoons$ gas. The value of the equilibrium gas pressure is independent of volume of liquid or gas. Increased surface area of the liquid will increase the rate at which equilibrium can be attained but not the value at equilibrium. A larger surface area will allow more molecules to enter the gaseous phase but at the same time, if equilibrium exists, more will be condensing to the liquid phase.

12.52 If the beverage is cold enough, water in the air immediately surrounding the glass will be cooled sufficiently to condense it on the vessel. The temperature needs to be below the temperature at which the air around the glass would just be at the equilibrium vapor pressure of water. At the lower temperature some of the water vapor must condense.

12.53 A humidity of 100% implies air saturated with H_2O vapor which exerts a partial pressure equal to the vapor pressure of water at that temperature. Vapor pressure increases with temperature; therefore, the amount of water in a given volume of saturated air also increases.

12.54 As the warm moist air rises, the temperature begins to fall. The decrease in temperature brings about a lowering of the average K.E. rendering more molecules capable of condensing. This increased condensation is in the form of rain. Often the cooling of air as it is forced to rise over a mountain is sufficient to cause the air to exceed 100% humidity or to have a water vapor pressure that exceeds the allowed value at the new temperature. When this happens, rain frequently falls.

12.55 The critical temperature is the highest temperature at which the substance can exist as a liquid regardless of the pressure. The critical pressure is the vapor pressure of a substance at its critical temperature.

12.56 The boundary between liquid and vapor disappears at temperatures above the critical temperature.

12.57 Below its critical temperature, -267.8°C (Table 12.2)

12.58 On a cool day the CO_2 is below its critical temperature (31°C or 88°F) and a separate liquid phase exists.

12.59 The water is able to evaporate more quickly if the saturated atmosphere above the coffee is slowly blown away. This increased rate of evaporation allows the coffee to cool more quickly.

12.60 If solid I_2 has a higher vapor pressure than does solid NaCl at the same temperature, one can conclude that the strength of attractive forces in solid I_2 must be less than those in solid NaCl. This verifies what we already know; I_2 molecules attract each other with very weak forces while NaCl particles are attracted by their neighbors via strong ionic forces.

12.61 The vapor pressure of a solid increases with increased temperature.

12.62 The process is known as freeze-drying.

12.63 Le Chatelier's principle states that "when a system in a state of dynamic equilibrium is disturbed by some outside influence that upsets the equilibrium, the system responds by undergoing a change in a direction that reduces the disturbance and, if possible, brings the system back to equilibrium."

12.64 (a) Increasing the temperature will cause a shift to the right.

(b) Increasing the temperature will cause a shift to the right.

12.65 It will shift the equilibrium to the left since the solid will occupy much less space.

12.66 That temperature at which the vapor pressure of a liquid is equal to the atmospheric pressure is known as the boiling point of the liquid. The boiling point of a liquid at one standard atmosphere, 760 torr, is referred to as its normal boiling point.

12.67 **~88°C**

12.68 (From Figure 12.17) **~73°C** (From Table 11.2) **between 73 and 75°C**

12.69 The stronger the intermolecular attractive forces, the more difficult it is for molecules to break away from the liquid and enter the gaseous state. To make a compound with strong intermolecular forces boil, one has to provide more kinetic energy in the form of a higher temperature.

12.70 For solid $\rightarrow$ liquid, ΔH is the molar heat of fusion.

For liquid $\rightarrow$ solid, ΔH is the molar heat of crystallization.
These two have the same magnitude but have opposite signs.

12.71 The temperature at which solid $\rightleftharpoons$ liquid exists is called either the freezing point or the melting point.

12.72 The only difference in freezing point and melting point depends on whether you imagine approaching it from high or low temperatures.

12.73 Fusion is the process of melting.

12.74 From solid $\rightarrow$ liquid only small changes in separation take place and only small increases in potential energy occur. When a liquid evaporates, large changes in intermolecular distances occur with correspondingly larger potential energy changes.

12.75 Crystalline solids have highly regular, symmetrical shapes. They possess faces that intersect each other at characteristic interfacial angles.

12.76 An amorphous solid has particles arranged in a chaotic fashion. Amorphous solids lack the internal order found in crystals.

12.77 A lattice is a regular or repetitive pattern of points or particles. A unit cell is the grouping of particles that is repeated throughout the solid and generates the entire lattice. We can create an infinite number of chemical structures by simply varying the chemical environment about each point in a lattice.

12.78 With the use of the Bragg equation, $2d \sin \theta = n\lambda$, we can find the spacing between the successive layers of particles in a crystal: d is the spacing between successive layers, θ is the angle at which X rays enter and leave, λ is the wavelength of the X rays and n is an integer.

12.79 The quantities a, b & c, corresponding to the edge lengths of the cell and α, β & γ, corresponding to the angles at which the edges intersect one another, describe a particular lattice.

12.80 See Figure 12.30 (a), (b), and (c).

12.81 (a) Its lattice arrangement is a face-centered cubic lattice of chloride ions with sodium ions at the center of each edge and in the center of the unit cell.

(b) Cl^- 8 corners x 1/8Cl^- per corner = 1 Cl^-

6 faces x 1/2Cl^- per face = 3 Cl^-

Total = 4 Cl^-

Na^+ 12 edges x 1/4Na^+ per edge = 3 Na^+

1 center x 1 Na^+ per center = 1 Na^+

4 Na^+

Na_4Cl_4 or **4 Formula Units per Unit Cell**

(c) **No.** The cations and anions would need to have the same magnitude of charge if all the corners and faces are to be occupied (i. e., anion/cation ratio of one-to-one).

12.82 (a) $2d \sin \theta = n\lambda$ $n = 1$

$2 \times 1{,}000 \times \sin \theta = 1 \times 229$ $\sin \theta = 229/2{,}000 = 0.1145$ $\theta =$ **6.57°**

(b) $2 \times 200 \times \sin \theta = 1 \times 229$ $\sin \theta = 229/400 = 0.5725$ $\theta =$ **34.9°**

12.83 $2d \sin \theta = n\lambda$ $n = 1$ $\lambda = 141$ pm

(a) $2d \sin 20.0° = 1 \times 141$

$2 \times d \times 0.342 = 141$ $d = 141/0.684 =$ **206 pm**

(b) $2d \sin 27.4° = 1 \times 141$ $d = 141/0.920 =$ **153 pm**

(c) $2d \sin 35.8° = 1 \times 141$ $d = 141/1.170 =$ **121 pm**

12.84 If $n = 1$, $2 \times 200 \sin \theta = 1 \times 141$, $\theta = 20.6°$

If $n = 2$, $2 \times 200 \sin \theta = 2 \times 141$, $\theta = 44.8°$

If $n = 3$, $2 \times 200 \sin \theta = 3 \times 141$, $\sin \theta = 1.06$, impossible, $\sin \theta \leq 1$

Answers: **20.6°** and **44.8°**

12.85 (Draw a picture of the unit cell). In the body-centered cubic unit cell there is a sphere at each corner and in the very center. The corner spheres each touch the center sphere but they do not touch each other. The distance from one corner to the corner diagonally through the unit cell is 4 radii or 4 r. From the unit cell two right triangles can be constructed. The first involves 2 edges (E) and the diagonal across the face (DF). The other involves the unit cell diagonal (DC), an edge (E), and the diagonal across the face (DF). In the first triangle we can use the known edge length (E) to calculate the length of the diagonal across the face of the unit cell (DF). Then in the second triangle we can use the now known length of the diagonal across the face (DF) and the edge length (E) to calculate the unit cell diagonal (DC). Knowing the unit cell diagonal (DC) is 4r, we then have the radius of the atom.

First Triangle $(E)^2 + (E)^2 = (DF)^2$

$(288.4)^2 + (288.4)^2 = (DF)^2$ DF = 407.86 pm

Second Triangle $(E)^2 + (DF)^2 = (DC)^2$

$(288.4)^2 + (407.86)^2 = (DC)^2$ $249{,}500 = (DC)^2$

DC = 499.5 pm 4r = 499.5 pm **r = 124.9 pm**

12.86 A diagonal (DF) drawn across the face of the unit cell forms a right triangle with 2 edges (E) as the two sides of the triangle. The diagonal (DF) is 4 radii.

$(DF)^2 = (E)^2 + (E)^2 = (407.86)^2 + (407.86)^2$

DF = 576.80 = 4r **r = 144.20 pm**

12.87 (See the discussion with Problem 12.86)

$(4r)^2 = (DF)^2 = (E)^2 + (E)^2$ $(4 \times 143)^2 = 2(E)^2$

$E^2 = 163{,}600$ E (or length) = **404 pm**

12.88 (See the discussion with Problem 12.85)

$(E)^2 + (E)^2 = (DF)^2$

$(412.3)^2 + (412.3)^2 = DF^2$ DF = 583.1

$(E)^2 + (DF)^2 = (DC)^2$ $(412.3)^2 + (583.1)^2 = (DC)^2$

DC = 714.1 DC = 2r (Cs^+) + 2r (Cl^-)

714.1 = 2r (Cs^+) + (2 x 181 pm) r (Cs^+) = **176 pm**

12.89 Unit cell edge length = 658 pm

658 pm = 2r (Rb^+) + 2r (Cl^-) 658 pm = 2r(Rb^+) + (2 x 181 pm)

r (Rb^+) = 148 pm **148 pm** or **1.48 angstroms**

12.90 For a discussion of the body-centered unit cell, see the solution to Problem 12.85. For a discussion of the face-centered unit cell, see the solution to Problem 12.86. For the simple cubic E = 2 radii or 2r. The body-centered unit cell has 2 atoms per unit cell. The face-centered unit cell has 4 atoms per unit cell (See Problem 12.81). The simple cubic unit cell has 1 atom per unit cell.

(a) Volume = $(E)^3 = (2r)^3 = (2 \times 144 \text{ pm})^3 = 2.39 \times 10^7 \text{ pm}^3$

$$\frac{107 \text{ g Ag}}{\text{mol}} \times \frac{1 \text{ mol}}{6.022 \times 10^{23} \text{atoms}} \times \frac{1 \text{ atom}}{\text{unit cell}} \times \frac{1 \text{ unit cell}}{2.39 \times 10^7 \text{pm}^3} \times \frac{(10^{10}\text{pm})^3}{\text{cm}^3}$$

$$= \mathbf{7.50 \ g/cm^3}$$

(b) $(E)^2 + (E)^2 = (DF)^2$ and $(E)^2 + (DF)^2 = (DC)^2$

$(DC)^2 = (4r)^2 = (4 \times 144 \text{ pm})^2 = 3.32 \times 10^5 \text{ pm}^2$

$3.32 \times 10^5 \text{ pm}^2 = (E)^2 + (E)^2 + (E)^2$ E = 333 pm

Volume = $E^3 = (333)^3 = 3.68 \times 10^7 \text{ pm}^3$

$$\frac{107 \text{ g Ag}}{\text{mol}} \times \frac{1 \text{ mol}}{6.022 \times 10^{23} \text{atoms}} \times \frac{2 \text{ atoms}}{\text{unit cell}} \times \frac{1 \text{ unit cell}}{3.68 \times 10^7 \text{pm}^3} \times \frac{(10^{10}\text{pm})^3}{\text{cm}^3}$$

$$= \mathbf{9.74 \ g/cm^3}$$

(c) $(E)^2 + (E)^2 = (DF)^2 = (4r)^2$ $2(E)^2 = (4 \times 144 \text{ pm})^2$

E = 407 pm Volume = $(E)^3 = 6.76 \times 10^7 \text{ pm}^3$

$$\frac{107 \text{ g Ag}}{\text{mol}} \times \frac{1 \text{ mol}}{6.022 \times 10^{23} \text{atoms}} \times \frac{4 \text{ atoms}}{\text{unit cell}} \times \frac{1 \text{ unit cell}}{6.76 \times 10^7 \text{pm}^3} \times \frac{(10^{10}\text{pm})^3}{\text{cm}^3}$$

$$= \mathbf{10.6 \ g/cm^3}$$

The density calculated for the **face-centered cubic** structure (part c) is closest to the actual density

12.91 Volume of sphere = $4/3\pi\, r^3 = 4/3\pi\, (50\text{ pm})^3 = 5.236 \times 10^5\text{ pm}^3$
(a) volume of primitive cubic unit cell (See Problem 12.90).

$V = (2r)^3 = (2 \times 50\text{ pm})^3 = 1.0 \times 10^6\text{ pm}^3$

Vacant space = $1.0 \times 10^6\text{ pm} - 5.24 \times 10^5\text{ pm}^3 = 4.8 \times 10^5\text{ pm}^3$

$\frac{4.8 \times 10^5}{1.0 \times 10^6} \times 100\% =$ **48% vacant**

(b) For a discussion of the body-centered cubic unit cell see Problem 12.85.
$(4r)^2 = (DC)^2 = (E)^2 + (E)^2 + (E)^2$ $(4 \times 50\text{ pm})^2 = 3(E)^2$

$E = 115\text{ pm}$ $V = (115\text{ pm})^3 = 1.52 \times 10^6\text{ pm}^3$ (2 spheres per unit cell)

Vacant space = $1.52 \times 10^6\text{ pm}^3 - (2 \times 5.24 \times 10^5\text{ pm}^3) = 4.7 \times 10^5\text{ pm}^3$

$\frac{4.7 \times 10^5}{1.52 \times 10^6} \times 100\% =$ **31% vacant**

(c) For a discussion of the face-centered cubic unit cell see Problem 12.86.
$(4r)^2 = (DF)^2 = (E)^2 + (E)^2$ $(4 \times 50)^2 = 2(E)^2$

$E = 141\text{ pm}$

$V = (141\text{ pm})^3 = 2.80 \times 10^6\text{ pm}^3$ (Four spheres per unit cell)

Vacant space = $2.80 \times 10^6 - (4 \times 5.24 \times 10^5) = 7.04 \times 10^5\text{ pm}^3$

$\frac{7.04 \times 10^5}{2.80 \times 10^6} \times 100\% =$ **25.1% vacant**

12.92 The Br^- forms a face-centered cubic unit cell. (See Problem 12.86).
$(4r)^2 = (DF)^2 = (E)^2 + (E)^2$ $(4r)^2 = (550\text{ pm})^2 + (550\text{ pm})^2$

$r = 194\text{ pm}$ = radius of Br^- Edge = (2 x radius of Br^-) + (2 x radius Li^+)

$550\text{ pm} = (2 \times 194\text{ pm}) + 2r(Li^+)$ **r = 81pm**

The 81 pm represents the maximum radius of the Li^+. If it is smaller, it simply will not fill all the available space between the bromide ions.

12.93 Volume of the cube = $(412.3\ \text{pm})^3 = 7.009 \times 10^7\ \text{pm}^3$

Atoms per unit cell: if face-centered = 4CsCl (See 12.85)

if body-centered = 2CsCl

Calculated density of the face-centered unit cell =

$$\frac{4 \text{ formula units CsCl}}{7.009 \times 10^7 \text{pm}^3} \times \frac{168.4 \text{ g}}{\text{mole}} \times \frac{1 \text{ mole}}{6.022 \times 10^{23} \text{ form. units}} \times \frac{(10^{10}\text{pm})^3}{\text{cm}^3}$$

$= \mathbf{15.96\ g/cm^3}$

Not very close to the known value of 3.99 g/cm³

Calculated density of the body-centered unit cell =

$$\frac{2 \text{ formula units CsCl}}{7.009 \times 10^7 \text{pm}^3} \times \frac{168.4 \text{ g}}{\text{mole}} \times \frac{1 \text{ mole}}{6.022 \times 10^{23} \text{ form. units}} \times \frac{(10^{10}\text{pm})^3}{\text{cm}^3}$$

$= \mathbf{7.979\ g/cm^3}$ **Not very close to the known value of 3.99 g/cm³**
(Calculation of the density of the primitive cubic cell yields 3.99 g/cm³, the value of CsCl.)

12.94 $$\frac{2 \text{ Na atoms}}{\text{unit cell}} \times \frac{22.99 \text{ g Na}}{\text{mol}} \times \frac{1 \text{ mol}}{6.022 \times 10^{23} \text{atoms}} \times \frac{1 \text{ cm}^3}{0.97 \text{ g}} = \frac{7.87 \times 10^{-23} \text{ cm}^3}{\text{unit cell}}$$

Edge = $(7.87 \times 10^{-23}\ \text{cm}^3)^{1/3} = 4.3 \times 10^{-8}$ cm or **0.43 nm**

12.95 vol/unit cell = $[546.26\ \text{pm} \times (10^{-10}\ \text{cm/pm})]^3 = 1.6300 \times 10^{-22}\ \text{cm}^3$

$$\frac{1.6300 \times 10^{-22} \text{ cm}^3}{\text{unit cell}} \times \frac{3.180 \text{ g}}{\text{cm}^3} \times \frac{1 \text{ mol}}{78.08 \text{ g}} \times \frac{6.022 \times 10^{23} \text{ form. units}}{\text{mol}}$$

= **3.998 formula units/unit cell ≈ 4**

12.96 (See the solution to Problem 12.81)

$$\frac{\text{vol}}{\text{unit cell}} = \frac{4 \text{ formula units}}{\text{unit cell}} \times \frac{58.44 \text{ g}}{\text{mol}} \times \frac{1 \text{ mol}}{6.022 \times 10^{23} \text{ form. units}}$$

$$\frac{1 \text{ cm}^3}{2.165 \text{ g}} \times \frac{(10^{10} \text{ pm})^3}{\text{cm}^3} = \frac{1.793 \times 10^8 \text{ pm}^3}{\text{unit cell}}$$

Edge length = $(1.793 \times 10^8 \text{ pm}^3)^{1/3} = 563.9$ pm

Edge length = 2 x radius Cl^- + 2 x radius Na^+

563.9 pm = (2 x 181 pm) + 2 radius Na^+ radius Na^+ = **101 pm**

12.97 Please see Table 12.5.

12.98 (a) molecular (b) molecular
(c) metallic (d) ionic
(e) molecular (f) ionic
(g) molecular

12.99 (a) molecular (b) ionic
(c) ionic (d) metallic
(e) covalent (f) molecular
(g) ionic

12.100 $SnCl_4$ is likely, based on the data given, to form a molecular solid, while $SnCl_2$ would form an ionic solid.

12.101 Boron is probably a covalent solid based on the data given.

12.102 Paraffin is a molecular solid.

12.103 OsO_4 is probably a molecular solid.

12.104 Calcite is probably an ionic solid.

12.105 Prepare a heating curve like that in Figure 12.38 and give the actual values of T_f, T_b, ΔH_{fus} and ΔH_{vap}.

12.106 The temperature cannot rise above the melting point as long as there is liquid in contact with solid.

12.107 Glass is an amorphous solid. The disorientation of the molecules gives rise to a melting point range.

12.108 When water is cooled, it reaches 32°F (0°C) at which it may be present as liquid water until further heat is extracted. As additional heat is removed, freezing will begin. The answer to the question is yes.

12.109 All the heat that is added is used to increase the potential energy of the molecules being converted to the gas phase. Therefore, the average kinetic energy and the temperature of all the molecules (both gaseous and liquid) remains constant until all the molecules have become gaseous.

12.110 Prepare a diagram like that in Figure 12.41. Position the triple point at 150 torr and 20°C. Have the solid-liquid line pass through 760 torr and 25°C and the liquid-gas line pass through 760 torr and 95°C with both intersecting at the triple point.

12.111

At 22°C

State	Pressure (torr)
vapor	up to 160
vapor-liquid	160
liquid	160 to 250
solid-liquid	250
solid	250 to 1,000

At 10°C

State	Pressure (torr)
vapor	up to 75
solid-vapor	75
solid	75 to 1,000

12.112 Draw a curve like that in Figure 12.38. Indicate melting point (at 25°C), boiling point (at 95°C), heat of fusion and heat of vaporization.

12.113 The density of the solid is greater than that of the liquid because the solid-liquid line leans to the right (connecting the triple point, 20° - 150 torr, and the melting point, 25° - 760 torr).

12.114 Solid, vapor, solid, vapor, solid, liquid.

12.115 The triple point of I_2 occurs at a temperature above that to which it was heated and at a pressure above atmospheric pressure.

12.116 Increases in pressure lead to the production of the more dense phase.
(a) Therefore, increases in pressure will cause the production of more liquid in the case of H_2O. The melting point will be lowered.
(b) Increases in pressure will cause the production of more solid in the case of CO_2. The melting point will be increased.

12.117 For (b) boiling point, (c) heat of vaporization, (d) surface tension, (f) heat of sublimation, and (g) critical temperature to be arranged from highest value to lowest value, the compounds would follow the order: propylene glycol, isopropyl alcohol, acetone, methyl ethyl ether, and butane. For (a) vapor pressure and (e) rate of evaporation, the order would be reversed with butane having the highest value.

12.118 Joules required equals the heat needed for five steps: (1) warming the ice, (2) melting the ice, (3) warming the water, (4) boiling the water, and (5) warming the steam.

(1) heat = $(2.05\ J\ g^{-1}\ {}^\circ C^{-1})(1.00\ mol \times 18.02\ g\ mol^{-1})(20^\circ C) = 739\ J$
(2) heat = $(5.98 \times 10^3\ J\ mol^{-1})(1.00\ mol) = 5.98 \times 10^3\ J$
(3) heat = $(4.18\ J\ g^{-1}\ {}^\circ C^{-1})(1.00\ mol \times 18.02\ g\ mol^{-1})(100^\circ C) = 7.53 \times 10^3\ J$
(4) heat = $(40.6 \times 10^3\ J\ mol)(1.00\ mol) = 4.06 \times 10^4\ J$
(5) heat = $(2.01\ J\ g^{-1}\ {}^\circ C^{-1})(1.00\ mol \times 18.02\ g\ mol^{-1})(20^\circ C) = 724\ J$

Total = **5.56×10^4 J**

12.119 AY + YB = extra distance travelled by more penetrating ray = $n\lambda$; XY = d

$$\sin\theta = \frac{YB}{XY} = \frac{YB}{d} \text{ and } \sin\theta = \frac{AY}{XY} = \frac{AY}{d} \text{ or } YB = d\sin\theta \text{ and } AY = d\sin\theta$$

therefore, $AY + YB = d\sin\theta + d\sin\theta = n\lambda$

$2d\sin\theta = n\lambda$

12.120 Body-centered unit cell = 2 atoms/unit cell

Vol. of the unit cell = $(288.4\ pm)^3 = 23{,}990{,}000\ pm^3 = 2.399 \times 10^{-23}\ cm^3$

$$\frac{51.996\ g\ Cr}{1\ mol\ Cr} \times \frac{cm^3}{7.19\ g} \times \frac{1\ unit\ cell}{2.399 \times 10^{-23}\ cm^3} \times \frac{2\ atoms}{unit\ cell}$$

= 6.03×10^{23} atoms/mole

13 PHYSICAL PROPERTIES OF COLLOIDS AND SOLUTIONS

13.1 Pure substances, unlike mixtures, have constant composition.

13.2 Suspensions have dispersed particles of greater than 1,000 nm in size. Fine sand suspended in water and snow being blown about through the air are two examples of suspensions.

13.3 **From 1nm to 1,000 nm**

(a) **1×10^{-6} mm to 1×10^{-3} mm**

(b) **3.9×10^{-8} in. to 3.9×10^{-5} in.**

13.4 Filtration, centrifugation, and settling under the influence of gravity are methods for separation of a suspension of a solid in a liquid.

13.5 By spinning a sample in a centrifuge, the centrifugal force thus produced behaves as a very powerful artificial gravity and drives the suspended particles to the bottom of the container.

13.6 The scattering of light by colloidal dispersions allows one to view from the side a focused beam of light as it passes through the dispersions. The scattering of light by colloidal dispersions is known as the Tyndall effect. Only colloidal dispersions show the Tyndall effect because the dispersed particles in solution are too small to deflect light and the dispersed particles in suspensions will not deflect the light but will instead block out the light.

13.7

	Dispersing phase	Dispersed phase	Kind of colloid
(a) Styrofoam	solid	gas	solid foam
(b) cream	liquid	liquid	emulsion
(c) lard	solid	liquid	solid emulsion
(d) jelly	liquid	solid	gel, sol
(e) liq. rubber cement	liquid	solid	sol, gel

13.8 Emulsifying agents stabilize emulsions by preventing the dispersed particles from sticking to each other when they collide.

13.9 Test for the Tyndall effect.

13.10 An electrical charge on the surface of dispersed particles in a colloidal dispersion will stabilize the dispersion. A colloid stabilized in this way can be coagulated by removal or neutralization of the electrical charge.

13.11 First AgCl forms and begins to form dispersed particles. The excess Cl^- adsorbs on the AgCl particles and acts to prevent them from growing too large to remain suspended. As more Ag^+ is added, the electrical charge is neutralized by the formation of more AgCl and the particles grow in size and will settle out of solution.

13.12 A gas dispersed in a gas. Because gases will mix on the molecular level.

13.13 Solid, liquid and gaseous solutions are possible.

13.14 The dispersed particles in a solution are smaller than those of colloidal dispersions or suspensions, on the order of the size of molecules or ions.

13.15 Substitutional solid solutions exist when atoms, molecules, or ions of the solute replace particles of the solvent in the crystalline lattice. Brass is an example.

13.16 Interstitial solid solutions exist when solute particles fit into spaces between "solvent" particles in the host lattice. Tungsten carbide is an example.

13.17 An interstitial solid solution would be more likely to exist under the conditions described in this question.

13.18 Mole fraction = moles solute/total moles of all species in the solution.

Mole percent = 100% x mole fraction.

Weight fraction = mass of a particular solute/total mass of all components.

Weight percent = 100% x weight fraction.

Molarity = moles solute/volume of solution (in liters).

Molality = moles solute/kilograms of solvent.

13.19 All concentration units are ratios (fractions).

13.20 45.0 g $C_3H_5(OH)_3$ = 45.0 g/92.1 g/mol = 0.489 mol $C_3H_5(OH)_3$

100.0 g H_2O = 100.0 g/18.02 g/mol = 5.549 mol H_2O

$$X_{glycerin} = \frac{0.489 \text{ mol}}{0.489 \text{ mol} + 5.549 \text{ mol}} = \mathbf{0.0810}$$

$$\omega_{glycerin} = \frac{45.0 \text{ g}}{45.0 \text{ g} + 100.0 \text{ g}} = \mathbf{0.310}$$

weight % glycerin = 0.310 x 100% = **31.0%**

$$m_{glycerin} = \frac{0.489 \text{ mol glycerin}}{0.100 \text{ kg } H_2O} = \mathbf{4.89\ m}$$

13.21 45.0 g of benzene = 45.0 g/78.06 g/mol = 0.576 mol benzene

80.0 g of toluene = 80.0 g/92.15 g/mol = 0.868 mol toluene

(a) $\text{weight \% toluene} = \frac{80.0 \text{ g}}{80.0 \text{ g} + 45.0 \text{ g}} \times 100\% = \mathbf{64.0\%\ toluene}$

$\text{weight \% benzene} = \frac{45.0 \text{ g}}{80.0 \text{ g} + 45.0 \text{ g}} \times 100\% = \mathbf{36.0\%\ benzene}$

(b) $X_{toluene} = \frac{0.868 \text{ mol toluene}}{0.868 \text{ mol} + 0.576 \text{ mol}} = \mathbf{0.601}$

$X_{benzene} = \frac{0.576 \text{ mol benzene}}{0.868 \text{ mol} + 0.576 \text{ mol}} = \mathbf{0.399}$

(c) $m = \frac{0.576 \text{ mol benzene}}{0.0800 \text{ kg toluene}} = \mathbf{7.20\ m}$

13.22 $\text{Weight of 1 L soln.} = \frac{1.107 \text{ g}}{\text{mL}} \times \frac{1{,}000 \text{ mL}}{\text{L}} = 1{,}107 \text{ g/L}$

Weight of water in 1 L soln. = 1,107 g - 121.8 g = 985 g H_2O

Moles of $Zn(NO_3)_2$ in 1 L soln. = 121.8 g/189.4 g/mol = 0.6431 mol

(continued)

13.22 (continued)

(a) Weight % $Zn(NO_3)_2 = \dfrac{121.8\ g}{1{,}107\ g} \times 100\% =$ **11.00%**

(b) m = 0.6431 mol/0.985 kg H_2O = **0.653 m**

(c) X of $Zn(NO_3)_2 = \dfrac{0.6431\ mol}{0.6431\ mol + (985\ g/18.02\ g/mol)} =$ **0.0116**

(d) M = 0.6431 mol/1 L soln. = **0.6431 M**

13.23 40.0 mol of H_2O ⇔ 40.0 mol x 18.02 g/mol or 721 g

0.30 mol of $CuCl_2$ ⇔ 0.30 mol x 134.5 g/mol or 40 g

(a) X of $CuCl_2$ = 0.30/(0.30 + 40.0) = **0.0074**

(b) m of $CuCl_2$ = 0.30 mol/0.721 kg H_2O = **0.42 m**

(c) weight % $CuCl_2$ = [40 g/(40 + 721) g] x 100% = **5.3%**

13.24 Calculate the mass in grams of $CHCl_3$ in one kilogram of solution.

$$1.00\ \text{kg soln.} \times \frac{1{,}000\ g}{kg} \times \frac{12.4\ g\ CHCl_3}{10^6\ g\ soln.} \Leftrightarrow 1.24 \times 10^{-2}\ g\ CHCl_3$$

(a) Weight % $CHCl_3$ = (1.24×10^{-2} g/1,000 g) x 100% = **1.24×10^{-3}%**

(b) For a very dilute solution like the one in this problem, its density will be 1.00 g/mL (See Question 13.29). Therefore, 1.00 kg of solution will occupy 1.00 L.

$$\text{M of } CHCl_3 = \frac{1.24 \times 10^{-2} g/119.5\ g\ mol^{-1}}{1.00\ L} = \mathbf{1.04 \times 10^{-4}\ M}$$

13.25 $X_{alcohol} = 0.250 = \dfrac{0.250\ mol\ alcohol}{0.250\ mol\ alc. + ?\ mol\ H_2O}$

? mol H_2O= 0.750 mol

(continued)

13.25 (continued)

$$\text{wt. \% alc.} = \frac{0.250 \text{ mol x } 60.1 \text{ g mol}^{-1}}{(0.250 \text{ mol x } 60.1 \text{ g mol}^{-1})+(0.750 \text{ mol x } 18.02 \text{ g mol}^{-1})} \text{ x } 100\%$$

$$= \frac{15.02 \text{ g}}{28.54 \text{ g}} \text{ x } 100\% = \mathbf{52.6\,\%}$$

$$\text{m (alc).} = \frac{0.250 \text{ mol alc.}}{0.01352 \text{ kg } H_2O} = \mathbf{18.5\ m}$$

13.26 9.6 g $NaHCO_3$ = 9.6 g/84.01 g mol^{-1} = 0.11 mol

100 g H_2O = 100 g/18.02 g mol^{-1} = 5.55 mol

$$X(NaHCO_3) = \frac{0.11}{0.11 + 5.55} = \mathbf{0.019} \qquad m(NaHCO_3) = \frac{0.11 \text{ mol}}{0.100 \text{ kg}} = \mathbf{1.1\ m}$$

13.27 $6.25 \text{ m} = \dfrac{6.25 \text{ mol NaCl}}{1.000 \text{ kg } H_2O}$

$$X_{NaCl} = \frac{6.25 \text{ mol NaCl}}{6.25 \text{ mol NaCl} + (1{,}000 \text{ g } H_2O/\ 18.02 \text{ g mol}^{-1})} = \frac{6.25}{6.25 + 55.49} = \mathbf{0.101}$$

$$\omega_{NaCl} = \frac{6.25 \text{ mol NaCl x } 58.44 \text{ g mol}^{-1}}{(6.25 \text{ mol NaCl x } 58.44 \text{ g mol}^{-1}) + 1{,}000 \text{ g } H_2O} = \frac{365 \text{ g}}{365 \text{ g} + 1{,}000 \text{ g}} = \mathbf{0.267}$$

13.28 $14.0\%\ Na_2CO_3 = \dfrac{14.0 \text{ g } Na_2CO_3}{14.0 \text{ g } Na_2CO_3 + (100 - 14.0) \text{ g } H_2O}$

$$\text{X of } Na_2CO_3 = \frac{14.0 \text{ g } Na_2CO_3/106 \text{ g mol}^{-1}}{(14.0 \text{ g } Na_2CO_3/106 \text{ g mol}^{-1}) + 86.0 \text{ g } H_2O/18.02 \text{ g mol}^{-1}}$$

$$= \frac{0.132}{0.132 + 4.77} = \mathbf{0.0269}$$

$$\text{m of } Na_2CO_3 \text{ soln} = \frac{0.132 \text{ mol } Na_2CO_3}{0.0860 \text{ kg } H_2O} = \mathbf{1.53\ m}$$

13.29 (a) $M = \dfrac{150 \times 10^{-3}\ \text{g}/24.3\ \text{g mol}^{-1}}{1.00\text{L}} = \mathbf{0.00617\ M\ Mg^{2+}}$

(b) $m = \dfrac{150 \times 10^{-3}\ \text{g}/24.3\ \text{g mol}^{-1}}{1\ \text{kg} - 150 \times 10^{-6}\text{kg}} = \dfrac{6.173 \times 10^{-3}\ \text{mol}}{0.99985\ \text{kg}} = \mathbf{0.00617\ m\ Mg^{2+}}$

Note: For very dilute solutions molarity and molality are equal to 2 or 3 significant figures.

13.30 $m = \dfrac{96.0\ \text{g}/98.1\ \text{g mol}^{-1}}{4.0\ \text{g}\ H_2O \times 1\ \text{kg}/1{,}000\ \text{g}} = \mathbf{2.4 \times 10^2\ m\ H_2SO_4}$

$$X_{\text{sulfuric acid}} = \frac{96.0\ \text{g}/98.1\ \text{g mol}^{-1}}{(96.0\ \text{g}/98.1\ \text{g mol}^{-1}) + (4.0\ \text{g}/18.02\ \text{g mol}^{-1})}$$

$$= \frac{0.9786}{0.9786 + 0.2220} = \mathbf{0.82}$$

$$X_{H_2O} = \frac{0.2220}{0.9786 + 0.2220} = \mathbf{0.18}$$

13.31 $2.25\ m\ NH_4NO_3\ \text{soln.} = \dfrac{2.25\ \text{mol}\ NH_4NO_3}{1.00\ \text{kg}\ H_2O}$

$$\text{Weight \%}\ NH_4NO_3 = \frac{2.25\ \text{mol}\ NH_4NO_3 \times 80.06\ \text{g mol}^{-1}}{2.25\ \text{mol} \times 80.06\ \text{g mol}^{-1} + 1{,}000\ \text{g}\ H_2O} \times 100\%$$

$$= \frac{180.1}{180.1 + 1{,}000} = \mathbf{15.3\%\ NH_4NO_3}$$

$$X_{NH_4NO_3} = \frac{2.25\ \text{mol}\ NH_4NO_3}{2.25\ \text{mol}\ NH_4NO_3 + 1{,}000\ \text{g}\ H_2O/18.02\ \text{g mol}^{-1}}$$

$$= \frac{2.25}{2.25 + 55.49} = \mathbf{0.0390}$$

$$X_{H_2O} = \frac{55.49}{2.25 + 55.49} = \mathbf{0.9610}$$

13.32 $X_{benzene} = 0.240$ $X_{chloroform} = 1 - 0.240 = 0.760$

(a) mole percent $CHCl_3 = 0.760 \times 100\% =$ **76.0 mol %**

(b) $$m\,(C_6H_6) = \frac{0.240 \text{ mol } C_6H_6}{0.760 \text{ mol } CHCl_3 \times 119.4 \text{ g mol}^{-1} \times 1 \text{ kg}/1{,}000 \text{ g}}$$

$= 2.64 \text{ mol kg}^{-1} =$ **2.64 m C_6H_6**

(c) $$m\,(CHCl_3) = \frac{0.760 \text{ mol } CHCl_3}{0.240 \text{ mol } C_6H_6 \times 78.12 \text{ g mol}^{-1} \times 1 \text{ kg}/1{,}000 \text{ g}}$$

$= 40.5$ mol/kg $=$ **40.5 m $CHCl_3$**

(d) Weight % C_6H_6

$$= \frac{0.240 \text{ mol} \times 78.12 \text{ g/mol}}{(0.240 \text{ mol} \times 78.12 \text{ g/mol}) + (0.760 \text{ mol} \times 119.3 \text{ g/mol})} \times 100\%$$

$$= \frac{18.749}{18.749 + 90.668} \times 100\% = \mathbf{17.1\,\%}$$

Weight % $CHCl_3$ = 100.0% - 17.1% = 82.9%

13.33 $$m = \frac{222.6 \text{ g}/62.08 \text{ g mol}^{-1}}{0.2000 \text{ kg } H_2O} = \mathbf{17.93 \text{ m}}$$

$$M = \frac{222.6 \text{ g}/62.08 \text{ g mol}^{-1}}{(200.0 \text{ g} + 222.6 \text{ g})(1 \text{ mL}/1.072 \text{ g})(1 \text{ L}/1{,}000 \text{ mL})} = 9.096 \frac{\text{mol}}{\text{L}} = \mathbf{9.096 \text{ M}}$$

13.34 4.03 M ethylene glycol is 4.03 mol ethylene glycol in 1.000 L of solution. The 1.000 L of solution weighs 1.045 g/mL x 1000 mL or 1045 g. Of the 1045 grams of solution, 4.03 mol ethylene glycol x 62.08 g mol^{-1} (or 250 g) would be ethylene glycol and 1045 - 250 (or 795) grams would be water.

Weight % ethylene glycol $= \frac{250}{1045} \times 100\% =$ **23.9 %**

$$X_{ethylene\ glycol} = \frac{4.03 \text{ mol ethylene glycol}}{4.03 \text{ mol} + (795 \text{ g } H_2O/18.02 \text{ g mol}^{-1})} = \mathbf{0.0837}$$

$$m = \frac{4.03 \text{ mol}}{0.795 \text{ kg}} = \mathbf{5.07 \text{ m}}$$

13.35 1 kg of H_2O will contain the number of moles of the ions listed in the table. That number of moles will weigh: (0.566 mol x 35.45 g mol^{-1}) + (0.486 mol x 22.99 g mol^{-1}) + (0.055 mol x 24.3 g mol^{-1}) + (0.029 mol x 96.1 g mol^{-1}) + (0.011 mol x 40.1 g mol^{-1}) + (0.011 mol x 39.1 g mol^{-1}) + (0.002 mol x 61 g mol^{-1}) = 36.35 g of ions

$$\text{Mass of Cl}^- = \frac{0.566 \text{ mol Cl}^-}{1 \text{ kg H}_2\text{O}} \times \frac{35.45 \text{ g Cl}^-}{\text{mol}} \times \frac{1 \text{ kg H}_2\text{O}}{(1{,}000 \text{ g H}_2\text{O} + 36.35 \text{ g ions})}$$

$$\times \frac{1024 \text{ g soln.}}{\text{L}} \times 3.78 \text{ L soln.} = \mathbf{74.9 \text{ g Cl}^-}$$

$$\text{Mass of Na}^+ = \frac{0.486 \text{ mol Na}^+}{1 \text{ kg H}_2\text{O}} \times \frac{22.99 \text{ g Na}^+}{\text{mol}} \times \frac{1 \text{ kg H}_2\text{O}}{(1{,}000 \text{ g H}_2\text{O} + 36.35 \text{ g ions})}$$

$$\times \frac{1024 \text{ g soln.}}{\text{L}} \times 3.78 \text{ L soln.} = \mathbf{41.7 \text{ g Na}^+}$$

Mass of Mg^{2+} = 0.055 x 24.3 x 3.73 = **5.0 g Mg^+**
Mass of SO_4^{2-} = 0.029 x 96.1 x 3.73 = **10 g SO_4^{2-}**
Mass of Ca^{2+} = 0.011 x 40.1 x 3.73 = **1.6 g Ca^{2+}**
Mass of K^+ = 0.011 x 39.1 x 3.73 = **1.6 g K^+**
Mass of HCO_3^- = 0.002 x 61 x 3.73 = **0.5 g HCO_3^-**

Total Mass of ions = **135.3 g** or **135 g**

13.36 $$50.0 \text{ g soln.} \times \frac{10.0 \text{ g Na}_2\text{CO}_3}{100 \text{ g soln.}} \times \frac{286.2 \text{ g Na}_2\text{CO}_3\cdot10\text{H}_2\text{O}}{106.0 \text{ g Na}_2\text{CO}_3}$$

$$\Leftrightarrow \mathbf{13.5 \text{ g Na}_2\text{CO}_3\cdot10\text{H}_2\text{O}}$$

13.37 Since the gas molecules are in constant, random motion, they mix in a short period of time. The random motion of the gas molecules will yield greater disorder if possible and, in this case, that will result in spontaneous mixing.

13.38 The tendency for a system to spontaneously move toward that state which has greater disorder is what is occurring during the formation of an alcohol-water solution.

13.39 In the case of the mixing of hexane and salt, the salt has very strong ionic forces that would need to be broken before it could be distributed throughout the hexane. The tendency toward disorder alone cannot overcome the strong attraction within the solid salt. Therefore, a salt-hexane solution cannot be formed.

13.40 Since "like dissolves like," small alcohols will dissolve in water. In an alcohol-water solution there are intermolecular attractions between the OH group of the alcohol and the water molecules. As the size of the alcohol is increased from one carbon atom to two carbon atoms, etc., the OH group in the alcohol becomes a smaller and smaller portion of the molecule and the effect of the interaction between it and water will have less and less impact on the solubility of the entire molecule.

13.41 Ammonia is quite soluble because it can form solute-solvent forces of attraction with water (hydrogen bonds) that are comparable to the strength of the solvent-solvent forces of attraction in water. H_2 and O_2 are incapable of forming such strong solute-solvent forces of attraction with water.

13.42 NH_3. Because it is capable of forming hydrogen bonds with the water.

13.43 Substances that exhibit similar intermolecular attractive forces tend to be soluble in one another.

13.44 The ions of the salt increase the effective polarity of the solvent and, therefore, decrease the solubility of solutes that are not as polar.

13.45 Methyl alcohol molecules surround H_2O molecules resulting in micelle-like particles that are soluble in the gasoline.

13.46 The ion is said to be hydrated when it is surrounded by water molecules. When the solute particle becomes surrounded by molecules of the solvent, it is referred to as being solvated; hydration is a special case of solvation.

13.47 When dissolving an ionic compound the water molecules surround the ions "insulating" their charges from each other.

13.48 Micelles are collections of fatty acid anions which are colloidal in size. Soaps are fatty acid salts which when dissolved become separated fatty acid anions and metal cations. The hydrophobic end of the fatty acid anion will dissolve in each other or in grease or oil forming a layer with the hydrophilic part of the fatty acid anion pointing toward the solvent; this makes the oil more soluble in water and allows it to be carried away with the water.

13.49 They do not form precipitates with the cations found in hard water.

13.50 The amount of energy that is absorbed or released when a substance enters solution is called the heat of solution. A negative ΔH_{soln} value means the solution temperature will increase.

13.51 Positive

13.52 Acetone molecules are attracted to water molecules more strongly than they are attracted to other acetone molecules because between two acetone molecules hydrogen bonding cannot occur. However, there will be hydrogen bonding between the hydrogens of water (since they are connected directly to an oxygen) and the pairs of electrons on the oxygen of acetone.

13.53 When hexane dissolves in ethanol, some of the hydrogen bonding between ethanol molecules will be disrupted. The energy needed to break hydrogen bonds in ethanol is not regained when ethanol and hexane are mixed since no strong forces exist between ethanol and hexane.

13.54 In an ideal gas there are no intermolecular attractions. An ideal solution is one in which the solute-solute, solute-solvent and solvent-solvent interactions are all the same (but they are not zero).

13.55 The heat of solution is equal to the difference between the lattice energy and the hydration energy. If the lattice energy is greater, a net input of energy is required; therefore, the solution process is endothermic.

13.56 The dominant energy effect when gases dissolve in a liquid is caused by the solvation of the gas molecules, which is exothermic.

13.57 $AlCl_3$ probably has a large value for its hydration energy.

13.58 Lattice energy

13.59 $\Delta H_{soln.}(AlCl_3) = -321\ kJ/mol$

$$\frac{-321\ kJ}{mol} \times 10.0\ g \times \frac{1\ mol}{133.3\ g} = -24.1\ kJ \text{ or } \mathbf{24.1\ kJ\ liberated}$$

13.60 $\Delta H_{soln.}(NH_4NO_3) = 26\ kJ/mol$

$$\frac{26\ kJ}{mol} \times 115\ g \times \frac{1\ mol}{80.1\ g} \times \frac{10^3\ J}{kJ} \times \frac{1\ cal}{4.184\ J} = \mathbf{8900\ cal} \text{ or } \mathbf{8.9\ kcal}$$

13.61 KI would become more soluble with an increase in temperature. Addition of heat favors an endothermic process. Using Le Chatelier's principle, heat is a reactant in this process and solution is a product. Increased heat will yield increased solution.

13.62 In <u>fractional crystallization</u>, an impure product is dissolved in a small amount of hot solvent (in which it is more soluble than the impurities). The solution is then cooled. As the solution cools, the pure product separates from the mixture, leaving the impurities behind.

13.63 Since the solution process for a gas in a liquid is nearly always exothermic, the reverse reaction is endothermic and gases are less soluble as the temperature is raised.

13.64 Increasing the pressure will increase the rate at which molecules leave the gas and enter the solution. This will continue until equilibrium is reestablished, at which time the concentration of the solute in the solution will have increased.

13.65 Pressure only has an appreciable effect on equilibria where sizable volume changes occur. When a liquid or solid dissolves in a liquid, only very small changes in volume occur.

13.66 $\frac{5.00 \times 10^{-2} \text{ g/30.0 g/mol}}{6.56 \times 10^{-2} \text{ g/30.0 g/mol}} = \text{mol ratio} \qquad \frac{C_1}{P_1} = \frac{C_2}{P_2}$

$$P_2 = P_1 \times \frac{C_2}{C_1} = P_1 \times \text{mol ratio}$$

$$P_{ethane} = 751 \text{ torr} \times \frac{5.00 \times 10^{-2} \text{ g}}{6.56 \times 10^{-2} \text{ g}} = \mathbf{572 \text{ torr}}$$

13.67 $p_g = C_g/k_g = 0.0478 \text{ g L}^{-1}/6.50 \times 10^{-5} \text{ g L}^{-1} \text{ torr}^{-1} = 735 \text{ torr}$

$P_{total} = p_g + p_{water} = 735 + 23.8 = \mathbf{759 \text{ torr}}$

13.68 $C_1 = k \times p_1$ or $k = C_1/p_1 = \text{constant}$

$k = (2.09 \times 10^{-4}/0.968) \text{ g L}^{-1} \text{ atm}^{-1} = 2.16 \times 10^{-4} \text{ g L}^{-1} \text{ atm}^{-1}$

$C_2 = k \times p_2 = 2.16 \times 10^{-4} \text{ g L}^{-1} \text{ atm}^{-1} \times 1{,}000 \text{ atm} = \mathbf{0.216 \text{ g/L}}$

13.69 $(5.34 \times 10^{-5} \text{ g L}^{-1} \text{ torr}^{-1})(0.20 \times (760 - 24)\text{torr}) = 7.86 \times 10^{-3} \text{ g/L}$

$7.86 \times 10^{-3} \text{ g/L} \times 1\text{L} = \mathbf{7.9 \times 10^{-3} \text{ g}}$

13.70 The vapor pressure of a solvent depends on the fraction of the total number of molecules at the surface of the solution that are solvent molecules; i.e., the mole fraction of the solvent.

13.71 When the vapor pressure of a mixture is greater than that predicted, it is said to exhibit a positive deviation from Raoult's law; conversely, when a solution gives a lower vapor pressure than we would expect from Raoult's law, it is said to show a negative deviation.

13.72 ΔH_{soln} for positive deviations is endothermic; whereas, ΔH_{soln} for negative deviations is exothermic.

13.73 Raoult's Law states that $P_{soln.} = X_{solvent} P°_{solvent}$

$P°_{solvent} = 93.4$ torr

$$X_{solvent} = \frac{1{,}000 \text{ g benzene}/78.06 \text{ g mol}^{-1}}{(1{,}000 \text{ g benzene}/78.06 \text{ g mol}^{-1}) + (56.4 \text{ g } C_{20}H_{42}/283 \text{ g mol}^{-1}}$$

$$= \frac{12.81}{13.01} = 0.985$$

$P_{soln.} = 0.985 \times 93.4$ torr $=$ **92.0 torr**

13.74 $P_{soln.} = X_{solvent} P°_{solvent}$ $\qquad 130 = X_{solvent}\, 160$

$X_{solvent} = 0.812$ $\qquad X_{glycerol} = 1 - 0.812 =$ **0.188**

13.75 $P_T = X_A P°_A + X_B P°_B$

$$P_T = \frac{25.0 \text{ g}/100.1 \text{ g mol}^{-1}}{(25.0 \text{ g}/100.1 \text{ g mol}^{-1}) + (35.0 \text{ g}/114.1 \text{ g mol}^{-1})} \times 791 \text{ torr}$$

$$+ \frac{35.0 \text{ g}/114.1 \text{ g mol}^{-1}}{(25.0 \text{ g}/100.1 \text{ g mol}^{-1}) + (35.0 \text{ g}/114.1 \text{ g mol}^{-1})} \times 352 \text{ torr}$$

$$= \frac{0.250}{0.250 + 0.307} \times 791 \text{ torr} + \frac{0.307}{0.250 + 0.307} \times 352 \text{ torr}$$

$= (0.449 \times 791$ torr$) + (0.551 \times 352$ torr$) =$ **549 torr**

13.76 60.0 g benzene x 1 mol/78.1 g = 0.768 mol benzene

40.0 g toluene x 1 mol/92.1 g = 0.434 mol toluene

$P_{soln} = X_A P°_A + X_B P°_B = \{[0.768/(0.768 + 0.434)] \times 93.4 \text{ torr}\} +$

$\{[0.434/(0.434 + 0.768)] \times 26.9 \text{ torr}\}$

$= (0.639 \times 93.4$ torr$) + (0.361 \times 26.9$ torr$) =$ **69.4 torr**

13.77 $P_{soln.} = X_A P^\circ_A + X_B P^\circ_B$

$$137 \text{ torr} = \frac{400 \text{ g}/154 \text{ g mol}^{-1}}{(400 \text{ g}/154 \text{ g mol}^{-1}) + (43.3 \text{ g}/? \text{ mol}^{-1})} \times 143 \text{ torr}$$

$$+ \frac{43.3 \text{ g}/\ ? \text{ mol}^{-1}}{(400 \text{ g }/154 \text{ g mol}^{-1}) + (43.3 \text{ g }/? \text{ g mol}^{-1})} \times 85 \text{ torr}$$

$$137 = \frac{(2.597)(143)}{2.597 + 43.3/?} + \frac{(43.3/?)(85)}{2.597 + 43.3/?}$$

$$2.597 + 43.3\ /? = \frac{371.4}{137} + \frac{3681/?}{137} \qquad 43.3/? - \frac{3681/?}{137} = 0.1139$$

$5{,}932/? - 3{,}681/? = 15.604$ $\qquad$ **? = 144 g/mol**

13.78 (a) 511 torr = $X_{chloroform}$ x 526 torr

$X_{chloroform} = 0.971$ $\qquad$ $X_{unkn.} = 1 - 0.971 =$ **0.029** $= X_{solute}$

(b) $$X_{solute} = \frac{\text{moles solute}}{\text{moles solute + moles } CHCl_3} \qquad 0.029 = \frac{x}{x+1}$$

$0.029\text{ x} + 0.029 = \text{x}$ $\qquad$ x = **0.030 mol solute**

(c) 0.030 mol = 8.3 g/? M.M. $\qquad$ M.M. = **277** or **280 (2 sign. figs.)**

13.79 $$X_{water} = \frac{100 \text{ g}/18.02 \text{ g mol}^{-1}}{(100 \text{ g}/18.02 \text{ g mol}^{-1}) + (150 \text{ g}/92.11 \text{ g mol}^{-1})} = 0.773$$

91.1 torr = 0.773 P°_{water} $\qquad$ P°_{water} = 118 torr

118 torr = **55°C** (From Table 11.2)

13.80 When a mixture of two liquids is boiled, the vapor is always richer in the more volatile component. The process of successive condensations and boilings to produce fractions even richer in the more volatile component is known as fractional distillation.

13.81 Approximately three times for Figure 13.22 and 4 or 5 times for Figure 13.23.

13.82 Sketch a diagram like that in Figure 13.22. It should have pure CCl_4 as the left axis and pure C_6H_6 as the right one. The curves start at the boiling point of CCl_4 (76.8°C) at the left axis and meet again at the boiling point of benzene (80.1°C) at the right axis.

13.83 Draw a diagram like that in Figure 13.23. If water is on the left and butyl alcohol on the right, the lines will start at 100°C (boiling point of water), decrease to 92.4°C (boiling point of the azeotrope, which is a mole fraction of water of 0.716), and then increase to 117.8°C at the right axis (the boiling point of butyl alcohol). These substances show positive deviation from ideality, i. e., minimum-boiling azeotrope.

13.84 Colligative properties are properties that depend on the relative number of particles of solute and solvent instead of on their specific chemical nature.

13.85 In the presence of a solute, the rate of freezing at a particular temperature is decreased because fewer solvent particles are in contact with the solid. The rate of melting, however, is the same since no solute is incorporated in the solid solvent to re-establish equilibrium. The temperature must be lowered so that the solvent freezes faster from the solution and the solvent melts more slowly from the solid until the rates of these two processes become equal.

13.86 $\Delta T_b = K_b\, m$

$$\text{B.P.}_{\text{soln.}} - 100.0°\text{C} = (0.51°\text{C m}^{-1})\left(\frac{55.0\text{ g}/92.11\text{ g m}^{-1}}{0.250\text{ kg H}_2\text{O}}\right)$$

$\text{B.P.}_{\text{soln.}} - 100.0°\text{C} = (0.51)(2.388) = 1.2°\text{C}$ $\quad$ $\text{B.P.}_{\text{soln.}} =$ **101.2°C**

$\Delta T_f = K_f\, m = (1.86°\text{C m}^{-1})(2.388\text{ m}) = 4.44°\text{C}$

F.P. = 0.00°C - 4.44°C = **-4.44°C**

13.87 $\Delta T_f = K_f\, m$

$$0.307°\text{ C} = (5.12°\text{C m}^{-1})\left(\frac{3.84\text{ g/MM}}{0.500\text{ kg benzene}}\right)$$

MM = **128** Empirical formula is given as C_4H_2N

Therefore, the molecular formula based upon calculated MM must be twice the empirical formula: $\mathbf{C_8H_4N_2}$

13.88 $\Delta T_f = K_f m$ $\quad 0.744^\circ C = (1.86^\circ C\ m^{-1})\left(\frac{16.9\ g/MM}{0.250\ kg\ H_2O}\right)$

$0.1000 = 16.9/MM \qquad MM = 169.0$

moles C = 169 g x (57.2/100) x (1 mol/12.01 g) = 8.05 mol

moles H = 169 g x (4.77/100) x (1 mol/1.01 g) = 7.98 mol

moles O = 169 g x (38.1/100) x (1 mol/16.0 g) = 4.02 mol

Emp. Form. = C_2H_2O Emp. FM = 42 Therefore, Mol. Form. = $\mathbf{C_8H_8O_4}$

13.89 $\Delta Tf = Kf\ m$ $\quad 0.750^\circ C = (1.86^\circ C\ m^{-1})\left(\frac{?\ g/180.2\ g\ mol^{-1}}{0.150\ kg\ H_2O}\right)$

? = **10.9 g** added

$$\Delta T_b = K_b\ m = (0.51^\circ C\ m^{-1})\left(\frac{10.9\ g/180.2\ g\ mol^{-1}}{0.150\ kg}\right) = 0.21^\circ C$$

B.P. = 100.00 + 0.21 = **100.21°C**

13.90 $\Delta T_f = K_f m$ $\quad 2.47^\circ C = (1.86^\circ C\ m^{-1})(?\ m)$

? m = 1.328 m , $\Delta T_b = K_b m$, $\Delta T_b = (0.51^\circ C\ m^{-1})(1.33\ m)$

$\Delta T_b = 0.68^\circ C$ $\qquad$ B.P. = 100.00 + 0.68 = **100.68°C**

13.91 0.075 = fraction dissociated, (assume $MX \rightarrow M^+ + X^-$)

1.000 - 0.075 = fraction undissociated = 0.925

Total = [(0.925) x 0.100 m] + [(2 x 0.075) x 0.100 m] = 0.108 m

$\Delta T_f = (1.86^\circ C\ m^{-1})\ (0.108\ m) = 0.201^\circ C$

F.P. = 0.000°C - 0.201°C = **-0.201°C**

13.92 If we select a volume of 1.00 L of each, we have: 1.000 $L_{ethyl.}$ x 1.113 g/mL x 1,000 mL/L x 1 mol/62.08 g = 17.93 mol ethylene glycol and 1.000 L H_2O x 1.000 g/mL x 1,000 mL/L x 1kg/1,000 g = 1.000 kg water.

$\Delta T_f = K_f m$ $\Delta T_f = (1.86°C\ m^{-1})(17.93\ mol/1.00\ kg) = 33.3°C$

F.P. = 0.0°C - 33.3°C = **-33.3°C**

°F = (9/5)°C + 32.0 = [(9/5)(-33.3)] + 32.0 = **-28.0°F** This is the temperature at which the first crystals will form. At -34°C all the solution might not be solidified; therefore, this solution might give some protection to -34°C but its protection at that temperature would be minimal if the solution were to act as an ideal solution.

13.93 In dialysis small ions, small molecules, and solvent are allowed to pass through a membrane, but in osmosis only solvent is allowed through the membrane.

13.94 A semipermeable membrane is a film that restricts the passage of solute through it while allowing passage of solvent.

13.95 Solutions that have the same osmotic pressure are called isotonic solutions. During intravenous feeding, the solute concentration must be carefully controlled to prevent excessive movement of fluid into or out of the cells.

13.96 $\pi = MRT$

$$\pi = \left(\frac{5.0\ g/342\ g\ mol^{-1}}{1\ L}\right)(0.0821\ L\ atm\ mol^{-1}\ K^{-1})(298\ K)(760\ torr\ atm^{-1})$$

π = **270 torr** (2 sign. figures)

13.97 $\pi = MRT$ $$\frac{3.74\ torr}{760\ torr\ atm^{-1}} = \frac{(0.400\ g/MM)}{1.00\ L}(0.0821\ L\ atm\ mol^{-1}K^{-1})(300\ K)$$

MM = **2,000** (3 sign. figures)

13.98 $\Delta T_f = m K_f = 0.10\ m\ MgSO_4 \times \frac{2\ m\ ions}{1\ m\ salt} \times 1.86°C\ m^{-1} = 0.37°C$

F.P. = 0.00°C - 0.37°C = **-0.37°C** (if completely dissociated)

13.99 $\Delta T_f = m K_f = 0.10\ m\ CaCl_2 \times \frac{3\ m\ ions}{1\ m\ salt} \times 1.86°C\ m^{-1} = 0.56°C$

F.P. = 0.00°C - 0.56 = **-0.56°C** (if completely dissociated)

13.100 It suggests that there is less than one mole of particles in the solution for each mole expected to have been present in the solution, i. e., an electrolyte is not completely dissociated.

13.101 $\pi = MRT = 2 \times 0.010\ M \times (0.0821\ atm\ M^{-1}\ K^{-1})(298\ K)(760\ torr/atm)$

$= \mathbf{370\ torr}$ (2 sign. figures)

13.102 From Problem 13.35: total m = 1.160 m.

$M = (1.160\ mol/1000\ g)(1.024\ g/mL)(1000\ mL/L) = 1.188\ M$

$\pi = 1.188\ M \times 0.0821\ atm\ M^{-1}\ K^{-1} \times 298\ K = 29.07\ atm = \mathbf{29.1\ atm}$

Greater than 29.1 atm

13.103 $\Delta T_f = m\ K_f$

$\Delta T_f = (0.10\ m)(1.86°C\ m^{-1}) = 0.19°C$ (calculated as a nonelectrolyte solution)

$$i = \frac{(\Delta T_f)\ \text{measured}}{(\Delta T_f)\ \text{calculated as nonelectrolyte}}$$

i = 1.21 (from Table 13.6)

$$1.21 = \frac{(\Delta T_f)\ \text{measured}}{0.19}$$

$\Delta T_f = 0.23°C$ F.P. = 0.00 - 0.23 = **-0.23°C**

13.104 $MgSO_4$. The greater the degree of attraction between ions, the less it dissociates. In the case of $MgSO_4$ it agrees with predictions; there is a doubly charged cation and a doubly charged anion.

13.105 $Al_2(SO_4)_3$

13.106 KCl, i factor = **2**; $NiCl_2$, i factor = **3**; $Al_2(SO_4)_3$, i factor = **5**.

13.107 Calculated freezing point: $\Delta T_f = (0.130\ m)(1.86°C\ m^{-1})$

$\Delta T_f = 0.242$ calculated as nonelectrolyte

$$i = \frac{(\Delta T_f)\ \text{measured}}{(\Delta T_f)\ \text{calculated}} = \frac{0.72}{0.242} = 2.98 = 3.0$$

This i value is consistent with: $Hg_2Cl_2(s) + H_2O \rightarrow Hg_2^{2+}(aq) + 2Cl^-(aq)$ (in which the mercury(I) remains a dimer).

13.108 First calculate, from the freezing point data, the molality of the benzene solution.

From Table 13.5 $K_{f(benzene)} = 5.12°C\ m^{-1}$

$\Delta T_f = mK_f$ or $5.50°C - 2.63°C = m(5.12°C\ m^{-1})$

$m = 0.561\ mol\ kg^{-1}$

Next calculate the mole fraction of benzene (C_6H_6) in the solution.

$$X_{benzene} = \frac{(1\ kg\ benzene \times 1000\ g\ kg^{-1} \times 1\ mol/78.1\ g)}{(0.561\ mol) + (1000\ g \times 1\ mol/78.1\ g)} = \frac{12.80}{0.561 + 12.80} = 0.958$$

From the mole fraction the vapor pressure can be calculated:

$P_{soln} = X_{benzene(solvent)}\ P^{o}_{benzene}$ $\qquad P_{soln} = (0.958)(93.4\ torr) = \mathbf{89.5\ torr}$

13.109 Assuming that ethylene glycol is nonvolatile, which it is very close to being at the temperatures in this problem:

$$\Delta T_b = mK_b = \left(\frac{10.0\ g/62.08\ g\ mol^{-1}}{0.0900\ kg}\right)0.51 = 0.913°C$$

B.P. = **100.91°C**

$$\Delta T_f = mK_f = \left(\frac{10/62.08}{0.0900}\right)1.86 = 3.33°C$$

F.P. = **-3.33°C**

$$V.P. = 760\ torr\left(\frac{90.0\ g/18.01\ g\ mol^{-1}}{(90.0\ g/18.01\ g\ mol^{-1}) + 10.0\ g/62.08\ g\ mol^{-1}}\right)$$

$$= 760\ torr\left(\frac{4.997}{4.997 + 0.161}\right) = \mathbf{736\ torr}$$

14 CHEMICAL THERMODYNAMICS

14.1 **Thermo** implies heat; **dynamics** implies movement or change.

14.2 **System & Surroundings** - By **system** we mean that particular portion of the universe upon which we wish to focus our attention. Everything else we call the **surroundings.**
Isothermal change - change occurring at constant temperature.
Adiabatic change - a change that occurs without heat transfer between system and surroundings.
State function - a quantity whose value for a system in a particular state is independent of the system's prior history.

14.3 Pressure-volume work and electrical work.

14.4 A **spontaneous change** is one that will take place by itself, without outside aid.

14.5 A **reversible process** is one that can be made to reverse its direction by the smallest change in an opposing force such as pressure or temperature.

14.6 Isothermal: immerse the system in a vat of water kept at a constant temperature by a thermostat. Adiabatic: keep the system in an insulated container.

14.7 The first law of thermodynamics states that if a system undergoes some series of changes that ultimately brings it back to its original state, the net energy change is zero.

14.8 The ΔE is the difference between the heat that is added to the system as it passes from the initial to the final state and the work done by the system upon its surroundings. E represents all of the energies, K.E. + P.E.; whereas, ΔE represents $E_{final} - E_{initial}$. ΔE, therefore, is only dependent on the energy the molecules have finally minus what they had initially, regardless of the path.

14.9 A "perpetual motion machine" is a device that could run forever without a net consumption of energy. Such a machine is not possible because it would violate the first law of thermodynamics by creating energy to run the machine.

14.10 We cannot calculate E because of the very nature of E, which represents the K.E. as well as the P.E. For instance, because there is no stationary reference point in the universe, we cannot measure absolute velocity; therefore, we cannot know K.E. We cannot account for all attractive and repulsive forces felt by the particles; thus we cannot determine P.E.

14.11 (a) There would be no change in temperature. There are no attractive forces between the molecules in an ideal gas, so there would be no change in PE when the molecules move further apart.
(b) Since gases cool on expansion, the average kinetic energy of the molecules must decrease. Therefore, ΔE is positive, q is positive, $w = 0$. Heat would have to be supplied to keep the temperature of the system constant (isothermal).

14.12 Because during an isothermal expansion or compression there is no change in the K.E. or P.E. of the material. K.E. remains constant because T remains constant. P.E. is zero because in an ideal gas there are no intermolecular attractive forces.

14.13 The extra P.E. can't just disappear. It shows up as an increase in the heat of reaction.

14.14 $P\Delta V$ has the units Joules, when P is in pascals and ΔV is in m^3, which is in energy units rather than other units. The common units of $P\Delta V$ are L x atm. The conversion value 101.3 J/L•atm is commonly used to convert from liter•atmospheres to joules.

14.15 ΔE is the heat of reaction at constant volume, whereas ΔH is the heat of reaction at constant pressure. Most reactions that are of interest to us take place at constant P, not constant V.

14.16 First step: $\Delta V = V_2 - 10.0\text{ L}$

$$V_2 = 10.0\text{ L} \times \frac{15.0\text{ atm}}{7.50\text{ atm}} = 20.0\text{ L}$$

$\Delta V = 20.0\text{ L} - 10.0\text{ L} = 10.0\text{ L}$

$w = -P\Delta V = -7.50\text{ atm} \times (20.0\text{ L} - 10.0\text{ L}) = \mathbf{-75.0\text{ L atm}}$

$\mathbf{\Delta T = 0,\ \Delta E_{system} = 0,\ \Delta E_{surroundings} = 0,\ q = -w = 75.0\text{ L atm}}$

Second Step: $\Delta V = V_3 - 20.0\text{ L}$

$$V_3 = 20.0\text{ L} \times \frac{7.50\text{ atm}}{1.00\text{ atm}} = 150\text{ L}$$

$\Delta V = 150\text{ L} - 20.0\text{ L} = 130\text{ L}$

$w = -P\Delta V = -1.00\text{ atm} \times 130\text{ L} = \mathbf{-130\text{ L atm}}$

$\mathbf{\Delta T = 0,\ \Delta E_{system} = 0,\ \Delta E_{surroundings} = 0,\ q = -w = 130\text{ L atm}}$

14.17 $w = -P\Delta V$ $\qquad V_2 = 50.0\ m^3 \times \frac{200\ kPa}{100\ kPa} = 100\ m^3$

$w = -100\ kPa \times (100\ m^3 - 50.0\ m^3) = -5{,}000\ kPa\ m^3 = \mathbf{-5.0 \times 10^3\ kJ}$

$\mathbf{-w = q = 5.0 \times 10^3\ kJ}$

14.18 $w = -3.00\ atm\ (0.250\ L - 0.500\ L) = 0.750\ L\ atm$

$w = 0.750\ L\ atm \times \frac{101.3\ J}{1\ L\ atm} \times \frac{1\ kJ}{1{,}000\ J} = \mathbf{0.0760\ kJ}$

$\mathbf{q = 12.6\ kJ}$

$\Delta E_{system} = q + w = 12.6\ kJ + (0.0760\ kJ) = \mathbf{12.7\ kJ}$

$\Delta E_{surr.} = \mathbf{-12.7\ kJ}$

14.19 (a) $\mathbf{q = 35\ J,\ \ w = 40\ J,}$ $\qquad \Delta E_{system} = q + w = 35\ J + (+40J) = \mathbf{75\ J}$
(b) $\mathbf{q_{surr.} = -35\ J,\ \ w_{surr.} = -40\ J,\ \ \Delta E_{surr.} = -75\ J}$

14.20 ΔE and ΔH differ by the term $P\Delta V$. For reactions occurring in open containers, the volume change for reactions involving just liquids and solids is very small. Therefore, the values of ΔE and ΔH are nearly identical for these reactions.

14.21 In reaction (a) (only) the moles of gaseous material is increasing as the reaction goes from left to right. The increase in number of moles of gas will result in an increase in volume at constant pressure and temperature. Therefore, for the reaction, $P\Delta V$ will be positive and ΔE will be a larger negative value than ΔH according to the equation: $\Delta H = \Delta E + P\Delta V$.

14.22 Atomization energy is the energy needed to break all bonds and to reduce the gaseous molecule entirely to gaseous atoms.

$H_2O(g) \rightarrow 2H(g) + O(g) \qquad \Delta H = \Delta H_{atom}$

14.23 The calculated ΔH_f° is usually from average bond energies.
Tabulated bond energies are averages as the bond energies for each kind of bond differ slightly for different compounds.

14.24 $H_2(g) + 2C(s) \rightarrow C_2H_2(g) \qquad \Delta H_f^o$

$H_2(g) \rightarrow 2H(g) \qquad \Delta H_1$

$2C(s) \rightarrow 2C(g) \qquad \Delta H_2$

$2H(g) + 2C(g) \rightarrow C_2H_2(g) \qquad \Delta H_3$

$\Delta H_f^\circ = \Delta H_1 + \Delta H_2 + \Delta H_3$
$= 2(218\ kJ) + 2(715\ kJ) + [2(-415\ kJ) + (-833\ kJ)] = \mathbf{203\ kJ/mol}$

14.25 $3H_2(g) + 6C(s) \rightarrow C_6H_6(g)$

$3H_2(g) \rightarrow 6H(g)$ ΔH_1

$6C(s) \rightarrow 6C(g)$ ΔH_2

$6H(g) + 6C(g) \rightarrow C_6H_6(g)$ ΔH_3

ΔH_3 = formation of 3 (C=C), 3(C-C), and 6(H-C) bonds.

$\Delta H_f^\circ = \Delta H_1 + \Delta H_2 + \Delta H_3$

= 6(218 kJ) +6(715 kJ) + [3(-607 kJ) +3(-348 kJ) +6(-415 kJ)] = **243 kJ/mol**

Resonance energy = 243 kJ -82.8 kJ =**160 kJ**. Compounds that have resonance structures are much more stable than predicted by non-resonance structures.

14.26 $3C(s) + 3H_2(g) \rightarrow CH_3CHCH_2$ $\Delta H_f^\circ = ?$

$3H_2(g) \rightarrow 6H(g)$ ΔH_1

$3C(s) \rightarrow 3C(g)$ ΔH_2

$6H(g) + 3C(g) \rightarrow CH_3CHCH_2$ ΔH_3

ΔH_3 = formation of 6 (C-H), 1 (C-C), and 1 (C=C) bond.

$\Delta H_f^\circ = \Delta H_1 + \Delta H_2 + \Delta H_3$

= 6(218 kJ) + 3(715 kJ) + [6(-415 kJ) + 1(-348 kJ) + 1(-607 kJ)] = **8 kJ/mol**

14.27 $3C(s) + 4H_2(g) \rightarrow CH_3CH_2CH_3$ $\Delta H_f^\circ = ?$

$4H_2(g) \rightarrow 8H(g)$ ΔH_1

$3C(s) \rightarrow 3C(g)$ ΔH_2

$8H(g) + 3C(g) \rightarrow CH_3CH_2CH_3$ ΔH_3

ΔH_3 = formation of 8 (C-H) and 2 (C-C) bonds

$\Delta H_f^\circ = \Delta H_1 + \Delta H_2 + \Delta H_3$

= 8(218 kJ) + 3(715 kJ) + [8(-415 kJ) + 2(-348 kJ)] = **-127 kJ mol^{-1}**

Literature value is -104 kJ mol^{-1} (Table 6.1)

14.28 A spontaneous change is one that takes place without assistance. A nonspontaneous change requires an outside force to drive the reaction.

14.29 Many spontaneous reactions give off energy. However, there are some instances where energy is absorbed during a spontaneous reaction. Both the change in energy (heat content) and in degree of randomness (entropy) must be considered when discussing the likelihood of a reaction being spontaneous.

14.30 In any process there is a natural tendency or drive toward increased randomness because a highly random distribution of particles represents a condition of higher statistical probability than an ordered one. A system's entropy is proportional to its statistical probability.

14.31 ΔS must be positive and the product of $T\Delta S$ must be greater than ΔH.

14.32 energy/temperature; e.g., J/K

14.33 (a), (b), and (d)

14.34 (a) negative (b) positive (c) positive (d) negative (e) negative (f) negative (g) positive (h) negative

14.35 (a) positive (b) negative (c) positive (d) positive (e) negative

14.36 (a) $\approx$ zero (b) positive (c) negative (d) positive (e) negative

14.37 During any spontaneous change, there is always an increase in the entropy of the universe.

14.38 ΔG must be negative which means ΔH must be negative and ΔS must be positive (if independent of temperature).

14.39 Because of the amount of energy required to reverse the spontaneous distribution of the pollutants,i.e., must counteract the enormous entropy increase.

14.40 $\Delta G = \Delta H - T\Delta S$ (Note the temperature in the equation).

14.41 Between 10,000 and 20,000 atm. It is not theoretically possible to change graphite to diamond at 1 atm.

14.42 The third law of thermodynamics states that the entropy of any pure crystalline substance at absolute zero is equal to zero.

14.43 There is perfect order (zero randomness) in a pure crystalline substance at absolute zero. A mixture would have a positive entropy at 0 K because of the random distribution of particles throughout the mixture.

14.44 For both a and b, Δn = -0.5 moles. Both reactions involve only gases. Therefore, one can not predict the extent of change of ΔS in either reaction without performing some calculations.

(a) $\Delta S = (1 \text{ mol} \times 256 \text{ J}^{-1} \text{ K}^{-1}) - (1/2\text{mol} \times 205 \text{ J mol}^{-1} \text{ K}^{-1})$
$- (1 \text{ mol} \times 248 \text{ J mol}^{-1} \text{ K}^{-1}) = \mathbf{-94\ J\ K^{-1}}$

(b) $\Delta S = (1 \text{ mol} \times 213.6 \text{ J mol}^{-1} \text{ K}^{-1}) - (1 \text{ mol} \times 197.9 \text{ J mol}^{-1} \text{ K}^{-1})$
$- (0.5 \text{ mol} \times 205 \text{ J mol}^{-1} \text{ K}^{-1}) = \mathbf{-87\ J\ K^{-1}}$

Therefore, reaction (a) is accompanied by the greater entropy change.

14.45 (a) C(s) (graphite) + $2Cl_2(g) \rightarrow CCl_4(\ell)$

$\Delta S = (214.4\ J\ K^{-1}) - [5.69\ J\ K^{-1} + (2 \times 223.0)\ J\ K^{-1}]$

$\Delta S_f° =$ **-237.3 J K^{-1}**

(b) $Mg(s) + O_2(g) + H_2(g) \rightarrow Mg(OH)_2(s)$

$\Delta S = (63.1\ J\ K^{-1}) - [32.5\ J\ K^{-1} + 205.0\ J\ K^{-1} + 130.6\ J\ K^{-1}]$

$\Delta S_f° =$ **-305.0 J K^{-1}**

(c) $Pb(s) + S(s) + 2O_2(g) \rightarrow PbSO_4(s)$ $\Delta S_f° =$ **-358 J K^{-1}**

(d) $Na(s) + 1/2H_2(g) + C(s) + 3/2O_2(g) \rightarrow NaHCO_3(s)$

$\Delta S_f° =$ **-274 J K^{-1}**

(e) $1/2N_2(g) + 3/2H_2(g) \rightarrow NH_3(g)$ $\Delta S_f° =$ **-99.2 J K^{-1}**

14.46 Since entropy values have a reference point it is possible to tabulate absolute values, but enthalpy has no reference point that is known to have a value of zero.

14.47 (a) $\Delta S° = [S°_{Al_2O_3} + 2S°_{Fe}] - [2S°_{Al} + S°_{Fe_2O_3}]$

= [51.0 J/K + 2(27.3) J/K] - [2(28.3) J/K + 87.4 J/K]

= **-38.4 J/K**

(b) $\Delta S° = [S°_{SiO_2} + 2S°_{H_2O}] - [S°_{SiH_4} + 2S°_{O_2}]$

= [41.8 J/K + 2(188.7) J/K] - [205 J/K + 2(205.0) J/K]

= **-196 J/K**

(c) $\Delta S° = [S°_{CaSO_4}] - [S°_{CaO} + S°_{SO_3}]$

= (107 J/K) - (39.8 J/K + 256 J/K)

= **-189 J/K**

(d) $\Delta S° = [S°_{Cu} + S°_{H_2O}] - [S°_{CuO} + S°_{H_2}]$

= (33.15 J/K + 188.7 J/K) - (42.6 J/K + 130.6)

= **48.6 J/K**

(e) $\Delta S° = [S°_{C_2H_6}] - [S_{C_2H_4} + S°_{H_2}]$

= (230 J/K) - (220 J/K + 130.6 J/K)

= **-121 J/K**

14.48 (a) $\Delta S° = [S°_{C_2H_4}] - [\ S°_{C_2H_2} + S°_{H_2}]$

= (220 J/K) - (201 J/K + 130.6 J/K)

= **-112 J/K**

(b) $\Delta S° = [S°_{H_2SO_4}] - [\ S°_{SO_3} + S°_{H_2O}]$

= (157 J/K) - (256 J/K + 70 J/K)

= **-169 J/K**

(continued)

14.48 (continued)

(c) $\Delta S° = [S°_{MgCl_2 \cdot 2H_2O}] - [S°_{Mg(OH)_2} + 2(S°_{HCl})]$
$= (180 J/K) - [63.1\ J/K + (2 \times 186.7)\ J/K]$
$=$ **-256 J/K**

(d) $\Delta S° = [S°_{CO} + S°_{H_2O}] - [S°_{CO_2} + S°_{H_2}]$
$= (197.9\ J/K + 188.7\ J/K) - (213.6\ J/K + 130.6\ J/K)$
$=$ **42.4 J/K**

(e) $\Delta S° = [10S°_{N_2} + 3S°_{CO_2} + 4S°_{H_2O}] - [10S°_{N_2O} + S°_{C_3H_8}]$
$= [(10 \times 191.5) J/K + (3 \times 213.6) J/K + (4 \times 188.7) J/K]$
$- [(10 \times 220.0) J/K + 269.9\ J/K] =$ **840.7 J/K**

14.49 $\Delta G°$ = sum of ΔG_f's of products - sum of ΔG_f's of reactants

(a) $2Al(s) + Fe_2O_3(s) \rightarrow Al_2O_3(s) + 2Fe(s)$

$$\Delta G° = [1\ mol \times \frac{(-1577\ kJ)}{mol} + 2\ mol\ (0)] - [2\ mol\ (0) + 1\ mol \times \frac{(-741.0\ kJ)}{mol}]$$

$= -1577\ kJ + 741.0\ kJ =$ **-836 kJ**

(b) $SiH_4(g) + 2O_2(g) \rightarrow SiO_2(s) + 2H_2O(g)$
$\Delta G° = [1\ (-856\ kJ) + 2\ (-228\ kJ)] - [1\ (+52.3\ kJ) + 2(0)] =$ **-1364 kJ**

(c) $CaO(s) + SO_3(g) \rightarrow CaSO_4(s)$ $\Delta G° =$ **-346 kJ**

(d) $CuO(s) + H_2(g) \rightarrow Cu(s) + H_2O(g)$ $\Delta G° =$ **-101 kJ**

(e) $C_2H_4(g) + H_2(g) \rightarrow C_2H_6(g)$ $\Delta G° =$ **-101 kJ**

14.50 (a) $\Delta G° = [+68.2\ kJ] - [+209\ kJ + 0] =$ **-141 kJ**
(b) $\Delta G° = [-689.9\ kJ] - [-370\ kJ + (-237\ kJ)] =$ **-83 kJ**
(c) $\Delta G° = [-1118\ kJ] - [-833.9\ kJ + 2(-95.4)\ kJ] =$ **-93 kJ**
(d) $\Delta G° = [(-137\ kJ) + (-228\ kJ)] - [-395\ kJ + 0] =$ **30 kJ**
(e) $\Delta G° = [10(0) + 3(-395)\ kJ + 4(-228) kJ] - [10(104)\ kJ + (-23)\ kJ]$
$=$ **-3114 kJ**

14.51 $\Delta G° = [6 \times \Delta G_f°(CO_2) + 6 \times \Delta G_f°(H_2O)] - [1 \times \Delta G_f°(glu) + 6 \times \Delta G_f°(O_2)]$
$= [6 \times (-395\ kJ) + 6 \times (-237\ kJ)] - [1 \times (-910.2\ kJ) + 6 \times (0)] =$ **-2882 kJ**
or -2880 kJ (only 3 sign. figs)

14.52 (a) $Pb(s) + PbO_2(s) + 2H_2SO_4(\ell) \rightarrow 2PbSO_4(s) + 2H_2O(\ell)$
$\Delta G° = [2(-811.3\ kJ) + 2(-237\ kJ)] - [1(0) + 1(-219\ kJ) + 2(-689.9\ kJ) =$ **-498 kJ**

(b) $CH_4(g) + 4Cl_2(g) \rightarrow CCl_4(\ell) + 4HCl(g)$
$\Delta G° = [1(-65.3) + 4(-95.4)] - [1(-50.6) + 4(0)] =$ **-396.3 kJ**

(c) $13N_2O(g) + C_4H_{10}(g) \rightarrow 13N_2(g) + 4CO_2(g) + 5H_2O(g)$
$\Delta G° = [13(0) + 4(-395) + 5(-228)] - [13(104) + 1(-17.0)] =$ **-4055 kJ**
or -4.06×10^3 kJ (only 3 sign. fig.)

14.53 $3CaCO_3(s) \rightarrow 3CaO(s) + 3CO_2(g)$ $\quad \Delta G_1 = 3(+130\ kJ)$
$3CaO(s) + 2H_3PO_4(\ell) \rightarrow Ca_3(PO_4)_2(s) + 3H_2O(\ell)$ $\quad \Delta G_2 = 1(-512\ kJ)$

$3CaCO_3(s) + 2H_3PO_4(\ell) \rightarrow Ca_3(PO_4)_2(s) + 3CO_2(g) + 3H_2O(\ell)$
$\Delta G_3 = \Delta G_1 + \Delta G_2$
$= 3(130\ kJ) + (-512\ kJ)$
$\mathbf{\Delta G_3 = -122\ kJ}$

14.54 $CO(g) + 1/2O_2(g) \rightarrow CO_2(g)$ $\quad \Delta G_1 = 1/2(-516\ kJ)$
$Mn_2O_3(s) \rightarrow 2MnO(s) + 1/2O_2$ $\quad \Delta G_2 = -1/2(-312\ kJ)$

$Mn_2O_3(s) + CO(g) \rightarrow 2MnO(s) + CO_2(g)$ $\quad \Delta G_3 = \Delta G_1 + \Delta G_2$
$= 1/2(-516\ kJ) - (1/2)(-312\ kJ)$
$=$ **-102 kJ**

14.55 Advantage - maximum work is obtained.
Disadvantage - the change takes forever to occur.

14.56 Maximum amount of useful work at Standard Temperature (25°C) and Standard Pressure (1 atm) = $\Delta G°$.
For $C_3H_8(g) + 5O_2(g) \rightarrow 3CO_2(g) + 4H_2O(g)$
$\Delta G° = [3 \times \Delta G_f°(CO_2) + 4 \times \Delta G_f°(H_2O)] - [1 \times \Delta G_f°(C_3H_8) + 5 \times \Delta G_f°(O_2)]$
$= [3 \times (-395\ kJ) + 4 \times (-228\ kJ)] - [1 \times (-23) + 5 \times (0)] =$ **-2074 kJ**
The maximum useful work is the work that would be available under ideal, reversible conditions. A real process does not follow a reversible path. Therefore, we always get less than this maximum amount of work in any real process that uses the above reaction.

14.57 See Figure 14.14 (c) The position of the equilibrium favors products.

14.58 Refer to Figure 14.13. Once a minimum free energy has been achieved by the system, the composition of the system can no longer change since such a change involves going "uphill" on the free energy curve. At the minimum, both reactants and products possess the same free energy and, therefore, $\Delta G = 0$.

14.59 At equilibrium at constant pressure: $0 = \Delta H - T\Delta S$ or $\Delta S = \frac{\Delta H}{T}$

$\Delta S_{vap} = 40.7 \times 10^3$ J mol^{-1}/373 K = **109 J mol^{-1} K^{-1}**
$\Delta S_{fus} = 6.02 \times 10^3$ J mol^{-1}/273 K = **22.1 J mol^{-1} K^{-1}**
Both should be positive since both processes increase the randomness of the system. One would expect vaporization to have a greater increase in randomness than does melting and the above values verify that expectation.

14.60 $Br_2(\ell) \rightleftharpoons Br_2(g)$ $\Delta H = +30.9 - (0) = +30.9$ kJ
$\Delta S = 245.4 - 152.2 = +93.2$ J K^{-1} $T\Delta S = \Delta H$ (constant pressure)
$T = \Delta H/\Delta S = 30{,}900$ J/93.2 J K^{-1} = **332 K** or 332-273 = **59°C**

14.61 $\Delta G°$ determines the position of equilibrium between reactants and products. Whether we start with pure reactants or pure products, some reaction will occur (accompanied by a free energy decrease) until equilibrium is reached.

14.62 The sign and magnitude of $\Delta G°$ tells us where the reaction is going or something about the position of equilibrium. A large $\Delta G°$ indicates a reaction that is far from equilibrium and one in which a considerable change will be observed.

14.63 There is none.

14.64 Whether the reaction is spontaneous ($\Delta G = -$) or not, and how fast the reaction occurs.

14.65 See Figure 14.14.

14.66 (a) $1/2N_2(g) + O_2(g) \rightarrow NO_2(g)$ **$\Delta G°$ = +51.9 kJ/mol (No)**
(b) $2HNO_3(\ell) + Ag(s) \rightarrow AgNO_3(s) + NO_2(g) + H_2O(\ell)$
$\Delta G°$ = -57 kJ/mol (Yes, but costly)
(c) $2NH_3(g) + 3O_2(g) \rightarrow NO_2(g) + NO(g) + 3H_2O(g)$
$\Delta G°$ = -511 kJ/mol (Yes)
(d) $CuO(s) + NO(g) \rightarrow NO_2(g) + Cu(s)$ **$\Delta G°$ = +92 kJ/mol (No)**
(e) $NO(g) + 1/2O_2(g) \rightarrow NO_2(g)$ **$\Delta G°$ = -34.9 kJ/mol (Yes)**
(f) $9H_2O(g) + 7N_2O(g) \rightarrow 6NH_3(g) + 8NO_2(g)$
$\Delta G°$ = 1637 kJ/8 mol NO_2 (No)

14.67 First calculate the moles present in 4.00 L of $C_4H_{10}(g)$ at 25°C and 1 atm.

PV = nRT   n = [(1 atm)(4.00 L)]/[(0.0821 L atm mol^{-1} K^{-1})(298 K)]

n = 0.163 (if 1 atm is assumed to have 3 significant figures)

Calculate ΔG° for: $C_4H_{10}(g) + 13/2O_2(g) \rightarrow 4CO_2(g) + 5H_2O(\ell)$

ΔG° = [4(-395) kJ + 5(-237)kJ] -[1(-17.0)kJ + 13/2(0)]
= -2750 kJ per mole C_4H_{10}

For 0.163 mole there would be 2750 x 0.163 or **448 kJ** of useful work.

14.68 $C_8H_{18}(\ell) + 25/2O_2(g) \rightarrow 8CO_2(g) + 9H_2O(\ell)$

$\Delta G^\circ_{rxn} = [8\Delta G_f^o(CO_2) + 9\Delta G_f^o(H_2O)] - [1\Delta G_f^o(C_8H_{18}) + 25/2\Delta G_f^o(O_2)$

= [8(-395 kJ) + 9(-237 kJ)] - [1(12.8 kJ) + 25/2(O)]

= (-3160) + (-2133) - (12.8) = -5306 kJ

or -5310 kJ[mol(C_8H_{18})]$^{-1}$

$$\frac{-5310 \text{ kJ}}{\text{mol } C_8H_{18}} \times \frac{1 \text{ mol}}{114 \text{ g}} = 46.6 \text{ kJ g}^{-1}$$

46.6 kJ g^{-1} x 35.3% effic. = 16.4 kJ g^{-1}

$$2.00 \times 10^4 \text{ kJ} \times \frac{1 \text{ g}}{16.4 \text{ kJ}} = \mathbf{1.22 \times 10^3 \text{ g needed}}$$

15 CHEMICAL EQUILIBRIUM IN GASEOUS SYSTEMS

15.1 In a dynamic equilibrium, products are constantly changing to reactants and reactants are constantly changing to products but the overall effect is no net change in concentration.

15.2 (a) $\frac{[NO]^2}{[N_2][O_2]} = K_c$ (b) $\frac{[NO_2]^2}{[NO]^2[O_2]} = K_c$ (c) $\frac{[H_2S]^2}{[H_2]^2[S_2]} = K_c$

(d) $\frac{[NO_2]^4[O_2]}{[N_2O_5]^2} = K_c$ (e) $\frac{[POCl_3]^{10}}{[P_4O_{10}][PCl_5]^6} = K_c$

15.3 (a) $\frac{p_{NO}^2}{p_{N_2}p_{O_2}} = K_p$ (b) $\frac{p_{NO_2}^2}{p_{NO}^2 p_{O_2}} = K_p$ (c) $\frac{p_{H_2S}^2}{p_{H_2}^2 p_{S_2}} = K_p$

(d) $\frac{p_{NO_2}^4 p_{O_2}}{p_{N_2O_5}^2} = K_p$ (e) $\frac{p_{POCl_3}^{10}}{p_{P_4O_{10}} p_{PCl_5}^6} = K_p$

15.4 (a) $K_p = \dfrac{p_{CH_3OH}}{p_{CO}\,p_{H_2}^2}$, $K_c = \dfrac{[CH_3OH]}{[CO][H_2]^2}$

(b) $K_p = \dfrac{p_{CO_2}\,p_{H_2}}{p_{CO}\,p_{H_2O}}$, $K_c = \dfrac{[CO_2][H_2]}{[CO][H_2O]}$

(c) $K_p = \dfrac{p_{PCl_5}}{p_{PCl_3}\,p_{Cl_2}}$, $K_c = \dfrac{[PCl_5]}{[PCl_3][Cl_2]}$

(d) $K_p = \dfrac{p_{N_2}\,p_{H_2O}^4}{p_{NO_2}^2\,p_{H_2}^4}$, $K_c = \dfrac{[N_2][H_2O]^4}{[NO_2]^2[H_2]^4}$

(e) $K_p = \dfrac{p_{H_2O}^2\,p_{SO_2}^2}{p_{H_2S}^2\,p_{O_2}^3}$, $K_c = \dfrac{[H_2O]^2[SO_2]^2}{[H_2S]^2[O_2]^3}$

15.5 By convention. This simplifies tabulation of equilibrium constants by removing ambiguity.

15.6 Exp. 1 K_c = (0.23) (0.055)/0.0023 = 5.5
Exp. 2 K_c = (0.15) (0.37)/0.010 = 5.6
Exp. 3 K_c = (0.99) (0.47)/0.085 = 5.5
Exp. 4 K_c = (3.66) (1.50)/1.00 = 5.49
For all of these, K_c (average) = **5.5** (This is within experimental precision)

15.7 (a) $\dfrac{[HCl]^2}{[H_2][Cl_2]}$ or $\dfrac{p_{HCl}^2}{p_{H_2}\,p_{Cl_2}}$

(b) $\dfrac{[HCl]}{[H_2]^{1/2}[Cl_2]^{1/2}}$ or $\dfrac{p_{HCl}}{p_{H_2}^{1/2}\,p_{Cl_2}^{1/2}}$ K(a) would equal K(b) squared.

15.8 (a) $K_c = 1/(1.4 \times 10^7) = \mathbf{7.1 \times 10^{-8}}$

(b) $K_c = (1.4 \times 10^7)^2 = \mathbf{2.0 \times 10^{14}}$

15.9 $$K_p = \frac{P_{NO}^2}{P_{N_2} P_{O_2}} = \frac{P_{NO}^2 P_{H_2O}^3}{P_{NH_3}^2 P_{O_2}^{5/2}} \times \frac{P_{NH_3}^2}{P_{N_2} P_{H_2}^3} \times \frac{P_{H_2}^3 P_{O_2}^{3/2}}{P_{H_2O}^3}$$

$= (9 \times 10^{172})^{1/2} \times (9.1 \times 10^5) \times (8.6 \times 10^{79})^{-3/2}$

$= \mathbf{3 \times 10^{-28}}$

15.10 $$K_c = \frac{[ClF]^2[BrF_5]}{[BrF][ClF_3]^2} = \frac{[BrF_5]}{[BrF_3][F_2]} \times \frac{[ClF]^2[F_2]^2}{[ClF_3]^2} \times \frac{[BrF_3]}{[BrF][F_2]}$$

$= (8.6 \times 10^{35}) \times (7.8 \times 10^{12})^{-2} \times (7.3 \times 10^{27}) = \mathbf{1.0 \times 10^{38}}$

15.11 By looking at the size of the equilibrium constant, you can determine whether the reaction favors the forward or reverse reaction. If the number is much greater than one, the reaction will tend to proceed far toward completion. If, however, the K is much less than one, only small amounts of products will be present at equilibrium.

15.12 From the magnitude of K we can say that the tendency to proceed toward completion increases in the order (b) < (c) < (d) < (a).

15.13 Of the reactions given, reaction (a) will proceed the farthest toward completion if allowed to come to equilibrium. Reaction (b) will proceed the least toward completion.

15.14 One must be very careful when trying to compare the positions of equilibria of equations, such as those in this question, that have greatly different stoichiometry. As shown below, the stoichiometry of reactions (a) and (b) are similar and that (c) can be made to be more nearly like (a) and (b) if its coefficients are divided by four. Even then the answer must be rather tentative. For comparison, the equations are written without the compounds being identified.

(a) 2( ) + ( ) $\rightleftharpoons$ 2( ) $\quad K_p = \frac{p^2}{p^2 p} = \frac{p^2}{p^3} = 8.6 \times 10^{79}$

(continued)

15.14 (continued)

(b) ( ) + 3( ) $\rightleftharpoons$ 2( ) $K_p = \frac{p^2}{pp^3} = \frac{p^2}{p^4} = 9.1 \times 10^5$

(c) 4/4( ) + 5/4( ) $\rightleftharpoons$ 4/4( ) + 6/4( ) $K_p = \frac{pp^{3/2}}{pp^{5/4}} = \frac{p^{5/2}}{p^{9/4}} = \sqrt[4]{9 \times 10^{172}}$

$$\frac{p^{5/2}}{p^{9/4}} = 1.7 \times 10^{43}$$

In this comparison, one can conclude that reaction (a) will proceed farthest toward the right.

15.15 $\Delta G° = -RT \ln(1) = \mathbf{0}$

15.16 In Section 15.3 it is stated that for gases, the equilibrium constant is calculated from $\Delta G° = -RT \ln K_p$. **For gases K_p is calculated from $\Delta G°$.**

15.17 Equation 14.7 $\Delta G = \Delta H - T\Delta S$ or $\Delta G° = \Delta H° - T\Delta S°$
Equation 15.5 $\Delta G° = -RT \ln K_P$
When the two equations are combined and ln changed to log, one obtains:

$$\Delta H° - T\Delta S° = -2.303\ RT \log K_P$$

$$\log K_p = \frac{T\Delta S° - \Delta H°}{2.303\ RT} = \frac{T\Delta S°}{2.303\ RT} - \frac{\Delta H°}{2.303\ RT}$$

$$\log K_p = \frac{\Delta S°}{2.303\ R} - \frac{\Delta H°}{2.303\ R} \times \frac{1}{T}$$

(This is in the form $y = b + mx$, where $\log K_p$ corresponds to y and 1/T corresponds to x.)

Slope = m = $-\Delta H°/2.303\ R$ y intercept = $\Delta S°/2.303\ R$
Slope gives $\Delta H°$ **y intercept gives $\Delta S°$**

15.18 $\Delta G = -RT \ln K_P$ or $\Delta G° = -R(298\ K) \ln K_P$
$\Delta G°$ = sum of $\Delta G_f°$ products - sum of $\Delta G_f°$ reactants
(From Table 14.4) $\Delta G°$ = [2 mol (-273 kJ /mol) + 1 (0)] - [2 mol (-95.4 kJ/mol) + 1 (0)] = -355 kJ

-355×10^3 J (per mol) = $-(8.314\ J\ mol^{-1}\ K^{-1})(298\ K) \ln K_P$

$\mathbf{K_P = 2 \times 10^{62}}$

15.19 $\Delta G^\circ = -R\ (298\ K)\ \ln K_P$
ΔG° = sum of ΔG_f° products - sum of ΔG_f° reactants
ΔG° = [1 mol(86.8 kJ/mol) + 1 mol(-370 kJ/mol)] - [1 mol(-300 kJ /mol) + 1 mol(+51.9 kJ/mol)] = -35.1 kJ

-35.1×10^3 J (per mol) = $-(8.314\ J\ mol^{-1}K^{-1})(298\ K)\ \ln K_P$

K_P = **1×10^6**

15.20 (a) ΔG° = [1 mol(-773.6 kJ/mol) + 6 mol(-228 kJ/mol)] - [1 mol(-2222 kJ/mol)]
= **80 kJ**

(b) $\Delta G^\circ = -RT \ln K_P$
80×10^3 J (per mol) = $-(8.314\ J\ mol^{-1}\ K^{-1})(298K)\ \ln K_P$
$-32 = \ln K_P$ K_P = **1.3×10^{-14}**

15.21 $\Delta G = -RT \ln K_P$
-13.5×10^3 J (per mol) = $-(8.314\ J\ mol^{-1}\ K^{-1})(700\ K)\ \ln K_P$

$2.32 = \ln K_P$ K_P = **10**

15.22 395°C = 668 K $\Delta G = -RT \ln K_P$ $K_P = 4.56 \times 10^{-2}$
$\Delta G = -(8.314\ J\ mol^{-1}\ K^{-1})(668\ K)\ \ln (4.56 \times 10^{-2})$
$\Delta G = 1.71 \times 10^4$ J = **17.1 kJ**

15.23 527°C = 800 K $\Delta G = -RT \ln K_P$
$\Delta G = -(8.314\ J\ mol^{-1}\ K^{-1})(800\ K)\ \ln 5.10$

$\Delta G = -1.08 \times 10^4$ J = **-11 kJ**

15.24 ΔH° = [0 + 0] - [2 mol(-92.5 kJ/mol)] = +185 kJ

ΔS°=[1 mol(130.6 J $mol^{-1}\ K^{-1}$) + 1 mol(223.0 J $mol^{-1}\ K^{-1}$)] - [2 mol(186.7 J $mol^{-1}\ K^{-1}$)]
= - 19.80 J K^{-1}

$\Delta G = \Delta H - T\Delta S$ = 185,000 J - (773 K)(-19.80 J K^{-1})
$\Delta G = 2.00 \times 10^5$ J = **2.00×10^2 kJ**

$\Delta G = -RT \ln K_P$
2.00×10^5 J = $-(8.314\ J\ mol^{-1}\ K^{-1})(773\ K)\ \ln K_P$
$-31.1 = \ln K_P$ K_P = **3×10^{-14}**

15.25 $\Delta H° = [1 \text{ mol } (-84.5 \text{ kJ mol}^{-1})] - [1 \text{ mol } (51.9 \text{ kJ mol}^{-1}) + 1(0)] = -136.4 \text{ kJ}$

$\Delta S° = [1 \text{ mol } (230 \text{ J mol}^{-1} \text{ K}^{-1})] - [1 \text{ mol } (220 \text{ J mol}^{-1} \text{ K}^{-1}) + 1 \text{ mol } (130.6 \text{ J mol}^{-1} \text{ K}^{-1})]$
$= -121 \text{ J K}^{-1}$

$\Delta G = -RT \ln K_P = \Delta H - T\Delta S$

$\Delta H = T\Delta S - RT \ln K_P = T(\Delta S - R \ln K_P)$

$$\frac{\Delta H}{\Delta S - R \ln K_P} = T$$

$$\frac{-136.4 \times 10^3 \text{ J}}{(-120.6 \text{ J K}^{-1}) - (8.314 \text{ J mol}^{-1} \text{ K}^{-1})(\ln 1)} = T$$

$$T = \frac{-136.4 \times 10^3 \text{K}}{120.6 + 0} = \mathbf{1.13 \times 10^3 \text{ K}} \text{ or } 860°$$

15.26 $\Delta G = \Delta G° + RT \ln Q$

$$Q = \frac{3 \times 10^{-6} \text{atm}}{(2 \times 10^{-3} \text{atm})(1 \times 10^{-2} \text{ atm})^2} = 15$$

$\Delta G = (-13.5 \times 10^3 \text{ J}) + (8.314 \text{ J mol}^{-1} \text{ K}^{-1})(700 \text{ K})(\ln 15) = 2.3 \times 10^3 \text{ J}$

$\Delta G = +$; therefore, the system is **not at equilibrium.** The value of RT ln Q must decrease if ΔG is to be zero and if the system is to attain equilibrium. The value of RT ln Q will decrease if the **reaction proceeds spontaneously to the left.**

15.27 15.2 (a) and 15.4 (b), only

15.28 $K_P = K_c (RT)^{\Delta n}$ $\Delta n = \Delta$ moles of gases $= 0$

$\mathbf{K_P} = (4.05)[(0.0821 \text{ L atm mol}^{-1} \text{ K}^{-1})(773 \text{ K})]^0 = \mathbf{4.05}$

15.29 $K_P = K_c (RT)^{\Delta n}$

$K_P = (5.67 \text{ mol}^2/\text{L}^2)[(0.0821 \text{ L atm mol}^{-1} \text{ K}^{-1})(1773 \text{ K})]^2$

$\mathbf{K_P} = (5.67)(2.119 \times 10^4) \text{ atm}^2 = \mathbf{1.20 \times 10^5 \text{ atm}^2}$

15.30 $K_P = K_c (RT)^{\Delta n}$

$(6.5 \times 10^{-2} \text{ atm}^{-1}) = K_c [(0.0821 \text{ L atm mol}^{-1} \text{ K}^{-1})(373 \text{ K})]^{-1}$

$6.5 \times 10^{-2} = K_c (3.265 \times 10^{-2})$ $\mathbf{K_c = 2.0 \text{ L mol}^{-1}}$

15.31 At equilibrium, $K_c = [H_2O(g)]$, $K_p = p_{H_2O(g)}$

Thus, p_{H_2O} or $[H_2O]$ are constants that only change with temperature.

15.32 The concentrations of pure solids and liquids are invariant; they are constants that can be incorporated into K_p or K_c.

15.33 (a) $K_c = [CO_2(g)]$

(b) $K_c = \frac{[Ni(CO)_4(g)]}{[CO(g)]^4}$

(c) $K_c = \frac{[I_2(g)][CO_2(g)]^5}{[CO(g)]^5}$

(d) $K_c = \frac{[CO_2(g)]}{[Ca(HCO_3)_2(aq)]}$

(e) $K_c = [Ag^+(aq)][Cl^-(aq)]$

15.34 $PCl_3(g) + Cl_2(g) \rightleftharpoons PCl_5(g)$

(a) Addition of PCl_3 would drive reaction toward the products to compensate for the excess of PCl_3.

(b) Removal of Cl_2 would cause the reaction to proceed toward the reactants to re-establish equilibrium and make up for the loss of Cl_2.

(c) Removal of PCl_5 would cause the reaction to proceed toward the products to compensate for the loss of PCl_5.

(d) A decrease in the volume of the container would cause the pressure to increase, and the reaction would favor the side which has the least number of moles of gas, namely, the products.

(e) Addition of He (an inert gas), without a change in the size of container, would increase the pressure but not the partial pressures. Since the partial pressure values are what must yield the K_p value, there would be no effect on the position of equilibrium.

15.35 None of the above will effect the equilibrium constant for the reaction. The only change which will effect the equilibrium constant is a change in temperature.

15.36 (a) **decreased** (b) **increased** (c) **no change**
(d) **decreased** (e) **no change**

15.37 (a) **No change in the value of K** (b) **No change in the value of K**
(c) **No change in the value of K** (d) **The value of K will increase**
(e) **No change in the value of K**

15.38 (a) **increased** (b) **decreased** (c) **increased** (d) **no change**

15.39 (a) **decreased** (b) **increased** (c) **no change** (d) **decreased**

15.40 The answer to this question should be a sketch like that in Figure 15.2 with the top line being N_2, the middle line being H_2 and the bottom one being NH_3. Instead of the H_2 being added as in Figure 15.2, it would be N_2 being added in this answer.

15.41 (a) **no change** (b) **increased** (c) **decreased** (d) **increased**

15.42 $K_c = \dfrac{[PCl_5]}{[PCl_3][Cl_2]} = \dfrac{1}{K_c \text{ (calculated in Ex. 15.6)}} = \dfrac{1}{5.5} = \mathbf{0.18}$

15.43 $K_c = 5.5 \quad \Delta n = +1$
$K_P = K_c(RT)^{\Delta n} = (5.5 \text{ mol/L})[(0.0821 \text{ L atm mol}^{-1} \text{ K}^{-1})(298 \text{ K})]^1$
$K_P = (5.5)(2.447 \times 10^{+1} \text{ atm}) = \mathbf{1.3 \times 10^2 \text{ atm}}$

15.44 $Q = \dfrac{(0.30 \text{ M})(0.020 \text{ M})}{(0.040 \text{ M})(0.50 \text{ M})} = 0.30$
Q does not equal K_c. Therefore, **the system is not at equilibrium. Since the Q value is less than K_c, the reaction must proceed to the right if equilibrium is to be established.**

15.45 $K_c = \dfrac{[PCl_3][Cl_2]}{[PCl_5]} = \dfrac{[PCl_3](1.87 \times 10^{-1})}{1.29 \times 10^{-3}} = 33.3$

$[PCl_3] = \mathbf{2.30 \times 10^{-1} \text{ mol L}^{-1}}$

15.46 (a) $\mathbf{K_p} = \dfrac{p^2_{NO_2}}{p_{N_2O_4}} = \dfrac{(0.844 - 0.563)^2}{0.563} = \mathbf{1.40 \times 10^{-1}}$ **atm**

(b) $K_P = K_c(RT)^{\Delta n} \quad \Delta n = +1$
$1.40 \times 10^{-1} \text{ atm} = K_c\ [(0.0821 \text{ L atm mol}^{-1} \text{ K}^{-1})(298 \text{ K})]^1$
$\mathbf{K_c = 5.72 \times 10^{-3} \text{ mol L}^{-1}}$

(c) $\Delta G° = -RT \ln K_P = -(8.314 \text{ J mol}^{-1} \text{ K}^{-1})(298 \text{ K}) \ln(0.140)$
$\Delta G° = 4.87 \times 10^3 \text{ J mol}^{-1} = \mathbf{4.87 \text{ kJ mol}^{-1}}$

15.47 $K_c = \dfrac{[H_2][I_2]}{[HI]^2} = \dfrac{(1.0 \times 10^{-3} \text{ M})(2.5 \times 10^{-2} \text{ M})}{(2.2 \times 10^{-2} \text{ M})^2} = \mathbf{5.2 \times 10^{-2}}$

15.48 $K_p = \frac{p_{NOCl}^2}{p_{NO}^2 \, p_{Cl_2}} = \frac{(0.15 \text{ atm})^2}{(0.65 \text{ atm})^2 (0.18 \text{ atm})} = \mathbf{0.30 \text{ atm}^{-1}}$

15.49

	$2N_2O$ +	$3O_2$ ⇌	$4NO_2$
Init. Conc.	0.020	0.0560	0
Change	- 2X	-3X	+4X
Equil. Conc.	0.020 - 2X	0.0560 - 3X	0.020 or 4X

(a) $4X = 0.020$ $\quad X = 5.0 \times 10^{-3}$

$[N_2O] = 0.020 - 2X = 0.020 - 2(5.0 \times 10^{-3}) = \mathbf{0.010\ M}$

$[O_2] = 0.0560 - 3X = 0.0560 - 3(5.0 \times 10^{-3}) = \mathbf{0.041\ M}$

(b) $K_c = \frac{[NO_2]^4}{[N_2O]^2[O_2]^3} = \frac{(0.020)^4}{(0.010)^2(0.041)^3} = \mathbf{23\ L\ mol^{-1}}$

15.50

	SO_2 +	NO_2 ⇌	NO +	SO_3
Init. Conc.	0.0500	0.0500	0	0
Change	-X	-X	+X	+X
Equil. Conc.	0.0500 -X	0.0500 - X	+X	+X

$K_c = 85.0 = \frac{(X)(X)}{(0.0500 - X)^2}$

$\sqrt{85.0} = \frac{X}{(0.0500 - X)} = 9.22$

$X = 9.22\ (0.0500 - X)$

$X = 4.51 \times 10^{-2}$

$\mathbf{[NO] = [SO_3] = 0.0451\ M}$

$\mathbf{[SO_2] = [NO_2] = 0.0049\ M}$

15.51

	H_2 +	CO_2 ⇌	CO +	H_2O
Init. Conc.	0.200	0.200	0	0
Change	-X	-X	+X	+X
Equil. Conc.	0.200 -X	0.200 - X	+X	+X

$$K_c = \frac{(X)^2}{(0.200 - X)^2} = 0.771$$

$$\sqrt{0.771} = \frac{X}{(0.200 - X)}$$

$X = 0.0935$

$[H_2]$ = $[CO_2]$ = 0.106 M **[CO] = $[H_2O]$ = 0.0935 M**

15.52 $Q = \frac{(0.0150)(0.0100)}{(0.0100)(0.0200)} = 0.75$ $Q < K_c$, shifts to right

	SO_2 +	NO_2 ⇌	NO +	SO_3
Init. Conc.	0.0100	0.0200	0.0100	0.0150
Change	-X	-X	+X	+X
Equil. Conc.	0.0100 -X	0.0200 - X	0.0100 +X	0.0150 + X

$$K_c = 85.0 = \frac{(0.0100 + X)(0.0150 + X)}{(0.0100 - X)(0.0200 - X)}$$

$$85.0 = \frac{1.50 \times 10^{-4} + 2.5 \times 10^{-2} X + X^2}{2.00 \times 10^{-4} - 0.0300 X + X^2}$$

$84.0 X^2 - 2.575 X + 0.01685 = 0$
From the solution of this quadratic equation: X = 0.00947 (and, X = 0.0212; impossible)

$[SO_2]$ = 0.0005 M **$[NO_2]$ = 0.0105 M**
[NO] = 0.0195 M **$[SO_3]$ = 0.0245 M**

15.53 (a)

	$2CO_2$	$\rightleftharpoons$	$2CO$	+	O_2
Init. Conc.	1.0×10^{-3} M		0		0
Change	-2X		+2X		+X
Equil. Conc.	1.0×10^{-3} -2X		2X		X

$$K_c = 6.4 \times 10^{-7} = \frac{(2X)^2 X}{(1.0 \times 10^{-3} - 2X)^2}$$

If 2X is small compared to 1.0×10^{-3}, then:

$$6.4 \times 10^{-7} = \frac{(2X)^2 X}{(1.0 \times 10^{-3})^2} = \frac{4X^3}{(1.0 \times 10^{-3})^2}$$

$X = 5.4 \times 10^{-5}$ Check! Was 2X small compared to 1.0×10^{-3}? The value obtained for 2X was about 11% of the1.0×10^{-3}. This is about the limit allowed in most approximations. Using this value, the equilibrium concentrations are:

$[CO_2]$ = 8.9×10^{-4} M

[CO] = 1.1×10^{-4} M

$[O_2]$ = 5.4×10^{-5} M

To obtain a more precise solution, one would solve the above equation using a series of approximations. The above would be the first approximation. These values would be substituted into the equilibrium expression and the process would be repeated. The values after successive approximations are:

$[CO_2]$ = 9.0×10^{-4} M,

[CO] = 1.0×10^{-4} M,

$[O_2]$ = 5.1×10^{-5} M

(b) $\frac{(1.0 \times 10^{-3} - 9.0 \times 10^{-4})}{1.0 \times 10^{-3}} = \mathbf{0.10}$ = the fraction of CO_2 decomposed

15.54

	CO	+	Cl_2	$\rightleftharpoons$	$COCl_2$
Init. Conc.	0		0		0.020 M
Change	+X		+X		-X
Equil. Conc.	X		X		0.020 - X

$$K_c = 4.6 \times 10^9 = \frac{0.020 - X}{X^2}$$

Assume that X is very small compared to 0.020. Then:

$$4.6 \times 10^9 = \frac{0.020}{X^2} \qquad X = 2.1 \times 10^{-6}$$

[CO] = [Cl$_2$] = 2.1 x 10^{-6} M   [COCl$_2$] = 0.020 M

15.55 $K_P = P_{CO_2} \times P_{H_2O} = 0.25 \text{ atm}^2 = X^2$

$$P_{CO_2} = \sqrt{0.25 \text{ atm}^2} = \mathbf{0.50\ atm} = \mathbf{p_{CO_2}} = \mathbf{p_{H_2O}}$$

It is used in baking because it liberates gaseous CO_2 and H_2O that are trapped in the dough, thus causing it to rise.

15.56

	H_2	+	I_2	$\rightleftharpoons$	2HI
Init. Conc.	0.0100 M		0.0100 M		0.0740
Change	+X		+X		+0.050 M - 2X
Equil. Conc.	0.0100 M + X		0.0100 M + X		0.124 M - 2X

$$K_c = \frac{(0.074)^2}{(0.01 \times 0.01)} = 54.76 \qquad 54.76 = \frac{(0.124 - 2X)^2}{(0.0100 + X)^2}$$

$$\sqrt{54.76} = \frac{0.124 - 2X}{0.0100 + X} = 7.40 \qquad X = 5.3 \times 10^{-3}$$

[HI] = 0.124 - 2 x 5.3 x 10^{-3} = **0.113 M**

[H$_2$] = [I$_2$] = 0.0100 + 5.3 x 10^{-3} = **0.0153 M**

15.57

	$2NO_2 \rightleftharpoons$	N_2O_4
Init. Conc.	1.0 M	0
Change	- 2X	+X
Equil. Conc.	1.0 - 2X	X

$$7.5 = \frac{X}{(1.0 - 2X)^2}$$

$30X^2 - 31X + 7.5 = 0$

X = 0.3865 (the logical solution of the two solutions of the quadratic equation.)

$[NO_2]$ = 1.0 - 2 x 0.3865 = **0.23 M**

$[N_2O_4]$ = **0.39 M**

Double the size of the container!

	$2NO_2 \rightleftharpoons$	N_2O_4
Init. Conc.	0.50 M	0
Change	-2X	X
Equil. Conc.	0.50 - 2X	X

$$7.5 = \frac{X}{(0.50 - 2X)^2}$$

$30X^2 - 16X + 1.875 = 0$ X = 0.174

$[NO_2]$ = 0.50 - 2 x 0.174 = **0.15 M**

$[N_2O_4]$ = **0.17 M**

Yes! The larger container favors the NO_2 while the smaller container favored the N_2O_4. This is what one should expect based upon a knowledge of LeChatelier's Principle.

15.58

	CO	+	Cl_2	$\rightleftharpoons$	$COCl_2$
Init. Conc.	0.15 M		0.30 M		0
Change	-(0.15 - X)		- (0.15 - X)		+(0.15 - X)
Equil. Conc.	X		0.15 + X		0.15 - X

$$K_c = 4.6 \times 10^9 = \frac{0.15 - X}{(X)(0.15 + X)} \approx \frac{0.15}{X(0.15)}$$

$$X = \frac{0.15}{0.15 \times 4.6 \times 10^9} = 2.2 \times 10^{-10}$$

[CO] = 2.2 x 10^{-10} M

$[Cl_2]$ = 0.15 M + 2.2 x 10^{-10} M = **0.15 M**;

$[COCl_2]$ = 0.15 - 2.2 x 10^{-10} = **0.15 M**

15.59

	Initial Concentration	Change	Equilibrium Concentration
H_2	0.0200	- X	0.0200 - X
CO_2	0.0400	- X	0.0400 -X
CO	0	+X	X
H_2O	0	+X	X

$$K_c = 0.771 = \frac{(X)(X)}{(0.0200 - X)(0.0400 - X)} = \frac{X^2}{(8.00 \times 10^{-4}) - (0.0600\,X) + X^2}$$

$6.17 \times 10^{-4} - 4.63 \times 10^{-2}\,X + 0.771\,X^2 = X^2$

$.229\,X^2 + (4.63 \times 10^{-2})X - 6.17 \times 10^{-4} = 0$

$X = 1.25 \times 10^{-2}$

$[H_2]$ = 7.5 x 10^{-3} M **$[CO_2]$ = 2.75 x 10^{-2} M**

[CO] = 1.25 x 10^{-2} M **$[H_2O]$ = 1.25 x 10^{-2} M**

15.60 (a)

	Initial Concentration	Change	Equilibrium Concentration
BrF_3	0.0500	-(0.0500 - X)	X
F_2	0.0500	-(0.0500 - X)	X
BrF_5	0	+(0.0500 - X)	0.0500 - X ≈ 0.0500

$$K_c = 8.6 \times 10^{35} = \frac{(0.0500)}{(X)(X)} = \frac{0.0500}{X^2}$$

$X^2 = 5.81 \times 10^{-38}$

$X = 2.4 \times 10^{-19}$

$[BrF_3] = 2.4 \times 10^{-19}$ M

$[F_2] = 2.4 \times 10^{-19}$ M

$[BrF_5] = 0.0500 - (2.4 \times 10^{-19}) = 0.0500$ M

(b)

	Initial Concentration	Change	Equilibrium Concentration
BrF_3	0.100	-(0.100 - X)	X
F_2	0.200	-(0.100 - X)	0.100 + X ≈ 0.100
BrF_5	0	+(0.100 - X)	0.100 - X ≈ 0.100

$$K_c = 8.6 \times 10^{35} = \frac{(0.100)}{(X)(0.100)}$$

$X = 1.2 \times 10^{-36}$

$[BrF_3]$ = **1.2×10^{-36} M**

$[F_2] = 0.100 + 1.2 \times 10^{-36}$ = **0.100 M**

$[BrF_5] = 0.100 - 1.2 \times 10^{-36}$ = **0.100 M**

15.61

	N_2	+	O_2	$\rightleftharpoons$	2NO
Init. p	33.6		4.0		0
Change	-X		- X		+ 2X
Equil. p	33.6 - X		4.0 - X		2X

$$K_P = 4.8 \times 10^{-7} = \frac{(2X)^2}{(33.6 - X)(4.0 - X)} \approx \frac{(2X)^2}{(33.6)(4.0)}$$

$X = 4.0 \times 10^{-3}$

$p_{N_2} = 33.6$ atm $\quad p_{O_2} = 4.0$ atm $\quad$ $p_{NO} =$ **8.0×10^{-3} atm**

15.62 $\frac{P_iV_i}{T_i} = \frac{P_fV_f}{T_f}$

$$\frac{P_f}{P_i} = \frac{V_i\,T_f}{V_f\,T_i}$$

$\frac{V_i\,T_f}{V_f\,T_i}$ same for both gases

Therefore, $\frac{p_f(N_2)}{p_i(N_2)} = \frac{p_f(NO)}{p_i(NO)}$ $\qquad$ $\frac{0.80 \text{ atm}}{33.6 \text{ atm}} = \frac{p_f(NO)}{8.0 \times 10^{-3} \text{ atm}}$

$p_{f\,(NO)} =$ **1.9×10^{-4} atm**

16 ACID-BASE EQUILIBRIA IN AQUEOUS SOLUTIONS

16.1 $H_2O + H_2O \rightleftharpoons H_3O^+(aq) + OH^-(aq)$
"This is a very important equilibrium because it is present in any aqueous solution, regardless of what other reactions may also be taking place."

16.2 HCl and HNO_3 are the common strong monoprotic acids. The strong bases are the water soluble metal hydroxides of the metals of Group IA and of Group IIA from calcium down to barium.

16.3 $HCl(g) + H_2O(\ell) \rightarrow H^+(aq) + Cl^-(aq)$

or $\rightarrow H_3O^+(aq) + Cl^-(aq)$

$KOH(s) + H_2O(\ell) \rightarrow K^+(aq) + OH^-(aq)$

16.4 In pure water the $[H^+] = 1 \times 10^{-7}$. In the presence of an acid the dissociation of water is suppressed (Le Chatelier's principle) and the $[H^+]$ contributed from the H_2O is less than 10^{-7} M. Only in very dilute acid solutions or solutions of very weak acids must the dissociation of water be taken into account since only in these solutions will the concentration of H^+ contributed by water compare with that contributed by the acid.

16.5 (a) $[H^+] = \mathbf{1.0 \times 10^{-3}}$ **mol/L** $[OH^-] = \mathbf{1.0 \times 10^{-11}}$ **mol/L** pH = **3.00**

(b) $[H^+] = \mathbf{1.25 \times 10^{-1}}$ **mol/L** $[OH^-] = \mathbf{8.00 \times 10^{-14}}$ **mol/L** pH = **0.903**

(c) $[H^+] = \mathbf{3.2 \times 10^{-12}}$ **mol/L** $[OH^-] = \mathbf{3.1 \times 10^{-3}}$ **mol/L** pH = **11.49**

(continued)

16.5 (continued)

(d) $[H^+] = 4.2 \times 10^{-13}$ **mol/L** $[OH^-] = 2.4 \times 10^{-2}$ **mol/L** pH = **12.38**

(e) $[H^+] = 2.1 \times 10^{-4}$ **mol/L** $[OH^-] = 4.8 \times 10^{-11}$ **mol/L** pH = **3.68**

(f) $[H^+] = 1.3 \times 10^{-5}$ **mol/L** $[OH^-] = 7.7 \times 10^{-10}$ **mol/L** pH = **4.89**

(g) $[H^+] = 1.2 \times 10^{-12}$ **mol/L** $[OH^-] = 8.4 \times 10^{-3}$ **mol/L** pH = **11.92**

(h) $[H^+] = 2.1 \times 10^{-13}$ **mol/L** $[OH^-] = 4.8 \times 10^{-2}$ **mol/L** pH = **12.68**

16.6 $[H^+] = [OH^-] = \sqrt{2.42 \times 10^{-14}} = \mathbf{1.56 \times 10^{-7}}$ M

16.7 $pH = -\log [H^+]$; $pOH = -\log [OH^-]$;
$K_w = [H^+][OH^-]$ $\log K_w = \log [H^+] + \log [OH^-]$
$-\log K_w = -\log (1 \times 10^{-14}) = -\log [H^+] - \log [OH^-]$ $pK_w = 14 = pH + pOH$

16.8 (a) **acidic** (b) **basic** (c) **neutral** (d) **acidic** (e) **basic**

16.9 **(e)<(b)<(c)<(d)<(a)**

16.10 $pH = -\log[H^+] = -\log [1.56 \times 10^{-7}] =$ **6.807**

16.11 (a) $[H^+] = 0.050$ **mol/L** $[OH^-] = 2.0 \times 10^{-13}$ **mol/L**

(b) $[H^+] = 1.9 \times 10^{-6}$ **mol/L** $[OH^-] = 5.3 \times 10^{-9}$ **mol/L**

(c) $[H^+] = 1.0 \times 10^{-4}$ **mol/L** $[OH^-] = 1.0 \times 10^{-10}$ **mol/L**

(d) $[H^+] = 1.6 \times 10^{-8}$ **mol/L** $[OH^-] = 6.2 \times 10^{-7}$ **mol/L**

(e) $[H^+] = 1.1 \times 10^{-11}$ **mol/L** $[OH^-] = 9.1 \times 10^{-4}$ **mol/L**

(f) $[H^+] = 2.5 \times 10^{-13}$ **mol/L** $[OH^-] = 4.0 \times 10^{-2}$ **mol/L**

16.12 pOH = 14 - pH (a) **12.70** (b) **8.27** (c) **10.00** (d) **6.20**

(e) **3.06** (f) **1.39**

16.13 pH = - log [H^+]   pOH = - log[OH^-]

(a) pH = **3.00**   pOH = **11.00**

(b) pH = **0.903**   pOH = **13.097**

(c) pH = **11.49**   pOH = **2.51**

(d) pH = **12.38**   pOH = **1.62**

(e) pH = **3.68**   pOH = **10.32**

(f) pH = **4.89**   pOH = **9.11**

(g) pH = **11.92**   pOH = **2.08**

(h) pH = **12.68**   pOH = **1.32**

16.14 (a) $HOBr + H_2O \rightleftharpoons H_3O^+ + OBr^-$

(b) $HCN + H_2O \rightleftharpoons H_3O^+ + CN^-$

(c) $(CH_3)_3NH^+ + H_2O \rightleftharpoons H_3O^+ + (CH_3)_3N$

(d) $C_5H_5NH^+ + H_2O \rightleftharpoons H_3O^+ + C_5H_5N$

(e) $HCO_3^- + H_2O \rightleftharpoons H_3O^+ + CO_3^{2-}$

16.15 (a) $C_5H_5N + H_2O \rightleftharpoons C_5H_5NH^+ + OH^-$

(b) $CO_3^{2-} + H_2O \rightleftharpoons HCO_3^- + OH^-$

(c) $H_2PO_4^- + H_2O \rightleftharpoons H_3PO_4 + OH^-$

(d) $NO_2^- + H_2O \rightleftharpoons HNO_2 + OH^-$

(e) $C_6H_5NH_2 + H_2O \rightleftharpoons C_6H_5NH_3^+ + OH^-$

16.16 $K_a \times K_b = K_w = 1.0 \times 10^{-14}$

16.17 (d) < (c) < (a) < (b)

16.18 (c) < (d) < (b) < (a)

16.19 PH_2^- is a stronger base than HS^-.

16.20 CN^- is a stronger base than NO_2^-.

16.21 Ammonia

16.22 $SO_4^{2-} < C_2H_3O_2^- < HCO_3^- < OCl^- < NH_3$

16.23 $pK_a = -\log K_a = -\log 3.8 \times 10^{-9} = \mathbf{8.42}$

16.24 $pK_b = -\log K_b$

$K_b = \text{antilog}(-pK_b) = \text{antilog}(-3.84) = \mathbf{1.4 \times 10^{-4}}$

16.25 (a) $HC_7H_5O_2 \rightleftharpoons H^+ + C_7H_5O_2^-$

$K_a = \{[H^+][C_7H_5O_2^-]\}/[HC_7H_5O_2]$

(b) $N_2H_4 + H_2O \rightleftharpoons N_2H_5^+ + OH^-$

$K_b = \{[N_2H_5^+][OH^-]\}/[N_2H_4]$

(c) $HCHO_2 \rightleftharpoons H^+ + CHO_2^-$

$K_a = \{[H^+][CHO_2^-]\}/[HCHO_2]$

(d) $HC_8H_{11}N_2O_3 \rightleftharpoons H^+ + C_8H_{11}N_2O_3^-$

$K_a = \{[H^+][C_8H_{11}N_2O_3^-]\}/[HC_8H_{11}N_2O_3]$

(e) $C_5H_5N + H_2O \rightleftharpoons C_5H_5NH^+ + OH^-$

$K_b = \{[C_5H_5NH^+][OH^-]\}/[C_5H_5N]$

16.26 (a) $HNO_2 \rightleftharpoons H^+ + NO_2^-$ $K_a = 4.5 \times 10^{-4}$

	Init. conc.(M)	Change	Equil. Conc. (M)
H^+	~0	+X	X
NO_2^-	0	+X	X
HNO_2	0.30	-X	0.30 - X ≈ 0.30

$$\frac{(X)(X)}{(0.30 - X)} = 4.5 \times 10^{-4} = \frac{X^2}{0.30}$$

X = 1.2 x 10^{-2} mol/L = [H^+]

(b) $HF \rightleftharpoons H^+ + F^-$ $K_a = 6.5 \times 10^{-4}$

	Init. Conc.	Change	Equil. Conc.
H^+	~0	+X	X
F^-	0	+X	X
HF	1.0	-X	1.0 - X ≈ 1.0

$$\frac{X^2}{1.0 - X} \approx \frac{X^2}{1.0} = 6.5 \times 10^{-4}$$

X = [H^+] = 2.5 x 10^{-2} mol/L

(c) $K_a = 4.9 \times 10^{-10} = \frac{X^2}{0.025 - X}$ **X = [H^+] = 3.5 x 10^{-6} mol/L**

(d) $K_a = 1.5 \times 10^{-5} = \frac{X^2}{0.10 - X}$ **X = [H^+] = 1.2 x 10^{-3} mol/L**

(e) $K_a = 1.0 \times 10^{-5} = \frac{X^2}{0.050 - X}$ **X = [H^+] = 7.1 x 10^{-4} mol/L**

16.27 (a) $NH_3 + H_2O \rightleftharpoons NH_4^+ + OH^-$ $K_b = 1.8 \times 10^{-5}$

	Init. Conc.	Change	Equil. Conc.
NH_4^+	0	+X	X
OH^-	~0	+X	X
NH_3	0.15 M	-X	$0.15 - X \approx 0.15$

$$1.8 \times 10^{-5} = \frac{(X)(X)}{(0.15 - X)} \approx \frac{X^2}{0.15}$$

X = [OH⁻] = 1.6 x 10⁻³ mol/L

(b) $N_2H_4 + H_2O \rightleftharpoons N_2H_5^+ + OH^-$ $K_b = 1.7 \times 10^{-6}$

	Init. Conc.	Change	Equil. Conc.
$N_2H_5^+$	0	+X	X
OH^-	~0	+X	X
N_2H_4	0.20	-X	$0.20 - X \approx 0.20$

$$1.7 \times 10^{-6} = \frac{(X)(X)}{(0.20 - X)} \approx \frac{X^2}{0.20}$$

$X = [OH^-] = 5.8 \times 10^{-4}$ mol/L

(c) $K_b = 3.7 \times 10^{-4} = \frac{X^2}{0.80 - X}$ **$X = [OH^-] = 1.7 \times 10^{-2}$ mol/L**

(d) $K_b = 1.1 \times 10^{-8} = \frac{X^2}{0.35 - X}$ **$X = [OH^-] = 6.2 \times 10^{-5}$ mol/L**

(e) $K_b = 1.7 \times 10^{-9} = \frac{X^2}{0.010 - X}$ **$X = [OH^-] = 4.1 \times 10^{-6}$ mol/L**

16.28 $[OH^-][H^+] = 1.00 \times 10^{-14}$, $[OH^-] = 1.00 \times 10^{-14}/[H^+]$

(a) $[OH^-] = \mathbf{8.3 \times 10^{-13}}$ **mol/L** (b) $[OH^-] = \mathbf{4.0 \times 10^{-13}}$ **mol/L**

(c) $[OH^-] = \mathbf{2.9 \times 10^{-9}}$ **mol/L** (d) $[OH^-] = \mathbf{8.3 \times 10^{-12}}$ **mol/L**

(e) $[OH^-] = \mathbf{1.4 \times 10^{-11}}$ **mol/L**

16.29 $[H^+][OH^-] = 1 \times 10^{-14}$, $pH + (-\log [OH^-]) = 14$, $pH = 14 + \log [OH^-]$

(a) $pH = 14 + \log (1.6 \times 10^{-3}) = 14 - 2.8 = \mathbf{11.20}$

(b) pH = **10.76** (c) pH = **12.23** (d) pH = **9.79** (e) pH = **8.61**

16.30 $$K_a = \frac{[H^+][\text{anion}]}{[\text{undissoc. Acid}]} = \frac{(X)(X)}{(0.25 - X)}$$

$pH = -\log[H^+]$ $[H^+] = \text{antilog}(-pH) = \text{antilog}(-1.35)$

$[H^+] = 4.5 \times 10^{-2} = X$

$$K_a = \frac{(4.5 \times 10^{-2})^2}{0.25 - (4.5 \times 10^{-2})} = \frac{2.0 \times 10^{-3}}{0.21} = \mathbf{9.6 \times 10^{-3}}$$

16.31 $$K_a = \frac{[H^+][\text{anion}]}{[\text{undissoc. Acid}]} = \frac{(X)(X)}{(0.10 - X)}$$

$pH = 5.37 = -\log[H^+] = -\log(X)$ $X = 4.3 \times 10^{-6}$

$$K_a = \frac{(4.3 \times 10^{-6})^2}{0.10 - (4.3 \times 10^{-6})} = \frac{1.8 \times 10^{-11}}{0.10} = \mathbf{1.8 \times 10^{-10}}$$

16.32 $K_b = \frac{[\text{cation}][OH^-]}{[\text{unreacted base}]} = \frac{(X)(X)}{(0.10 - X)}$

pH = 14 - pOH = 14 - (-log[OH^-]) = 14 + log (X) = 8.75

$X = 5.6 \times 10^{-6}$

$$K_b = \frac{(5.6 \times 10^{-6})^2}{0.10 - (5.6 \times 10^{-6})} = \mathbf{3.1 \times 10^{-10}}$$

16.33 (a) $HCHO_2 \rightleftharpoons H^+ + CHO_2^-$ $K_a = 1.8 \times 10^{-4}$

	Init. Conc.	Change	Equil. Conc.
H^+	~0	+X	X
CHO_2^-	0	+X	X
$HCHO_2$	1.0	-X	1.0 - X

$$K_a = 1.8 \times 10^{-4} = \frac{(X)(X)}{1.0 - X} \approx \frac{X^2}{1.0}$$

$X = [H^+] = 1.3 \times 10^{-2}$ M $\quad$ % ionization = $\frac{1.3 \times 10^{-2}}{1.0} \times 100\% = \mathbf{1.3\ \%}$

(b) $K_a = 1.4 \times 10^{-5} = \frac{X^2}{0.010 - X}$ $\quad$ $X = [H^+] = [C_3H_5O_2^-]$

$X = 3.7 \times 10^{-4}$ $\quad$ % ionization = $\frac{3.7 \times 10^{-4}}{0.010} \times 100\% = \mathbf{3.7\ \%}$

(continued)

16.33 (continued)

(c) $K_a = 4.9 \times 10^{-10} = \frac{X^2}{0.025 - X}$ $X = 3.5 \times 10^{-6}$

$\%\text{ ionization} = \frac{3.5 \times 10^{-6}}{0.025} \times 100\% = \mathbf{0.014\%}$

(d) $K_a = 1.4 \times 10^{-5} = \frac{X^2}{0.35 - X}$ $X = 2.2 \times 10^{-3}$

$\%\text{ ionization} = \frac{2.2 \times 10^{-3}}{0.35} \times 100\% = \mathbf{0.63\%}$

(e) $K_a = 3.1 \times 10^{-8} = \frac{X^2}{0.50 - X}$ $X = 1.2 \times 10^{-4}$

$\%\text{ ionization} = \frac{1.2 \times 10^{-4}}{0.50} \times 100\% = \mathbf{0.024\%}$

(f) strong acid, assume **100% ionization**

16.34 (a) $K_a = 1.8 \times 10^{-5} = \frac{(X)(X)}{1.0 - X}$ $X = 4.2 \times 10^{-3}$

$\%\text{ ionization} = \frac{4.2 \times 10^{-3}}{1.0} \times 100\% = \mathbf{0.42\%}$

(b) $K_a = 1.8 \times 10^{-5} = \frac{(X)(X)}{0.10 - X}$ $X = 1.3 \times 10^{-3}$

$\%\text{ ionization} = \frac{1.3 \times 10^{-3}}{0.10} \times 100\% = \mathbf{1.3\%}$

(continued)

16.34 (continued)

(c) $K_a = 1.8 \times 10^{-5} = \frac{(X)(X)}{0.010 - X}$ $X = 4.2 \times 10^{-4}$

$$\% \text{ ionization} = \frac{4.2 \times 10^{-4}}{0.010} \times 100\% = \mathbf{4.2\%}$$

In the equilibrium represented by $HC_2H_3O_2 \rightleftharpoons H^+ + C_2H_3O_2^-$, collisions between ions would be expected to be less frequent in the more dilute solution. Therefore, on a molecular level we would expect the dilute solution to have a higher percent ionization. The above calculation verified this expectation.

16.35 $B + H_2O \rightleftharpoons HB^+ + OH^-$

$$K_b = \frac{[HB^+][OH^-]}{[B]}$$

From the pH, $[OH^-] = 2.5 \times 10^{-3}$
$[HB^+] = [OH^-] = 2.5 \times 10^{-3}$
$[B] = 0.012 - (2.5 \times 10^{-3})$

$$K_b = \frac{(2.5 \times 10^{-3})(2.5 \times 10^{-3})}{0.012 - (2.5 \times 10^{-3})} = \mathbf{6.6 \times 10^{-4}}$$

16.36 $HC_2H_3O_2 \rightleftharpoons H^+ + C_2H_3O_2^-$

	Init Conc.	H^+ added	Change	Equil. Conc.
H^+	~0	Y	-X	Y - X
$C_2H_3O_2^-$	1.0	0	-X	1.0 - X
$HC_2H_3O_2$	0	0	+X	X

$[H^+]$ from pH value $= 1.82 \times 10^{-5} = Y - X$

$$K_a = 1.8 \times 10^{-5} = \frac{[H^+][C_2H_3O_2^-]}{[HC_2H_3O_2]} = \frac{(1.82 \times 10^{-5})(1.0 - X)}{X}$$

(continued)

16.36 (continued)

$(1.8 \times 10^{-5})X = (1.82 \times 10^{-5}) - 1.82 \times 10^{-5} X$ $\quad X = 0.50$

$[H^+] = 1.82 \times 10^{-5}$ M, $[C_2H_3O_2^-] = 0.50$ M, $[HC_2H_3O_2] = 0.50$ M

Amount of HCl added = Y

Amount of H^+ remaining = Y - X = 1.82×10^{-5} M

$X = [HC_2H_3O_2] = 0.50$ M

$Y - (0.50) = 1.82 \times 10^{-5}$ M $\quad Y = 0.50$ M

$$\frac{0.50 \text{ mol HCl}}{\text{L}} \times 0.500 \text{ L} \times \frac{36.5 \text{ g}}{\text{mol}} = \mathbf{9.1 \text{ g HCl}}$$

16.37 $HC_4H_3N_2O_3 \rightleftharpoons H^+ + C_4H_3N_2O_3^-$

$K_a = 1.0 \times 10^{-5}$, $\quad$ M ($NaC_4H_3N_2O_3$) = init. conc. $C_4H_3N_2O_3^-$ =

$$10 \text{ mg} \times \frac{1 \text{ g}}{1000 \text{ mg}} \times \frac{1 \text{ mol}}{150 \text{ g}} \times \frac{1}{0.250 \text{ L}} = 2.67 \times 10^{-4} \text{ M}$$

	Init. Conc.	Change	Equil. Conc.
H^+	0.10	$-(2.67 \times 10^{-4} - X)$	~0.10
$C_4H_3N_2O_3^-$	2.67×10^{-4}	$-(2.67 \times 10^{-4} - X)$	X
$HC_4H_3N_2O_3$	0	$(2.67 \times 10^{-4} - X)$	$(2.67 \times 10^{-4} - X)$

$$1.0 \times 10^{-5} = \frac{(0.10)(X)}{(2.67 \times 10^{-4} - X)} \approx \frac{(0.10)(X)}{(2.67 \times 10^{-4})} \qquad X = 2.67 \times 10^{-8}$$

This shows that essentially 100% of the $C_4H_3N_2O_3^-$ is converted to barbituric acid.

16.38 $K_a = 1.4 \times 10^{-5} = \dfrac{(X)(X)}{(0.010 - X)}$ $\quad X = [H^+] = 3.7 \times 10^{-4}$ $\quad$ pH = **3.43**

16.39 $K_a = 1.8 \times 10^{-5} = \frac{(X)(X)}{(Y - X)}$ $X = [H^+] = 3.2 \times 10^{-3}$ M (from pH)

$$1.8 \times 10^{-5} = \frac{(3.2 \times 10^{-3})^2}{Y - (3.2 \times 10^{-3})}$$ $Y = [HC_2H_3O_2] =$ **0.57 M**

16.40 $K_b = \frac{(X)(X)}{Y - X}$ $K_b = 1.7 \times 10^{-6}$ $X = [OH^-] = 4.4 \times 10^{-4}$ M (from pH)

$$1.7 \times 10^{-6} = \frac{(4.4 \times 10^{-4})^2}{Y - (4.4 \times 10^{-4})}$$ $Y = [N_2H_4] =$ **0.11 M**

16.41 $K_a = \frac{(X)(X)}{0.010 - X}$ $X = [H^+] = 2.8 \times 10^{-5}$ M (from pH)

$$K_a = \frac{(2.8 \times 10^{-5})^2}{0.010 - (2.8 \times 10^{-5})} = \mathbf{7.8 \times 10^{-8}}$$

16.42 $HC_2H_3O_2 \rightleftharpoons H^+ + C_2H_3O_2^-$

	Init. Conc.	Change	Equil. Conc.
H^+	~0	+X	X
$C_2H_3O_2^-$	0	+X	X
$HC_2H_3O_2$	1.0×10^{-3}	-X	$(1.0 \times 10^{-3}) - X$

$$1.8 \times 10^{-5} = \frac{(X)(X)}{(1.0 \times 10^{-3}) - X}$$

$X^2 + (1.8 \times 10^{-5})X - 1.8 \times 10^{-8} = 0$ (Solve quadratic equation)

$X = 1.25 \times 10^{-4}$ or **$[H^+] = 1.3 \times 10^{-4}$ mol/L** This is the same answer, to two significant figures, that one would obtain by assuming $1.0 \times 10^{-3} - X \approx 1.0 \times 10^{-3}$.

16.43 $HCO_2H \rightleftharpoons CHO_2^- + H^+$ $K_a = 1.8 \times 10^{-4}$

	Init. Conc.	Change	Equil. Conc.
H^+	~0	+X	X
CHO_2^-	0	+X	X
$HCHO_2$	0.010	-X	0.010 - X

$$1.8 \times 10^{-4} = \frac{(X)(X)}{0.010 - X}$$

Solving this quadratic equation yields, $X = 1.3 \times 10^{-3}$ (To two significant figures, the same answer will be obtained if one assumed that $0.010\text{-}X \approx 0.010$.)

$[H^+] = 1.3 \times 10^{-3}$ M **$[CHO_2^-] = 1.3 \times 10^{-3}$ M**

$[HCHO_2] = 0.009$ M **$[OH^-] = 7.7 \times 10^{-12}$ M**

16.44 A buffer is any solution that contains both a weak acid and a weak base and has the property that the addition of small quantities of a strong acid are neutralized by the weak base while small quantities of a strong base are neutralized by the weak acid resulting in very little pH change.

(a) $NaCHO_2$ provides the weak conjugate base CHO_2^-; $HCHO_2$ is the weak acid.

(b) C_5H_5N provides the weak base; C_5H_5NHCl provides the weak acid $C_5H_5NH^+$.

(c) $NH_4C_2H_3O_2$ provides the weak base $C_2H_3O_2^-$ and the weak acid NH_4^+.

(d) $NaHCO_3$ provides HCO_3^-; HCO_3^- is both the weak acid and the weak base.

16.45 No. HCl is a strong acid and is completely dissociated. Cl^- is such a weak conjugate base that it cannot neutralize acids.

16.46 (a) When a strong acid is added, the HPO_4^{2-} will react to reduce its effect by forming $H_2PO_4^-$:

$$H^+(\text{added}) + HPO_4^{2-} \rightleftharpoons H_2PO_4^-$$

(b) When a strong base (OH^-) is added, the $H_2PO_4^-$ will react to reduce its effect by forming HPO_4^{2-} and H_2O:

$$H_2PO_4^- + OH^-(\text{added}) \rightleftharpoons HPO_4^{2-} + H_2O$$

16.47 (a) $HC_2H_3O_2 \rightleftharpoons H^+ + C_2H_3O_2^-$

	Init. Conc.	Change	Equil. Conc.
H^+	~0	+X	X
$C_2H_3O_2^-$	0.15 M	+X	$0.15 + X \approx 0.15$
$HC_2H_3O_2$	0.25	-X	$0.25 - X \approx 0.25$

$$K_a = 1.8 \times 10^{-5} = \frac{(X)(0.15)}{(0.25)}$$ **X = [H+] = 3.0 x 10-5 mol/L** ($X = [H^+] = 3.0 \times 10^{-5}$ mol/L)

(b) $1.8 \times 10^{-4} = \frac{(X)(0.50)}{(0.50)}$ **$X = [H^+] = 1.8 \times 10^{-4}$ mol/L**

(c) $4.5 \times 10^{-4} = \frac{(X)(0.40)}{(0.30)}$ **$X = [H^+] = 3.4 \times 10^{-4}$ mol/L**

(d) $1.8 \times 10^{-5} = \frac{(X)(0.15)}{(0.25)}$ $X = [OH^-] = 3.0 \times 10^{-5}$ mol/L

$[H^+] = 1.0 \times 10^{-14} / 3.0 \times 10^{-5} = 3.3 \times 10^{-10}$ mol/L

(e) $1.7 \times 10^{-6} = \frac{(X)(0.50)}{(0.30)}$ $X = [OH^-] = 1.0 \times 10^{-6}$ mol/L

$[H^+] = 1.0 \times 10^{-14}/1.0 \times 10^{-6} = 1.0 \times 10^{-8}$ mol/L

16.48 $K_a = 1.38 \times 10^{-4} = \frac{[H^+][C_3H_5O_3^-]}{[HC_3H_5O_3]}$ pH = 4.25 $[H^+] = 5.6 \times 10^{-5}$ M

$$\frac{[C_3H_5O_3^-]}{[HC_3H_5O_3]} = \frac{[NaC_3H_5O_3]}{[HC_3H_5O_3]} = \frac{K_a}{[H^+]} = \frac{1.38 \times 10^{-4}}{5.6 \times 10^{-5}} = 2.5$$

Reciprocal value = $\frac{[HC_3H_5O_3]}{[NaC_3H_5O_3]}$ = **0.40**

16.49 (a) $NH_3 + H_2O \rightleftharpoons NH_4 + OH^-$

$$K_b = 1.8 \times 10^{-5} = \frac{[NH_4^+][OH^-]}{[NH_3]} = \frac{(0.10)[OH^-]}{(0.10)} \qquad [OH^-] = 1.8 \times 10^{-5}$$

$[H^+] = 1.0 \times 10^{-14}/1.8 \times 10^{-5} = 5.6 \times 10^{-10}$ **pH = 9.26**

(b) $HC_2H_3O_2 \rightleftharpoons H^+ + C_2H_3O_2^-$ $K_a = 1.8 \times 10^{-5} = \frac{[H^+](0.40)}{(0.20)}$

$[H^+] = 9.0 \times 10^{-6}$ M **pH = 5.05**

(c) $N_2H_4 + H_2O \rightleftharpoons N_2H_5^+ + OH^-$ $K_b = 1.7 \times 10^{-6} = \frac{(0.10)[OH^-]}{(0.15)}$

$[OH^-] = 2.6 \times 10^{-6}$ **pH = 8.41**

(d) $HCl \rightarrow H^+ + Cl^-$ (not a buffer; 100% ionization) $[H^+] = 0.20$ **pH = 0.70**

16.50 $HC_2H_3O_2 \rightleftharpoons H^+ + C_2H_3O_2^-$

$$K_a = 1.8 \times 10^{-5} = \frac{(7.1 \times 10^{-6})[C_2H_3O_2^-]}{1.00} \qquad [C_2H_3O_2^-] = 2.5 \text{ M}$$

$$\frac{2.5 \text{ mol } C_2H_3O_2^-}{L} \times 1.00 \text{ L} \times \frac{1 \text{ mol } NaC_2H_3O_2}{\text{mol } C_2H_3O_2^-} \times \frac{82.1 \text{ g}}{\text{mol}} = \mathbf{210\ g}$$

16.51 pH = 10.00, $[H^+] = 1.0 \times 10^{-10}$ $[OH^-] = 1.0 \times 10^{-4}$

$NH_3 + H_2O \rightleftharpoons NH_4^+ + OH^-$

$$K_b = 1.8 \times 10^{-5} = \frac{[NH_4^+](1.0 \times 10^{-4})}{[NH_3]}$$

$$\frac{[NH_4^+]}{[NH_3]} = 0.18 \qquad \frac{[NH_3]}{[NH_4^+]} = \frac{1}{0.18} = \mathbf{5.6}$$

16.52 $HC_2H_3O_2 \rightleftharpoons H^+ + C_2H_3O_2^-$

	Init. Conc.	Change	Equil. Conc.
H^+	Y	-X	$Y - X = 1.0 \times 10^{-3}$
$C_2H_3O_2^-$	0.010	-X	0.010 - X
$HC_2H_3O_2$	0.010	+X	0.010 + X

$$K_a = 1.8 \times 10^{-5} = \frac{(1.0 \times 10^{-3})(0.010 - X)}{(0.010 + X)}$$

$X = 9.6 \times 10^{-3}$ $\quad$ $Y = 9.6 \times 10^{-3} + 1.0 \times 10^{-3} = 1.06 \times 10^{-2}$ M HCl

1.1×10^{-2} moles of HCl must be added

16.53 (a) Initial pH $\quad 1.8 \times 10^{-5} = \frac{[H^+][1.00]}{[1.00]}$ $\quad [H^+] = 1.8 \times 10^{-5}$

pH = 4.74

pH after the addition of 0.10 mol NaOH per 0.500 L or 0.20 M; pH = ? (Note: There will not be dilution during the addition of NaOH. The 0.20 M has been calculated using the volume of the buffer.)

	Init. Conc.	Effect of NaOH	Change	Equil. Conc.
$C_2H_3O_2^-$	1.00	+0.20	+X	$1.20 + X \approx 1.20$
H^+	~0		+X	X
$HC_2H_3O_2$	1.00	-0.20	-X	$0.80 - X \approx 0.80$

$$K_a = 1.8 \times 10^{-5} = \frac{(X)(1.20)}{(0.80)}$$

$X = [H^+] = 1.2 \times 10^{-5}$

pH = 4.92 $\quad$ **ΔpH** = 4.92 - 4.74 = **0.18**

(b) Initial pH = 4.74 $\quad$ pH after the addition of 0.20 M NaOH; equals ?

	Init. Conc.	Effect of NaOH	Change	Equil. Conc.
$C_2H_3O_2^-$	0.50	+0.20	+X	$0.70 + X \approx 0.70$
H^+	~0		+X	X
$HC_2H_3O_2$	0.50	-0.20	-X	$0.30 - X \approx 0.30$

(continued)

16.53 (continued)

$$K_a = 1.8 \times 10^{-5} = \frac{(X)(0.70)}{(0.30)} \qquad X = [H^+] = 7.7 \times 10^{-6}$$

pH = 5.11 $\quad$ **ΔpH** = 5.11 - 4.74 = **0.37**

(c) Initial pH = 5.11 (Same as final pH in Question 16.53 (b))
pH after the addition of the NaOH = ?

$$K_a = 1.8 \times 10^{-5} = \frac{(H^+)(0.90)}{(0.10)} \qquad -\log [H^+] = 5.70$$

ΔpH = 0.59

(d) Initial pH = 5.35
pH after the addition of 0.20 M NaOH = ?

	Init. Conc.	Effect of NaOH	Change	Equil. Conc.
$C_2H_3O_2^-$	0.80	+0.20	-X	1.0 - X
H^+	~0		-?	?
$HC_2H_3O_2$	0.20	-0.20	+X	0 + X

Since all of the $HC_2H_3O_2$ is consumed by the NaOH, the equilibrium must shift in the direction that tends to restore $HC_2H_3O_2$. This is hydrolysis of the $C_2H_3O_2^-$.

$$C_2H_3O_2^- + H_2O \rightarrow HC_2H_3O_2 + OH^-$$

$$K_{hy} = K_b (C_2H_3O_2^-) = \frac{K_w}{K_a} = \frac{[HC_2H_3O_2][OH^-]}{[C_2H_3O_2^-]}$$

	Init. Conc.	Change by Hydrolysis	Equil. Conc.
$HC_2H_3O_2$	0	+X	X
OH^-	~0	+X	X
$C_2H_3O_2^-$	1.0	-X	1.0 - X ≈ 1.0

$$\frac{1.0 \times 10^{-14}}{1.8 \times 10^{-5}} = \frac{(X)(X)}{1.0} \qquad X = [OH^-] = 2.4 \times 10^{-5}$$

pH = 9.38 $\quad$ **ΔpH** = 9.38 - 5.35 = **4.03**

(continued)

16.53 (continued)

(e) Initial pH = 5.70 (Same as final pH in part c)

After addition of NaOH, there will be: 1.00 M $C_2H_3O_2^-$, 0.00 M $HC_2H_3O_2$, and 0.10 M excess NaOH. The M of OH^- contributed by hydrolysis can be neglected when excess base is present. We will demonstrate below that its contribution is too small to change the pH.

	Init. Conc.	Change	Equil. Conc.
$C_2H_3O_2^-$	1.00	-X	1.00 - X ≈ 1.00
OH^-	0.10	+X	0.10 + X
$HC_2H_3O_2$	0	+X	X

$$K_{hy} = \frac{K_w}{K_a} = 5.56 \times 10^{-10} = \frac{(X)(0.10 + X)}{1.00}$$

$X = 5.6 \times 10^{-9}$ $[OH^-] = 0.10 + 5.6 \times 10^{-9} = 0.10$

pOH = 1.00 pH = 13.00 **ΔpH = 7.30**

16.54 Initial pH = ?

	Init. Conc.	Change	Equil. Conc.
$HCHO_2$	0.45 M	-X	0.45 - X ≈ 0.45 M
H^+	~0	+X	X
CHO_2^-	0.55 M	+X	0.55 + X ≈ 0.55 M

$$K_a = 1.8 \times 10^{-4} = \frac{(X)(0.55)}{0.45}$$ $X = [H^+] = 1.5 \times 10^{-4}$

Initial pH = 3.82 Final pH = ?

	Init. Conc.	Effect of HCl	Change	Equil. Conc.
$HCHO_2$	0.45 M	+ 0.10	-X	0.55 - X ≈ 0.55
H^+	~0		+X	X
CHO_2^-	0.55 M	-0.10	+X	0.45 + X ≈ 0.45

(continued)

16.54 (continued)

$K_a = 1.8 \times 10^{-4} = \frac{(X)(0.45)}{0.55}$ $X = [H^+] = 2.2 \times 10^{-4}$

Final pH = 3.66

ΔpH = -0.16

16.55 Initial pH = 3.82 Final pH = ?

	Init. Conc.	Effect of NaOH	Change	Equil. Conc.
$HCHO_2$	0.45 M	-0.20	-X	$0.25 - X \approx 0.25$
H^+	~0		+X	X
CHO_2^-	0.55 M	+0.20	+X	$0.75 + X \approx 0.75$

$K_a = 1.8 \times 10^{-4} = \frac{(X)(0.75)}{0.25}$ $X = [H^+] = 6.0 \times 10^{-5}$

Final pH = 4.22

ΔpH = 0.40

16.56 The pH does not change. In Exercise 16.49 (a) the concentration of NH_3 was 0.10 M and that of NH_4^+ was 0.10 M and the pH was 9.26. If that solution is diluted tenfold, the concentration of the NH_3 and NH_4^+ would each be 0.010 M. The calculated pH of the resulting solution is 9.26 or unchanged by the dilution of the buffered solution.

16.57 (a) **neutral (salt of a strong acid and a strong base)**
(b) **acidic (salt of a strong acid and a weak base)**
(c) **basic (salt of a strong base and a weak acid)**
(d) **acidic (salt of a strong acid and a weak base)**

16.58 $NaC_4H_7O_2$ would be the most basic.

$C_6H_5NH_3NO_3$ would be the most acidic.

16.59 (a) $C_2H_3O_2^- + H_2O \rightleftharpoons HC_2H_3O_2 + OH^-$

$$K_b = \frac{K_w}{K_a} = \frac{1.0 \times 10^{-14}}{1.8 \times 10^{-5}} = 5.6 \times 10^{-10}$$

$$K_b = \frac{[HC_2H_3O_2][OH^-]}{[C_2H_3O_2^-]} = \frac{(X)(X)}{(1.0 \times 10^{-3}) - X}$$

$X = 7.5 \times 10^{-7} = [OH^-]$ $\quad$ pOH = 6.12 $\quad$ **pH = 7.88**

(b) $NH_4^+ \rightarrow NH_3 + H^+$

$$K_a = \frac{K_w}{K_b} = 5.6 \times 10^{-10} = \frac{(X)(X)}{0.125 - X}$$

$X = [H^+] = 8.3 \times 10^{-6}$ $\quad$ **pH = 5.08**

(c) $CHO_2^- + H_2O \rightleftharpoons HCHO_2 + OH^-$

$$K_b = \frac{K_w}{K_a} = 5.6 \times 10^{-11} = \frac{(X)(X)}{0.10 - X}$$

$X = [OH^-] = 2.4 \times 10^{-6}$ $\quad$ pOH = 5.62 $\quad$ **pH = 8.38**

(d) $CN^- + H_2O \rightleftharpoons HCN + OH^-$

$$K_b = \frac{K_w}{K_a} = 2.0 \times 10^{-5} = \frac{(X)(X)}{0.10 - X}$$

$X = [OH^-] = 1.4 \times 10^{-3}$ $\quad$ **pH = 11.15**

(e) $NH_3OH^+ \rightleftharpoons NH_2OH + H^+$

$$K_a = \frac{K_w}{K_b} = 9.1 \times 10^{-7} = \frac{(X)(X)}{0.20 - X}$$

$X = [H^+] = 4.3 \times 10^{-4}$ $\quad$ **pH = 3.37**

16.60 $C_5H_5NH^+ \rightleftharpoons C_5H_5N + H^+$

$$K_a = \frac{1.0 \times 10^{-14}}{1.7 \times 10^{-9}} = \frac{(X)(X)}{0.10 - X} = \frac{X^2}{0.10}$$

$X = [C_5H_5N] = 7.7 \times 10^{-4}$

$$\textbf{\% of } C_5H_5NH^+ \textbf{ reacted} = \frac{7.7 \times 10^{-4}}{0.10} \times 100\% = \mathbf{0.77\ \%}$$

16.61 $Base^- + H_2O \rightleftharpoons H\text{-}Base + OH^-$ From pH: $[OH^-] = 2.2 \times 10^{-5}$

$$K_b = \frac{K_w}{K_a} = \frac{(2.2 \times 10^{-5})(2.2 \times 10^{-5})}{0.10 - (2.2 \times 10^{-5})} = 4.8 \times 10^{-9}$$

$$\frac{K_w}{K_a} = \frac{1.0 \times 10^{-14}}{K_a} = 4.8 \times 10^{-9}$$

$\mathbf{K_a = 2.1 \times 10^{-6}}$

16.62 $OCl^- + H_2O \rightleftharpoons HOCl + OH^-$

$$K_b = \frac{K_w}{K_a} = \frac{1.0 \times 10^{-14}}{3.1 \times 10^{-8}} = 3.2 \times 10^{-7}$$

$$K_b = \frac{[HOCl][OH^-]}{[OCl^-]} = \frac{(X)(X)}{0.67 - X} = 3.2 \times 10^{-7}$$

$X^2 = 0.67 \times 3.2 \times 10^{-7}$ $X = 4.6 \times 10^{-4} = [OH^-]$

pH = 10.66

16.63 $C_8H_{11}N_2O_3^- + H_2O \rightleftharpoons HC_8H_{11}N_2O_3 + OH^-$

$$K_b = \frac{K_w}{K_a} = \frac{1.0 \times 10^{-14}}{3.7 \times 10^{-8}} = 2.7 \times 10^{-7}$$

$$\frac{10\text{ mg}}{250\text{ mL}} \times \frac{1{,}000\text{ mL}}{\text{L}} \times \frac{1\text{ g}}{1{,}000\text{ mg}} \times \frac{1\text{ mol}}{206.2\text{ g}} = 1.9 \times 10^{-4}\text{ M}$$

$$K_b = 2.7 \times 10^{-7} = \frac{(X)(X)}{1.9 \times 10^{-4} - X} \approx \frac{X^2}{1.9 \times 10^{-4}}$$

$X = [OH^-] = 7.2 \times 10^{-6}$ **pH = 8.86**

16.64 $C_4H_3N_2O_3^- + H_2O \rightleftharpoons HC_4H_3N_2O_3 + OH^-$

$$K_b = \frac{K_w}{K_a} = \frac{1.0 \times 10^{-14}}{1.0 \times 10^{-5}} = 1.0 \times 10^{-9}$$

$$K_b = 1.0 \times 10^{-9} = \frac{(X)(X)}{0.0010 - X} \approx \frac{X^2}{0.0010}$$

$X = [HC_4H_3N_2O_3]$ = **1.0×10^{-6} M**

16.65 $C_7H_5O_2^- + H_2O \rightleftharpoons HC_7H_5O_2 + OH^-$

$$K_b = \frac{K_w}{K_a} = \frac{1.0 \times 10^{-14}}{6.5 \times 10^{-5}} = 1.5 \times 10^{-10}$$

$$K_b = \frac{[HC_7H_5O_2][OH^-]}{[C_7H_5O_2^-]} = \frac{(X)(X)}{(0.30 - X)} = 1.5 \times 10^{-10}$$

$$\frac{X^2}{0.30} \approx 1.5 \times 10^{-10}$$

$X = 6.7 \times 10^{-6}$ $[OH^-] = 6.7 \times 10^{-6}$ pOH = 5.17 **pH = 8.83**

16.66 (See Question 16.65)

$$K_b = \frac{[HC_7H_5O_2][OH^-]}{[C_7H_5O_2^-]} = \frac{X^2}{(0.020 - X)} \approx \frac{X^2}{0.020} = 1.6 \times 10^{-10}$$

$[OH^-] = 1.8 \times 10^{-6}$, pOH = 5.74, **pH = 8.26**

16.67 $CN^- + H_2O \rightleftharpoons HCN + OH^-$

$$K_b = \frac{K_w}{K_a} = \frac{1.0 \times 10^{-14}}{4.9 \times 10^{-10}} = 2.0 \times 10^{-5}$$

$$K_b = 2.0 \times 10^{-5} = \frac{(X)(X)}{0.0010 - X} \approx \frac{X^2}{0.0010}$$

$X = [OH^-] = 1.4 \times 10^{-4}$

Again, we see that the variable in the denominator is not negligible. Therefore, we may not drop the "X" value in the denominator. The following solution is by successive approximation; it can also be solved using the quadratic formula.

$$2.0 \times 10^{-5} = \frac{X^2}{0.0010 - X} \text{ or } \frac{X^2}{0.0010 - 1.4 \times 10^{-4}}$$

$X = [OH^-] = 1.3 \times 10^{-4}$ **pH = 10.11**

16.68 $H_2C_6H_6O_6 \rightleftharpoons H^+ + HC_6H_6O_6^-$

$$K_{a_1} = \frac{[H^+][HC_6H_6O_6^-]}{[H_2C_6H_6O_6]}$$

$HC_6H_6O_6^- \rightleftharpoons H^+ + C_6H_6O_6^{2-}$

$$K_{a_2} = \frac{[H^+][C_6H_6O_6^{2-}]}{[HC_6H_6O_6^-]}$$

16.69 $H_3C_6H_5O_7 \rightleftharpoons H_2C_6H_5O_7^- + H^+$ $\quad K_{a_1} = \dfrac{[H_2C_6H_5O_7^-][H^+]}{[H_3C_6H_5O_7]}$

$H_2C_6H_5O_7^- \rightleftharpoons HC_6H_5O_7^{2-} + H^+$ $\quad K_{a_2} = \dfrac{[HC_6H_5O_7^{2-}][H^+]}{[H_2C_6H_5O_7^-]}$

$HC_6H_5O_7^{2-} \rightleftharpoons C_6H_5O_7^{3-} + H^+$ $\quad K_{a_3} = \dfrac{[C_6H_5O_7^{3-}][H^+]}{[HC_6H_5O_7^{2-}]}$

16.70 $H_2SeO_3 \rightleftharpoons H^+ + HSeO_3^-$ $\quad K_{a_1} = 3 \times 10^{-3}$

$HSeO_3^- \rightleftharpoons H^+ + SeO_3^{2-}$ $\quad K_{a_2} = 5 \times 10^{-8}$

First Ionization

	Init. Conc.	Change	Equil. Conc.
H^+	~0.0	+X	X
$HSeO_3^-$	0.0	+X	X
H_2SeO_3	0.50	-X	0.50 - X

$$K_{a_1} = 3 \times 10^{-3} = \frac{(X)(X)}{0.50 - X}$$

Solved by successive approximations;

1) Assume 0.50 - X ≈ 0.50 and solve for X; X = 3.9×10^{-2} or 0.04
2) Assume 0.50 - X = 0.50 - 0.04 and solve for X; X = 3.7×10^{-2} or 0.04
3) Assume 0.50 - X = 0.50 - 0.04 No change from second approximation; therefore, X = 3.7×10^{-2}

(continued)

16.70 (continued)

Second Ionization

	Init. Conc.	Change	Equil. Conc.
H^+	3.7×10^{-2}	+X	$3.7 \times 10^{-2} + X$
SeO_3^{2-}	0	+X	X
$HSeO_3^-$	3.7×10^{-2}	-X	$3.7 \times 10^{-2} - X$

$$K_{a_2} = 5 \times 10^{-8} = \frac{(3.7 \times 10^{-2} + X)X}{(3.7 \times 10^{-2} - X)} \approx \frac{(3.7 \times 10^{-2})X}{3.7 \times 10^{-2}}$$

$X = 5 \times 10^{-8}$

$[H^+] = 0.037$ M = **0.04 M** pH = **1.4**

$[HSeO_3^-]$ = **0.04 M**

$[H_2SeO_3]$ = **0.46 M** $[SeO_3^{2-}]$ = **5×10^{-8} M**

16.71 (See Table 16.2) The second ionization will have no noticeable effect on the pH. We can ignore it. $H_2C_6H_6O_6 \rightleftharpoons H^+ + HC_6H_6O_6^-$

M.W. $(H_2C_6H_6O_6) = 176$

$$M\,(H_2C_6H_6O_6) = 0.500\text{g} \times \frac{1 \text{ mol}}{176 \text{ g}} \times \frac{1}{0.250 \text{ L}} = 1.14 \times 10^{-2}$$

$$K_a = 7.9 \times 10^{-5} = \frac{X^2}{1.14 \times 10^{-2} - X}$$

First approximation $X = [H^+] = 9.49 \times 10^{-4}$

Second approximation $X = [H^+] = 9.1 \times 10^{-4}$

pH = 3.04

16.72 Vitamin C = $H_2C_6H_6O_6$

$H_2C_6H_6O_6 \rightleftharpoons H^+ + HC_6H_6O_6^-$ $\qquad K_{a\,1} = 7.9 \times 10^{-5}$

$HC_6H_6O_6^- \rightleftharpoons H^+ + C_6H_6O_6^{2-}$ $\qquad K_{a\,2} = 1.6 \times 10^{-12}$

$$K_{a\,1} = \frac{(X)(X)}{0.050 - X} \approx \frac{X^2}{0.050} = 7.9 \times 10^{-5}$$

$X = 2.0 \times 10^{-3} = [H^+] = [HC_6H_6O_6^-]$

$0.050 - X = 0.050 - 0.002 = 0.048 = [H_2C_6H_6O_6]$

$$K_{a\,2} = \frac{(2.0 \times 10^{-3})Y}{2.0 \times 10^{-3}} = 1.6 \times 10^{-12}$$

$Y = [C_6H_6O_6^{2-}] = 1.6 \times 10^{-12}$

$[H^+] = [HC_6H_6O_6^-] = \mathbf{2.0 \times 10^{-3}\ M}$ $\qquad [OH^-] = \mathbf{5.0 \times 10^{-12}\ M}$

pH = **2.70** $\qquad [H_2C_6H_6O_6] = \mathbf{0.048\ M}$ $\qquad [C_6H_6O_6^{2-}] = \mathbf{1.6 \times 10^{-12}\ M}$

16.73 $H_2C_6H_6O_6 \rightleftharpoons H^+ + HC_6H_6O_6^-$ $\qquad K_{a_1} = 7.9 \times 10^{-5}$

$HC_6H_6O_6^- \rightleftharpoons H^+ + C_6H_6O_6^{2-}$ $\qquad K_{a_2} = 1.6 \times 10^{-12}$

$$M(H_2C_6H_6O_6) = 0.500\ g \times \frac{1\ mol}{176\ g} \times \frac{1}{0.200\ L} = 1.42 \times 10^{-2} M$$

<u>First Ionization</u>

	Init. Conc.	Change	Equil. Conc.
H^+	0.1 (from pH)	+X	$0.1 + X \approx 0.1$
$HC_6H_6O_6^-$	0	+X	+X
$H_2C_6H_6O_6$	1.42×10^{-2}	-X	$(1.42 \times 10^{-2}) - X \approx 1.42 \times 10^{-2}$

(continued)

16.73 (continued)

The second ionization will have no noticeable effect on the concentrations.

$$K_{a_1} = \frac{(0.1)X}{1.42 \times 10^{-2}} = 7.9 \times 10^{-5}$$

$X = 1.1 \times 10^{-5}$

$$\text{fraction dissociated} = \frac{1.1 \times 10^{-5}}{1.42 \times 10^{-2}} = \mathbf{7.9 \times 10^{-4}}$$

16.74 First Ionization of H_2CO_3

	Init. Conc.	Change	Equil. Conc.
H^+	1.0×10^{-3} (from pH)		$\approx 1.0 \times 10^{-3}$
HCO_3^-	0	+X	X
H_2CO_3	0.10	-X	$0.10 - X \approx 0.10$

$$K_{a_1} = \frac{(1.0 \times 10^{-3})(X)}{0.10} = 4.3 \times 10^{-7}$$

$X = [HCO_3^-] = 4.3 \times 10^{-5}$ M $= 4.3 \times 10^{-5}$ mol/L

Second Ionization

	Init. Conc.	Change	Equil. Conc.
H^+	1.0×10^{-3}	+Y	$1.0 \times 10^{-3} + Y \approx 1.0 \times 10^{-3}$
CO_3^{2-}	0	+Y	Y
HCO_3^-	4.3×10^{-5}	-Y	$4.3 \times 10^{-5} - Y \approx 4.3 \times 10^{-5}$

$$K_{a_2} = \frac{(1.0 \times 10^{-3})Y}{4.3 \times 10^{-5}} = 5.6 \times 10^{-11}$$

$Y = [CO_3^{2-}] = 2.4 \times 10^{-12}$ mol/L

16.75 First Ionization

	Init. Conc.	Change	Equil. Conc.
H^+	?	?	3.7×10^{-8} (from pH)
HCO_3^-	0	+X	X
H_2CO_3	2.6×10^{-2}	-X	2.6×10^{-2} - X

$$K_{a_1} = 4.3 \times 10^{-7} = \frac{(3.7 \times 10^{-8})X}{(2.6 \times 10^{-2} - X)}$$

$X = [HCO_3^-] = \mathbf{2.4 \times 10^{-2}}$ **mol/L**

Second Ionization

	Init. Conc.	Change	Equil. Conc.
H^+	--	--	3.7×10^{-8}
CO_3^{2-}	0	+X	X
HCO_3^-	2.4×10^{-2}	-X	2.4×10^{-2} - X

$$K_{a_2} = 5.6 \times 10^{-11} = \frac{(3.7 \times 10^{-8})X}{(2.4 \times 10^{-2} - X)} = \frac{(3.7 \times 10^{-8})X}{2.4 \times 10^{-2}}$$

$X = \mathbf{3.6 \times 10^{-5}}$ $\quad$ $[HCO_3^-] = 2.4 \times 10^{-2} - 3.6 \times 10^{-5} = \mathbf{2.4 \times 10^{-2}}$ **mol/L**

16.76 $C_6H_6O_6^{2-} + H_2O \rightleftharpoons HC_6H_6O_6^- + OH^-$

$$K_{b_1} = \frac{K_w}{K_{a_2}} = \frac{1.0 \times 10^{-14}}{1.6 \times 10^{-12}} = 6.2 \times 10^{-3}$$

$$K_{b_1} = 6.2 \times 10^{-3} = \frac{(X)(X)}{0.20 - X} \approx \frac{X^2}{0.20}$$

$X = [OH^-] = 3.5 \times 10^{-2}$

Oops! 0.20 - X is not ≈ 0.20! Must solve the following equation:

(continued)

16.76 (continued)

$$\frac{X^2}{0.20 - X} = 6.2 \times 10^{-3} \text{ or } \frac{X^2}{0.20 - 3.54 \times 10^{-2}} = 6.2 \times 10^{-3}$$

$X = [OH^-] = 3.2 \times 10^{-2}$ **pH = 12.51**

16.77 $SO_3^{2-} + H_2O \rightleftharpoons HSO_3^- + OH^-$

$$K_{b_1} = \frac{K_w}{K_{a_2}} = \frac{1.0 \times 10^{-14}}{1.0 \times 10^{-7}} = 1.0 \times 10^{-7}$$

$$K_{b_1} = \frac{X^2}{0.25 - X} \approx \frac{X^2}{0.25} = 1.0 \times 10^{-7}$$

$X = [OH^-] = 1.6 \times 10^{-4}$
The second hydrolysis (hydrolysis of the HSO_3^-) will yield negligible amounts of OH^-. **pH = 10.20**

16.78 $Na_2CO_3(s) + H_2O \rightarrow 2Na^+ + CO_3^{2-}$ (not an equilibrium)

$CO_3^{2-} + H_2O \rightleftharpoons HCO_3^- + OH^-$ $\quad K_{b_1} = \frac{K_w}{K_{a_2}}$

$HCO_3^- + H_2O \rightleftharpoons H_2CO_3 + OH^-$ $\quad K_{b_2} = \frac{K_w}{K_{a_1}}$

$H_2CO_3 \rightleftharpoons H_2O + CO_2$

	$CO_3^{2-} + H_2O$	$\rightleftharpoons$	HCO_3^-	+ OH^-
initial	.20		0	≈ 0
change	-X		+X	+X
equil.	.20 - X		X	X

(continued)

16.78 (continued)

$$K_{b_1} = \frac{1.0 \times 10^{-14}}{5.6 \times 10^{-11}} = \frac{X^2}{.20 - X} \qquad X = 5.9 \times 10^{-3}$$

	$HCO_3^- + H_2O$	$\rightleftharpoons$ H_2CO_3	+ OH^-
initial	5.9×10^{-3}	0	5.9×10^{-3}
change	-Y	+Y	+Y
equil	5.9×10^{-3} - Y	Y	5.9×10^{-3} + Y

$$K_{b_2} = \frac{1.0 \times 10^{-14}}{4.3 \times 10^{-7}} = \frac{Y(5.9 \times 10^{-3} + Y)}{5.9 \times 10^{-3} - Y} \qquad Y = 2.3 \times 10^{-8}$$

$[CO_3^{2-}]$ = 0.19 **$[HCO_3^-]$ = 5.9 X 10^{-3}**

[OH] = 5.9 X 10^{-3} **$[H_2CO_3]$ = 2.3 X 10^{-8}**

pH = 11.78 **$[H^+]$ = 1.7 x 10^{-12}**

16.79 Yes - due to hydrolysis

16.80 The endpoint occurs at the pH at which the indicator changes color, which may or may not occur at the pH that the equivalence point is reached.

16.81 (a) Since the titration involves a strong acid and a strong base, the **pH at the equivalence point will be 7.00.**

(b) $$\frac{0.0200 \text{ mol } H^+ \text{ (from } HNO_3)}{1{,}000 \text{ mL}} \times 15.0 \text{ mL} \times \frac{1 \text{ mol } OH^- \text{ (required)}}{1 \text{ mol } H^+ \text{ (available)}}$$

$$\times \frac{1{,}000 \text{ mL KOH (soln.)}}{0.0100 \text{ mol } OH^-}$$

= 30.0 mL KOH required to titrate to equivalence point

(continued)

16.81 (continued)

(c) mol H^+ available = $\frac{0.0200 \text{ mol } H^+}{1,000 \text{ mL}}$ x 15.0 mL = 3.00×10^{-4} mol H^+

mol OH^- available = $\frac{0.0100 \text{ mol } OH^-}{1,000 \text{ mL}}$ x 10.0 mL = 1.00×10^{-4} mol OH^-

Excess H^+ = $3.00 \times 10^{-4} - 1.00 \times 10^{-4}$ = 2.00×10^{-4} mol

$$\frac{2.00 \times 10^{-4} \text{ mol}}{15.0 \text{ mL} + 10.0 \text{ mL}} \times \frac{1,000 \text{ mL}}{\text{L}} = 8.00 \times 10^{-3} \text{ M}$$

pH = - log 8.00×10^{-3} = **2.10**

(d) mol H^+ available = $\frac{0.0200 \text{ mol } H^+}{1,000 \text{ mL}}$ x 15.0 mL = 3.00×10^{-4} mol H^+

mol OH^- available = $\frac{0.0100 \text{ mol } OH^-}{1,000 \text{ mL}}$ x 35.0 mL = 3.50×10^{-4} mol OH^-

Excess OH^- = $3.50 \times 10^{-4} - 3.00 \times 10^{-4}$ = 5.0×10^{-5} mol OH^- excess

$$\frac{5.0 \times 10^{-5} \text{ mol } OH^-}{15.0 \text{ mL} + 35.0 \text{ mL}} \times \frac{1,000 \text{ mL}}{\text{L}} = 1.0 \times 10^{-3} \text{ M } OH^-$$

pOH = 3.00 **pH = 11.00**

16.82 Titration Reaction: $NaOH + HC_4H_3N_2O_3 \rightarrow NaC_4H_3N_2O_3 + H_2O$

$$\frac{0.010 \text{ mol acid}}{\text{L}} \times 0.0250 \text{ L} \times \frac{1 \text{ mol base}}{1 \text{ mol acid}} \times \frac{\text{L (base)}}{0.020 \text{ mol base}}$$

= 0.0125 L (base) or 12.5 mL NaOH solution needed

$$\frac{0.010 \text{ mol acid}}{\text{L}} \times 0.0250 \text{ L} \times \frac{1 \text{ mol salt}}{1 \text{ mol acid}} = 2.5 \times 10^{-4} \text{ mol salt}$$

Concentration of the salt = $\frac{2.5 \times 10^{-4} \text{ mol}}{0.025 \text{ L} + 0.0125 \text{ L}}$ = 6.7×10^{-3} M

(continued)

16.82 (continued)

$$C_4H_3N_2O_3^- + H_2O \rightleftharpoons HC_4H_3N_2O_3 + OH^-$$

	Init. Conc.	Change	Equil. Conc.
$HC_4H_3N_2O_3$	0	+X	X
OH^-	~0	+X	X
$C_4H_3N_2O_3^-$	6.7×10^{-3}	-X	$6.7 \times 10^{-3} - X \approx 6.7 \times 10^{-3}$

$$K_b = \frac{K_w}{K_a} = \frac{1.0 \times 10^{-14}}{1.0 \times 10^{-5}} = \frac{X^2}{6.7 \times 10^{-3}}$$

$X = [OH^-] = 2.6 \times 10^{-6}$ pOH = 5.59 **pH = 8.41**

16.83 $HF + NaOH \rightleftharpoons NaF + H_2O$

(a) $\frac{0.200 \text{ mol HF}}{1{,}000 \text{ mL}} \times 50.0 \text{ mL} = 0.0100 \text{ mol HF}$

$\frac{0.100 \text{ mol NaOH}}{1{,}000 \text{ mL}} \times 5.0 \text{ mL} = 0.00050 \text{ mol NaOH}$

After the initial acid-base reaction, there will be 0.0100 - 0.0005 or 0.0095 mol HF and 0.0005 mol NaF or F^-. This is a buffer solution.

$$HF \rightleftharpoons H^+ + F^-$$

	Init. Conc.	Change	Equil. Conc.
H^+	~0	+X	X
F^-	0.00050 mol/0.055 L	+X	(0.00050/0.055) + X
HF	0.0095 mol/0.055 L	-X	(0.0095/0.055) - X

$$K_a = 6.5 \times 10^{-4} = \frac{(X)[(0.00050/0.055) + X]}{(0.0095/0.055) - X}$$

$X = [H^+] = 6.8 \times 10^{-3}$ M (from quadratic equation) **pH = 2.17**

(continued)

16.83 (continued)

(b) 0.0100 mol HF + 0.0050 mol NaOH yields 0.0050 mol excess HF and 0.0050 mol NaF. $HF \rightleftharpoons H^+ + F^-$

	Init. Conc.	Change	Equil. Conc.
H^+	~0	+X	X
F^-	0.0050 mol/0.100 L	+X	≈ 0.0050/0.100
HF	0.0050 mol/0.100 L	-X	≈ 0.0050/0.100

$$K_a = 6.5 \times 10^{-4} = \frac{(X)(0.050)}{(0.050)}$$

X =[H^+] = 6.5 x 10^{-4} **pH = 3.19** (Note: pH = pK_a at half neutralization)

(c) $0.0100 \text{ mol HF} \times \frac{1 \text{ mol NaOH}}{1 \text{ mol HF}} \times \frac{1 \text{ L NaOH}}{0.100 \text{ mol NaOH}} \times \frac{1{,}000 \text{ mL}}{\text{L}}$
= 100 mL NaOH solution required.

$$0.0100 \text{ mol HF} \times \frac{1 \text{ mol } F^- \text{ (or NaF)}}{1 \text{ mol HF}} = 0.0100 \text{ mol } F^-$$

$$\frac{0.0100 \text{ mol } F^-}{50.0 \text{ mL} + 100 \text{ mL}} \times \frac{1{,}000 \text{ mL}}{\text{L}} = 0.0667 \text{ M } F^-$$

$F^- + H_2O \rightleftharpoons HF + OH^-$

	Init. Conc.	Change	Equil. Conc.
F^-	0.0667	-X	0.0667 - X ≈ 0.0667
OH^-	~0	+X	X
HF	0	+X	X

$$K_b = \frac{K_w}{K_a} = \frac{1.0 \times 10^{-14}}{6.5 \times 10^{-4}} \approx \frac{(X)(X)}{0.0067}$$

X = [OH^-] = 1.0 x 10^{-6}

pOH = 6.00 **pH = 8.00**

[Note: If the concentration of OH^- produced by the dissociation of water is considered, the pH will be slightly less. That is, if the initial concentration of OH^- is taken as 1 x 10^{-7} rather than ≈ 0, a slightly different and more precise answer will be obtained.]

16.84 **Initial pH** $HC_4H_7O_2 \rightleftharpoons H^+ + C_4H_7O_2^-$

$K_a = 1.5 \times 10^{-5} = \frac{(X)(X)}{0.10 - X}$ $X = [H^+] = 1.2 \times 10^{-3}$ **pH = 2.92**

pH after adding 0.0010 mol NaOH;
0.0100 mol butyric acid plus 0.0010 mol NaOH

$1.5 \times 10^{-5} = \frac{(X)\{(0.0010/V) + X\}}{\{(0.0090/V) - X)\}}$ $X = [H^+] = 1.4 \times 10^{-4}$ **pH = 3.85**

pH after adding 0.0050 mol NaOH;

$1.5 \times 10^{-5} = \frac{(X)\{(0.0050/V) + X\}}{\{(0.0050/V) - X)\}}$ $X = [H^+] = 1.5 \times 10^{-5}$ **pH = 4.82**

pH after adding 0.0090 mol NaOH;

$1.5 \times 10^{-5} = \frac{(X)\{(0.0090/V) + X\}}{\{(0.0010/V) - X)\}}$ $X = [H^+] = 1.7 \times 10^{-6}$ **pH = 5.77**

pH after adding 0.010 mol NaOH;
0.010 mol NaOH and 0.010 mol $HC_4H_7O_2$ will form 0.010 mol $C_4H_7O_2^-$ and 0.010 mol Na^+ plus water. The $C_4H_7O_2^-$ will undergo hydrolysis.

$C_4H_7O_2^- + H_2O \rightleftharpoons HC_4H_7O_2 + OH^-$

$K_b = \frac{K_w}{K_a} = \frac{1.0 \times 10^{-14}}{1.5 \times 10^{-5}} = \frac{(X)(X)}{(0.010 \text{ mol}/0.10\text{L}) - X}$

$X = [OH^-] = 8.2 \times 10^{-6}$ **pH = 8.91**

pH after adding 0.011 mol NaOH;
0.010 mol $HC_4H_7O_2$ plus 0.011 mol NaOH will yield 0.010 mol $C_4H_7O_2^-$ and 0.001 mol of excess NaOH. The excess of the strong base will determine the pH.
0.001 mol NaOH = 0.001 mol OH^-.
0.001 mol OH^-/0.100 L = 0.01 M OH^- **pH = 12.0**
Plot a curve of pH vs. moles of NaOH added!
pH at equivalence point = 8.91
Indicator of choice: thymol blue or phenolphthalein

16.85 ?mol H^+ (initial) = 50.0 mL acid x $\frac{0.10 \text{ mol } H^+}{10^3 \text{ mL acid}}$ = **5.0×10^{-3} mol H^+**

If mol H^+ > mol OH^-, the **mol H^+ (XS) = mol H^+ - mol OH^-**
If mol OH^- > mol H^+, the **mol OH^- (XS) = mol OH^- - mol H^+**
(continued)

16.85 (continued)

$[H^+]$ = mol H^+/final total volume in liters

$[OH^-]$ = mol OH^-/final total volume in liters

point	mL base added:	mol base	M H^+	pH
1	0.00	0.00	0.10	1.00
2	10.00	0.0010	0.067	1.18
3	20.00	0.0020	0.043	1.37
4	30.00	0.0030	0.025	1.60
5	40.00	0.0040	0.011	1.95
6	45.00	0.0045	0.0053	2.28
7	49.00	0.0049	0.0010	3.00
8	50.00	0.0050	$1.0x10^{-7}$	7.00
9	51.00	0.0051	$1.0x10^{-11}$	11.00
10	55.00	0.0055	$2.1x10^{-12}$	11.68
11	60.00	0.0060	$1.1x10^{-12}$	11.96
12	70.00	0.0070	$6.0x10^{-13}$	12.22
13	80.00	0.0080	$4.3x10^{-13}$	12.36
14	90.00	0.0090	$3.5x10^{-13}$	12.46
15	100.00	0.0100	$3.0x10^{-13}$	12.52

These results will yield a plot very much like Figure 16.5 with an equivalence point at pH = 7.0 and 50.00 mL of NaOH added.

16.86

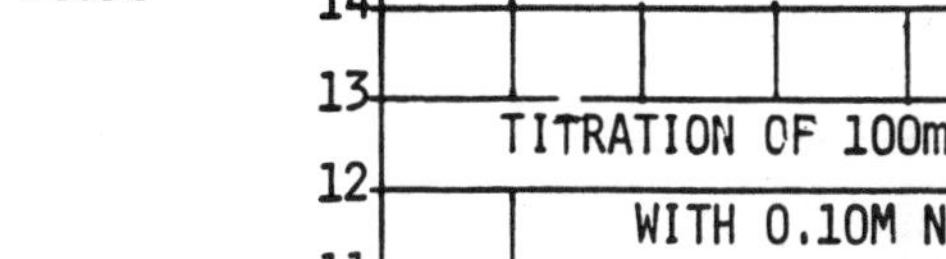

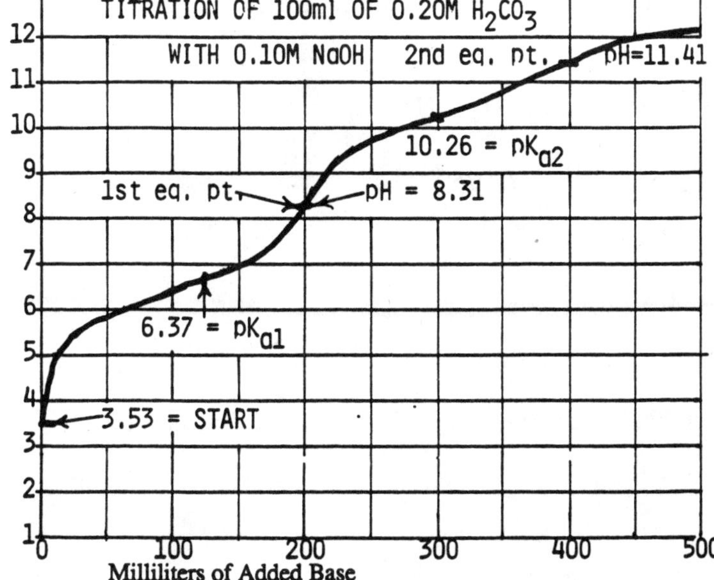

16.87 Acid and basic forms of an indicator differ in color (HIn $\rightleftharpoons$ $H^+ + In^-$). Therefore, depending on the pH range of your indicator, it will change to HIn in acid solution and In^- in basic solution. This color change ideally should correspond to the equivalence point of the titration. If too much indicator is added, it may interfere with the endpoint because it will react with the base in the titration.

16.88 Thymol Blue or Phenolphthalein
Congo Red pH range is too low. It would change color before the equivalence point is reached.

16.89 No. pH range for the color change is too low.

16.90 (a) HCN $K_a = 4.9 \times 10^{-10}$ If the concentration of the salt is about 0.10 M NaCN at the equivalence point, the pH at the equivalence point is:

$$\frac{K_w}{K_a} = \frac{1.0 \times 10^{-14}}{4.9 \times 10^{-10}} = \frac{X^2}{0.10}$$

$X = [OH^-] = 1.4 \times 10^{-3}$

pH = 11.15

In Table 16.5, **alizarin yellow** is the only suitable indicator shown.

(b) $C_6H_5NH_2$ $K_b = 3.8 \times 10^{-10}$
Assume that at the equivalence point the concentration of $C_6H_5NH_3Cl$ is approximately 0.10 M.

$$C_6H_5NH_3^+ \rightleftharpoons C_6H_5NH_2 + H^+$$

$$K_a = \frac{K_w}{K_b} = \frac{1.0 \times 10^{-14}}{3.8 \times 10^{-10}} = \frac{(X)(X)}{0.10 - X}$$

$X = [H^+] = 1.6 \times 10^{-3}$

pH = 2.8

The only suitable indicator listed in Table 16.5 is **thymol blue.**

16.91 Using the Henderson-Hasselbalch equation:

$$pH = pK_a + \log \frac{[anion]}{[acid]}$$

$$7.0 = -\log (1 \times 10^{-5}) + \log \frac{[anion]}{[acid]}$$

$$2.0 = \log \frac{[anion]}{[acid]} \qquad \frac{[anion]}{100} = [acid] \qquad 100 = \frac{[anion]}{[acid]}$$

The solution will be green.

16.92 (a) concentration of lactic acid $= 2.50\ g \times \frac{1\ mol}{90.0\ g} \times \frac{1}{0.500\ L} = 0.556\ M$

concentration of potassium lactate (lactate ion)

$$= 38.0\ g \times \frac{1\ mol}{128.1\ g} \times \frac{1}{0.500\ L} = 0.593\ M$$

$$pH = pK_a + \log\frac{[salt]}{[acid]} \text{ or } pK_a + \log\frac{[anion]}{[acid]}$$

$$pH = -\log(1.38 \times 10^{-4}) + \log\frac{0.593}{0.556} = 3.89$$

pH = 3.89

(b) concentration of lactic acid in solution before reaction

$$= 25.0\ g \times \frac{1\ mol}{90.0\ g} \times \frac{1}{0.550\ L} = 0.505\ M$$

Concentration of potassium lactate (lactate ion) in solution before reaction

$$= 38.0\ g \times \frac{1\ mol}{128.1\ g} \times \frac{1}{0.550\ L} = 0.540\ M$$

Concentration of HCl in solution before reaction

$$= 0.20\ M \times 0.050\ L \times \frac{1}{0.550\ L} = 0.018\ M$$

(continued)

16.92 (continued)

	$HC_3H_5O_3$	$\rightleftharpoons$	H^+	+	$C_3H_5O_3^-$
initial	0.505		0.018		0.540
change	+(1.8 x 10^{-2} - X)		-(0.018 - X)		-(0.018 - X)
equil.	0.523 - X		X		0.522 + X

$$K_a = \frac{(X)(0.522 + X)}{(0.523 - X)} \qquad 1.38 \times 10^{-4} = \frac{(X)(0.522)}{(0.523)}$$

X = $[H^+]$ = 1.34 x 10^{-4}

pH = 3.86

ΔpH = 3.86 - 3.89 = **-0.03**

(c) Quantities in solution before reaction:

$$\text{Lactic acid} = 25.0 \text{ g} \times \frac{1 \text{ mol}}{90.0 \text{ g}} \times \frac{1}{0.550 \text{ L}} = 0.505$$

Lactate ion = 0.540 M

$$\text{KOH} = 1.0 \text{ M} \times 0.0500 \text{ L} \times \frac{1}{0.550 \text{ L}} = 0.091 \text{ M}$$

	$C_3H_5O_3^- + H_2O$	$\rightleftharpoons$	$HC_3H_5O_3$	+	OH^-
initial	0.540 M		0.505 M		0.091 M
change	+(0.091 - X)		-(0.091 - X)		-(0.091 - X)
equil	0.631 - X		0.414 + X		+ X

$$K_b = \frac{K_w}{K_a} = \frac{1.00 \times 10^{-14}}{1.38 \times 10^{-4}} = \frac{(0.414 + X)(X)}{(0.631 - X)} \approx \frac{(0.414)X}{(0.631)}$$

X = $[OH^-]$ = 1.10 x 10^{-10}

pOH =9.96 pH = 4.04 ΔpH = 4.04 - 3.89 = **0.15**

16.93 Is HCO_3^- an acid or a base? It is both an acid and a base, but as which is it the stronger?

$K_a(HCO_3^-) = K_{a_2}(H_2CO_3) = 5.6 \times 10^{-11}$

$$K_b(HCO_3^-) = K_{hy}(HCO_3^-) = \frac{K_w}{K_{a_1}(H_2CO_3)} = \frac{1.0 \times 10^{-14}}{4.3 \times 10^{-7}} = 2.3 \times 10^{-8}$$

Since it is a stronger base than it is an acid, let's ignore its acid properties and calculate its pH as a base.

$$K_b = 2.3 \times 10^{-8} = \frac{[OH^-][H_2CO_3]}{0.50} = \frac{X^2}{0.50}$$

$X = [OH^-] = 1.1 \times 10^{-4}$

pH = 10.04

(The above assumption is not very accurate. The actual pH can be calculated by using the formula $\left[H^+\right] = \sqrt{K_{a_1} \times K_{a_2}}$ Then, $[H^+]$ would be calculated to be 4.91×10^{-9} and pH would be 8.31. The derivation of the equation $\left[H^+\right] = \sqrt{K_{a_1} \times K_{a_2}}$ requires the simultaneous consideration of both equilibria and is derived in the solution to Exercise 16.95. pH = 10 will be used in this solution. If you can derive the equation $\left[H^+\right] = \sqrt{K_{a_1} \times K_{a_2}}$, please use pH = 8.31.) Final pH = ?

	Init. Conc.	Effect of HCl	Change	Equil. Conc.
HCO_3^-	0.50	-0.05	+X	$0.45 + X \approx 0.45$ M
H^+	--	--	X	X
H_2CO_3	0	+0.05	-X	$0.05 - X \approx 0.05$ M

$$K_a = K_{a_1} = \frac{(X)(0.45)}{(0.05)} = 4.3 \times 10^{-7}$$

$X = [H^+] = 5 \times 10^{-8}$ pH = 7.3

ΔpH = 7.3 - 10.0 = **-2.7** (or 7.3 - 8.31 = -1.0)

16.94 $K_a = 1.8 \times 10^{-5} = \frac{[H^+][C_2H_3O_2^-]}{[HC_2H_3O_2]} = \frac{X^2}{0.50}$ $[H^+] = 5.62 \times 10^{-5}$

$$\frac{[C_2H_3O_2^-]}{[HC_2H_3O_2]} = 0.320 = \frac{\text{moles } C_2H_3O_2/V_T}{\text{moles } HC_2H_3O_2/V_T} = \frac{\text{moles } C_2H_3O_2^-}{\text{moles } HC_2H_3O_2}$$

Moles $C_2H_3O_2^-$ + moles $HC_2H_3O_2$ = initial number of moles of $NaC_2H_3O_2$ = (0.10 mole/L)(0.100 L) or 0.010 moles. Let X = moles $C_2H_3O_2^-$. Then, 0.010 - X will equal moles of $HC_2H_3O_2$.

$$\frac{X}{0.010 - X} = 0.320$$

$X = 2.43 \times 10^{-3}$ $0.010 - X = 7.57 \times 10^{-3}$ mole $HC_2H_3O_2$

From $HCl + C_2H_3O_2^- \rightarrow HC_2H_3O_2 + Cl^-$, 7.57×10^{-3} moles $HC_2H_3O_2$ required 7.57×10^{-3} moles HCl.

$$7.57 \times 10^{-3} \text{ mole HCl} \times \frac{1 \text{ L}}{6.0 \text{ mole HCl}} \times \frac{1{,}000 \text{ mL}}{\text{L}} = 1.26 \text{ mL} = \mathbf{1.3\ mL}$$

16.95 0.10 M NH_4^+ and 0.10 M NO_2^-

$$K_a(NH_4^+) = \frac{K_w}{K_b} = \frac{1.0 \times 10^{-14}}{1.8 \times 10^{-5}} = 5.6 \times 10^{-10}$$

$$K_b(NO_2^-) = \frac{K_w}{K_a} = \frac{1.0 \times 10^{-14}}{4.5 \times 10^{-4}} = 2.2 \times 10^{-11}$$

The values are too close to be able to neglect one or the other. You will need to consider both equations simultaneously. Consider what you already know.

(1) $NH_4^+ \rightleftharpoons NH_3 + H^+$ $K_a(NH_4^+) = \frac{K_w}{K_b}$

(2) $NO_2^- + H_2O \rightleftharpoons HNO_2 + OH^-$ $K_b(NO_2^-) = \frac{K_w}{K_a}$

(continued)

16.95 (continued)

(3) Initially, $[NH_4^+]^- = [NO_2^-]$ and this will probably not change significantly, since one is a very weak acid while the other is a very weak base.

(4) $[H^+][OH^-] = 1.0 \times 10^{-14}$ or $[OH^-] = 1.0 \times 10^{-14}/[H^+]$

Divide $K_a\ (NH_4^+)$ by $K_b\ (NO_2^-)$

$$\frac{K_a\ (NH_4^+)}{K_b\ (NO_2^-)} = \frac{[NH_3][H^+]/[NH_4^+]}{[HNO_2][OH^-]/[NO_2^-]}$$

Since the denominators are equal:

$$\frac{K_a\ (NH_4^+)}{K_b\ (NO_2^-)} = \frac{[NH_3][H^+]}{[HNO_2][OH^-]} = \frac{[NH_3][H^+]}{[HNO_2]\ K_w\ /\ [H^+]}$$

$$\frac{K_a\ (NH_4^+)}{K_b\ (NO_2^-)} = \frac{[NH_3][H^+]^2}{[HNO_2]\ K_w}$$

If $[NH_4^+] \approx [NO_2^-]$, then $[NH_3] \approx [HNO_2]$ and the equation becomes:

$$\frac{K_a\ (NH_4^+)}{K_b\ (NO_2^-)} = \frac{[H^+]^2}{K_w}$$

$$[H^+] = \sqrt{\frac{K_a\ (NH_4^+)\ K_w}{K_b\ (NO_2^-)}}$$

$$= \sqrt{K_a\ (NH_4^+)\ K_a\ (HNO_2)}$$

$$H^+ = \sqrt{5.6 \times 10^{-10} \times 4.5 \times 10^{-4}} = 5.0 \times 10^{-7}$$

pH = 6.30

16.96 **Acidic** pH = - log [H^+ (from acid) + H^+ (from water)]

= - log ($1.0 \times 10^{-8} + 1.0 \times 10^{-7}$) = **6.96**

A more precise answer to this question would require that one consider the effect that the H^+ from the acid would have on the autoionization of water.

	H^+ +	OH^- ⇌	H_2O
Before acid added	1.0×10^{-7}	1.0×10^{-7}	
Acid added	1.0×10^{-8}		
After acid reacts	$X + (1.0 \times 10^{-8})$	X	

$[X + (1.0 \times 10^{-8})]\,[X] = 1.0 \times 10^{-14}$ Solve this quadratic equation:

$X = 9.51 \times 10^{-8}$; $[H^+] = (9.51 \times 10^{-8}) + (1.0 \times 10^{-8}) = 1.05 \times 10^{-7}$:

pH = **6.98** It is very slightly acidic.

16.97

	HIO_3 ⇌	H^+ +	IO_3^-
before	0.20	≈0	0
change	-X	+X	+X
equil	0.20 - X	X	X

$K_a = 1.7 \times 10^{-1} = \frac{(X)(X)}{(0.20 - X)}$

(Use the quadratic formula)

X = 0.12

$[H^+] = 0.12$

$[IO_3^-] = 0.12$

$[HIO_3] = 0.08$

pH = **0.92**

16.98 $HC_2O_2Cl_3 \rightleftharpoons C_2O_2Cl_3^- + H^+$

	$HC_2O_2Cl_3$	$C_2O_2Cl_3^-$	H^+
before	0.60	0	≈0
change	-X	+X	+X
equil	0.60-X	X	X

$$K_a = 2.2 \times 10^{-1} = \frac{X^2}{(0.60 - X)}$$

For the quadratic formula $X^2 + 0.22\,X - 0.132 = 0$

$$X = \frac{-0.22 \pm \sqrt{(.22)^2 - 4(-0.132)}}{2} = 0.27$$

$[H^+] = 0.27$ **pH = 0.57**

By successive approximations

$$2.2 \times 10^{-1} = \frac{X^2}{.60} \qquad X = 0.36$$

$$2.2 \times 10^{-1} = \frac{X^2}{0.60 - 0.36} \qquad X = 0.23$$

$$2.2 \times 10^{-1} = \frac{X^2}{0.60 - 0.23} \qquad X = 0.29$$

$$2.2 \times 10^{-1} = \frac{X^2}{0.60 - 0.29} \qquad X = 0.26$$

$$2.2 \times 10^{-1} = \frac{X^2}{0.60 - 0.26} \qquad X = 0.27$$

$$2.2 \times 10^{-1} = \frac{X^2}{0.60 - 0.27} \qquad X = 0.27$$

$[H^+] = 0.27$ **pH = 0.57**

16.99 $PO_4^{3-} + H_2O \rightleftharpoons HPO_4^{2-} + OH^-$

	PO_4^{3-}	HPO_4^{2-}	OH^-
Before	1.0	0	≈0
change	-X	+X	+X
equil	1.0 - X	X	X

$$K_b = \frac{K_w}{K_{a\,3}} = \frac{1.0 \times 10^{-14}}{2.2 \times 10^{-12}} = \frac{(X)(X)}{1.0 - X}$$

$$4.55 \times 10^{-3} = \frac{X^2}{1.0 - X}$$

by successive approximations

$4.55 \times 10^{-3} = X^2 \quad X = 6.75 \times 10^{-2}$

$$4.55 \times 10^{-3} = \frac{X^2}{1.0 - (6.75 \times 10^{-2})} \quad X = 6.51 \times 10^{-2}$$

$$4.55 \times 10^{-3} = \frac{X^2}{1.0 - (6.5 \times 10^{-2})} \quad X = 6.5 \times 10^{-2}$$

$X = 6.5 \times 10^{-2} = [OH^-]$ pOH = 1.19

pH = 12.81

$NaOH \rightarrow Na^+ + OH^-$

$[OH^-]$ = **6.5×10^{-2} M = conc. of NaOH needed to give the same pH.**

17 SOLUBILITY AND COMPLEX ION EQUILIBRIA

17.1 The concentration of the solid is left out of the solubility equilibrium expression of a salt because it is not a variable and can be (and is) included within the constant K_{sp}.

17.2 (a) $Ag_2SO_3(s) \rightleftharpoons 2Ag^+(aq) + SO_3^{2-}(aq)$

$K_{sp} = [Ag^+]^2[SO_3^{2-}]$ (b) $K_{sp} = [Ca^{2+}][F^-]^2$

(c) $K_{sp} = [Fe^{3+}][OH^-]^3$ (d) $K_{sp} = [Mg^{2+}][C_2O_4^{2-}]$

(e) $K_{sp} = [Au^{3+}][Cl^-]^3$ (f) $K_{sp} = [Ba^{2+}][CO_3^{2-}]$

17.3 (a) $K_{sp} = [Pb^{2+}][F^-]^2$ (b) $K_{sp} = [Ag^+]^3[PO_4^{3-}]$

(c) $K_{sp} = [Fe^{2+}]^3[PO_4^{3-}]^2$ (d) $K_{sp} = [Li^+]^2[CO_3^{2-}]$

(e) $K_{sp} = [Ca^{2+}][IO_3^-]^2$ (f) $K_{sp} = [Ag^+]^2[Cr_2O_7^{2-}]$

17.4 $CuCl(s) \rightarrow Cu^+(aq) + Cl^-(aq)$

$K_{sp} = [Cu^+][Cl^-] = (1.0 \times 10^{-3})(1.0 \times 10^{-3}) = \mathbf{1.0 \times 10^{-6}}$

17.5 $PbCO_3(s) \rightarrow Pb^{2+}(aq) + CO_3^{2-}(aq)$

$K_{sp} = [Pb^{2+}][CO_3^{2-}] = (1.8 \times 10^{-7})(1.8 \times 10^{-7}) = \mathbf{3.2 \times 10^{-14}}$

17.6 $BaC_2O_4(s) \rightarrow Ba^{2+}(aq) + C_2O_4^{2-}(aq)$

$$\frac{0.0781\ g}{L} \times \frac{1\ mol}{225.4\ g} = \frac{3.47 \times 10^{-4}\ mol}{L}$$

$K_{sp} = [Ba^{2+}][C_2O_4^{2-}] = (3.47 \times 10^{-4})(3.47 \times 10^{-4}) = \mathbf{1.20 \times 10^{-7}}$

17.7 $CaCrO_4(s) \rightarrow Ca^{2+}(aq) + CrO_4^{2-}(aq)$

$K_{sp} = [Ca^{2+}][CrO_4^{2-}] = (1.0 \times 10^{-2})(1.0 \times 10^{-2}) = \mathbf{1.0 \times 10^{-4}}$

17.8 $PbI_2(s) \rightarrow Pb^{2+}(aq) + 2I^-(aq)$

$K_{sp} = [Pb^{2+}][I^-]^2 = (1.5 \times 10^{-3})(2 \times 1.5 \times 10^{-3})^2 = \mathbf{1.4 \times 10^{-8}}$

17.9 $$\frac{0.0981\ g}{0.200\ L} \times \frac{1\ mol}{245.2\ g} = \frac{2.00 \times 10^{-3}\ mol}{L}$$

$PbF_2(s) \rightarrow Pb^{2+}(aq) + 2F^-(aq)$

$K_{sp} = [Pb^{2+}][F^-]^2 = (2.00 \times 10^{-3})(2 \times 2.00 \times 10^{-3})^2 = \mathbf{3.20 \times 10^{-8}}$

17.10 $MgF_2(s) \rightarrow Mg^{2+}(aq) + 2F^-(aq)$

$$\frac{7.6 \times 10^{-2}\ g}{L} \times \frac{1\ mol}{62.3\ g} = \frac{1.2 \times 10^{-3}\ mol}{L}$$

$K_{sp} = [Mg^{2+}][F^-]^2 = (1.2 \times 10^{-3})(2 \times 1.2 \times 10^{-3})^2 = \mathbf{6.9 \times 10^{-9}}$

17.11 $$\frac{4.99\ g}{L} \times \frac{1\ mol}{312\ g} = \frac{1.6 \times 10^{-2}\ mol}{L}$$

$Ag_2SO_4(s) \rightarrow 2Ag^+(aq) + SO_4^{2-}(aq)$

$K_{sp} = [Ag^+]^2[SO_4^{2-}] = (2 \times 1.6 \times 10^{-2})^2(1.6 \times 10^{-2}) = \mathbf{1.6 \times 10^{-5}}$

17.12 pH = 9.50 $[OH^-]$ = antilog [-(14 - 9.50)] = 3.2×10^{-5}

$Ni(OH)_2(s) \rightarrow Ni^{2+}(aq) + 2OH^-(aq)$

$K_{sp} = [Ni^{2+}][OH^-]^2 = (1/2 \times 3.2 \times 10^{-5})(3.2 \times 10^{-5})^2 = \mathbf{1.6 \times 10^{-14}}$

17.13 $\frac{0.47 \text{ g } MgC_2O_4}{0.500 \text{ L}} \times \frac{1 \text{ mol}}{112 \text{ g}} = 8.4 \times 10^{-3}$ M

$MgC_2O_4(s) \rightarrow Mg^{2+}(aq) + C_2O_4^{2-}(aq)$

	Init. Conc.	Added	Equil. Conc.
Mg^{2+}	0	8.4×10^{-3}	8.4×10^{-3}
$C_2O_4^{2-}$	2.0×10^{-3}	8.4×10^{-3}	10.4×10^{-3}

$K_{sp} = (8.4 \times 10^{-3})(10.4 \times 10^{-3}) = \mathbf{8.7 \times 10^{-5}}$

17.14 (a) $AuCl_3(s) \rightarrow Au^{3+}(aq) + 3Cl^-(aq)$

	Init. Conc.	Change	Equil. Conc.
Au^{3+}	0	+X	X
Cl^-	0	+3X	3X

$K_{sp} = 3.2 \times 10^{-25} = (X)(3X)^3$ X = molar solubility = $\mathbf{3.3 \times 10^{-7}}$ **M**

(b) $Fe(OH)_2(s) \rightarrow Fe^{2+}(aq) + 2OH^-(aq)$

	Init. Conc.	Change	Equil. Conc.
Fe^{2+}	0	+X	X
OH^-	1×10^{-7}	+ 2X	2X

$K_{sp} = 2 \times 10^{-15} = (X)(2X)^2$ X = molar solubility = $\mathbf{8 \times 10^{-6}}$ **M**

(continued)

17.14 (continued)

(c) $BaSO_4(s) \rightarrow Ba^{2+}(aq) + SO_4^{2-}(aq)$

	Init. Conc.	Change	Equil. Conc.
Ba^{2+}	0	+X	X
SO_4^{2-}	0	+X	X

$K_{sp} = 1.5 \times 10^{-9} = (X)(X)$ X = molar solubility = **3.9 x 10⁻⁵ M**

(d) $Hg_2Cl_2(s) \rightarrow Hg_2^{2+}(aq) + 2Cl^-(aq)$

	Init. Conc.	Change	Equil. Conc.
Hg_2^{2+}	0	+X	X
Cl^-	0	+ 2X	2X

$K_{sp} = 2 \times 10^{-18}$ (From the Table in the Appendix) $= (X)(2X)^2$

X = molar solubility = **8×10^{-7} M**

(e) $CaF_2(s) \rightarrow Ca^{2+}(aq) + 2F^-(aq)$

	Init. Conc.	Change	Equil. Conc.
Ca^{2+}	0	+X	X
F^-	0	+2X	2X

$K_{sp} = 1.7 \times 10^{-10} = (X)(2X)^2$

X = molar solubility = **3.5×10^{-4} M**

(f) $MgC_2O_4(s) \rightarrow Mg^{2+}(aq) + C_2O_4^{2-}(aq)$

$K_{sp} = 8.6 \times 10^{-5} = (X)(X)$ X = molar solubility = **9.3×10^{-3} M**

17.15 $Mg(OH)_2(s) \rightarrow Mg^{2+}(aq) + 2OH^-(aq)$

	Init. Conc.	Change	Equil. Conc.
Mg^{2+}	0	+X	X
OH^-	1×10^{-7}	+2X	$1 \times 10^{-7} + 2X \approx 2X$

$K_{sp} = 7.1 \times 10^{-12} = (X)(2X)^2$ $X = 1.2 \times 10^{-4}$
$[OH^-] = 2.4 \times 10^{-4}$ **pH = 10.38**

17.16 $CaSO_4(s) \rightarrow Ca^{2+}(aq) + SO_4^{2-}(aq)$

$K_{sp} = 2 \times 10^{-4} = (X)(X)$ X = molar solubility = 1×10^{-2} M

$$\frac{1 \times 10^{-2} \text{ mol } CaSO_4}{L} \times 0.600 \text{ L} \times \frac{136 \text{ g}}{\text{mol}} = \mathbf{0.8 \text{ g } CaSO_4}$$

17.17 Volume of plaster = $\pi r^2 h = 3.14 \times (0.5 \text{ cm})^2(1.50 \text{ cm}) = 1.18 \text{ cm}^3$ (assuming that 0.5 cm has 3 significant figures)
Mass of plaster = $1.18 \text{ cm}^3 \times 0.97 \text{ g/mL} \times 1 \text{ mL/cm}^3 = 1.1$ g

$CaSO_4(s) \rightarrow Ca^{2+}(aq) + SO_4^{2-}(aq)$ $K_{sp} = 2. \times 10^{-4} = (X)(X)$

X = molar solubility = 1.4×10^{-2} mol/L

(1.4×10^{-2} mol/L) x 136 g/mol = 1.9 g/L

Liters required to dissolve 1.1 g: 1.1 g x 1L/1.9 g = 0.58 L or 580 mL

Time required: 0.58 L x 1d/2.0 L = **0.29 day or 7.0 hours**

17.18 $Al(OH)_3 \rightleftharpoons Al^{3+} + 3OH^-$
$K_{sp} = (X)(3X)^3 = 2 \times 10^{-33}$ $X = 2.9 \times 10^{-9}$ $3X = 8.7 \times 10^{-9} = 0.87 \times 10^{-8}$
($[OH^-]$ from the dissociation of water is 1×10^{-7} M or about 10 times this value; therefore we cannot neglect it.)
$K_{sp} = [Al^{3+}](1 \times 10^{-7})^3 = 2 \times 10^{-33}$
$[Al^{3+}] = 2 \times 10^{-12}$ M
2×10^{-12} mole of $Al(OH)_3$ dissolves per liter.

17.19 In a solution, a precipitate will form only if the mixture is supersaturated (i.e., when the value of the ion product exceeds the value of the K_{sp}).

17.20 (a) The ion concentrations are: 5.0×10^{-2} M Ag^+, 5.0×10^{-2} M NO_3^-, 1.0×10^{-3} M Na^+ and 1.0×10^{-3} M $C_2H_3O_2^-$. Possible salts include $AgNO_3$ (soluble -- no K_{sp} value), $NaC_2H_3O_2$ (soluble -- no K_{sp} value), $NaNO_3$ (soluble -- no K_{sp} value), $AgC_2H_3O_3$ (possible precipitate -- $K_{sp} = 2.3 \times 10^{-3}$). For $AgC_2H_3O_3$, ion product = $(5.0 \times 10^{-2})(1.0 \times 10^{-3}) = 5.0 \times 10^{-5}$ which is less than its K_{sp}. Therefore, **no precipitate will form.**

(b) 1.0×10^{-2} M Ba^{2+}, 2.0×10^{-2} M NO_3^-,2.0×10^{-2} M Na^+, and 2.0×10^{-2} M F^-. The possible salts include: $Ba(NO_3)_2$ (soluble -- no K_{sp} value), BaF_2 (possible precipitate -- $K_{sp} = 1.7 \times 10^{-6}$), NaF (soluble -- no K_{sp} value), $NaNO_3$ (soluble -- no K_{sp} value).
The only possible precipitate, BaF_2, has an ion product value = $(1.0 \times 10^{-2})(2.0 \times 10^{-2})^2 = 4.0 \times 10^{-6}$, which is greater than its K_{sp} value. **BaF_2 will precipitate.**

(c) The ion concentrations are: 9.3×10^{-3} M Ca^{2+}, 1.9×10^{-2} M Cl^-, 1.7×10^{-1} M Na^+, and 8.3×10^{-2} M SO_4^{2-}. The possible salts include: $CaCl_2$ (soluble), $CaSO_4$ (possible precipitate -- $K_{sp} = 2 \times 10^{-4}$), Na_2SO_4 (soluble), NaCl (soluble). For $CaSO_4$, the ion product value = $(9.3 \times 10^{-3})(8.3 \times 10^{-2}) = 7.7 \times 10^{-4}$, which is greater than its K_{sp} value. **$CaSO_4$ will precipitate.**

17.21 K_{sp} of $Fe(OH)_2 = 2 \times 10^{-15} = [Fe^{2+}][OH^-]^2 = (0.010)(X)^2$
$X^2 = 2 \times 10^{-15}/0.010 = 2 \times 10^{-13}$ $\quad X = [OH^-] = 4 \times 10^{-7}$
pOH = 6.4 **pH = 7.6** This is the minimum pH at which the value of the ion product will equal or exceed the solubility product constant.

17.22 Possible precipitation: $Ag^+ + Cl^- \rightarrow AgCl(s)$ $\quad K_{sp} = 1.7 \times 10^{-10}$

Ion	Init.Conc.	Conc. after Mix	Change	Equil. Conc.
Ag^+	0.20 M	0.10 M	- ~0.050 + Y	0.050 + Y
NO_3^-	0.20 M	0.10 M	none	0.10
H^+	0.10 M	0.050 M	none	0.050
Cl^-	0.10 M	0.050 M	- ~0.050 + Y	Y

$K_{sp} = 1.7 \times 10^{-10} = (0.050 + Y)(Y) \approx (0.050)\,Y$
$Y = 3.4 \times 10^{-9}$ $\quad [Ag^+] = 0.050 + 3.4 \times 10^{-9} =$ **0.050 M**
$[NO_3^-] = 0.10$ M $\quad$ **$[H^+] = 0.050$ M** $\quad$ **$[Cl^-] = 3.4 \times 10^{-9}$ M**

$$[OH^-] = \frac{1.0 \times 10^{-14}}{0.050} = 2.0 \times 10^{-13} \text{ M}$$

17.23 (a) The only possible precipitate is $CaCO_3$ ($K_{sp} = 9 \times 10^{-9}$).

Ion prod. = (0.025)(0.0050) = 1.2×10^{-4}

Since Ion Product > K_{sp}, **$CaCO_3$ will precipitate.**

(b) The only possible precipitate is $PbCl_2$ ($K_{sp} = 1.6 \times 10^{-5}$)
Ion Product = $(0.010)(0.060)^2 = 3.6 \times 10^{-5}$

Ion Product > K_{sp} **$PbCl_2$ will precipitate**

(c) The only possible precipitate is FeC_2O_4 ($K_{sp} = 2.1 \times 10^{-7}$)

Ion Product = $(1.5 \times 10^{-3})(2.2 \times 10^{-3}) = 3.3 \times 10^{-6}$

Ion Product > K_{sp} **FeC_2O_4 will precipitate.**

17.24 The common ion effect is the reduction of solubility caused by the presence of a common ion.

17.25 K_{sp} $(CaCO_3) = 9 \times 10^{-9} = [Ca^{2+}][CO_3^{2-}] = (X)(X + 0.50)$

X = **2×10^{-8} mol/L = molar solubility of $CaCO_3$ in 0.50 M Na_2CO_3**

17.26 K_{sp} $(AgCl) = 1.7 \times 10^{-10} = [Ag^+][Cl^-] = (X)(X + 0.060)$

X = **2.8×10^{-9} mol/L = molar solubility of AgCl in 0.020 M $AlCl_3$**

17.27 $K_{sp} = 1.6 \times 10^{-5} = [Pb^{2+}][Cl^-]^2 = (X)(2X + 0.060)^2$

$X = 4.4 \times 10^{-3}$ if 2X + 0.060 is assumed to be ≈ 0.060.
Not a good assumption.

Assume $2X + 0.060 \approx 2(4.4 \times 10^{-3}) + 0.060$ or 0.069.

Then $1.6 \times 10^{-5} = (X)(0.069)^2$ or $X = 3.6 \times 10^{-3}$

3.6×10^{-3} mol/L is the calculated molar solubility of $PbCl_2$ in 0.020 M $AlCl_3$

17.28 $K_{sp} = 1.9 \times 10^{-12} = [Ag^+]^2[CrO_4^{2-}] = (2X + 0.10)^2(X) \approx (0.10)^2(X)$

$X = 1.9 \times 10^{-10}$ mol Ag_2CrO_4/L in 0.10 M $AgNO_3$

17.29 $K_{sp} = 1.9 \times 10^{-12} = [Ag^+]^2[CrO_4^{2-}] = (2X)^2(X + 0.10) \approx (4X^2)(0.10)$

$X = 2.2 \times 10^{-6}$ mol Ag_2CrO_4/L in 0.10 M Na_2CrO_4

17.30 $K_{sp} = 1.7 \times 10^{-10} = [Ca^{2+}][F^-]^2 = (X)(2X + 0.010)^2 \approx (X)(0.010)^2$

$X = 1.7 \times 10^{-6}$ mol CaF_2/L in 0.010 M NaF

17.31 X = moles of NaF added per liter

K_{sp} (BaF_2) = $[Ba^{2+}][F^-]^2 = (6.8 \times 10^{-4})(2 \times 6.8 \times 10^{-4} + X)^2 = 1.7 \times 10^{-6}$

(from the solution of the quadratic equation) $X = 4.9 \times 10^{-2}$ mol NaF/L

$\frac{4.9 \times 10^{-2} \text{ mol}}{\text{L}} \times \frac{42 \text{ g}}{\text{mol}} = $ **2.1 g NaF**

17.32 $Fe(OH)_2 \rightleftharpoons Fe^{2+} + 2OH^-$
$[OH^-]$ = (from pH) 3.2×10^{-5}
$K_{sp} = 2 \times 10^{-15} = [Fe^{2+}](3.2 \times 10^{-5})^2$
$[Fe^{2+}] = 2 \times 10^{-6}$
Molar solubility of $Fe(OH)_2 = [Fe^{2+}] =$ **2×10^{-6} mole/L**

17.33 $Ca(OH)_2 \rightleftharpoons Ca^{2+} + 2OH^-$
(a) $K_{sp} = 6.5 \times 10^{-6} = (0.10 + X)(2X)^2$
$X = 4.0 \times 10^{-3}$ = molar solubility of $Ca(OH)_2$
= **4.0×10^{-3} mole/L**
(b) $K_{sp} = 6.5 \times 10^{-6} = (X)(2X + 0.10)^2$
$X = 6.5 \times 10^{-4}$ = molar solubility of $Ca(OH)_2$
= **6.5×10^{-4} mole/L**

17.34 Added HCl

$Mg(OH)_2(s) + 2HCl \rightarrow Mg^{2+} + 2Cl^- + 2H_2O$

0.00250 mol HCl will form 0.00125 mol Mg^{2+}

$[Mg^{2+}] = \frac{1.25 \times 10^{-3}\,mol}{1.025\,L} = 1.22 \times 10^{-3}\,M$

$Mg(OH)_2(s) \rightleftharpoons Mg^{2+} + 2OH^-$

$K_{sp} = 7.1 \times 10^{-12} = (1.22 \times 10^{-3} + X)(2X)^2 \approx (1.22 \times 10^{-3})(4X^2)$

$X = 3.8 \times 10^{-5}$

$[Mg^{2+}] = 1.22 \times 10^{-3} + 3.8 \times 10^{-5} =$ **1.3×10^{-3} M**

$[OH^-] = 2(3.8 \times 10^{-5}) = 7.6 \times 10^{-5}$ M **pH = 9.88**

17.35 M NaOH = (2.20 g/40.0 g/mol)/0.250 L = 0.220 M

	Init. Conc.	Change	Equil. Conc.
Na^+	0.220 M	-	- - -
OH^-	0.220 M	-2X	0.220 - 2X
Fe^{2+}	0.10 M	-X	0.10 - X
Cl^-	0.20 M	-	- - -

Assume that $X \approx 0.100$; then $[OH^-] \approx 0.020$ and $[Fe^{2+}] = Y$

$K_{sp} = 2 \times 10^{-15} = [Fe^{2+}][OH^-]^2 = (Y)(0.020)^2$

$Y = 5.0 \times 10^{-12}$ **$[Fe^{2+}] = 5 \times 10^{-12}$ M**

The amount $Fe(OH)_2$ formed will be: $[0.10\ mol/L - (5 \times 10^{-12}\ mol/L)] \times$ 0.250 L x 89.9 g/mol = **2.2 g $Fe(OH)_2$ precipitated**

17.36 (1.75 g NaOH/0.250 L)/40.0 g/mol = 0.175 M

	Init. Conc.	Change	Interm. Conc.	Assume	Equil. Conc.
OH^-	0.175 M	-2X	0.175 - 2X	$X \approx 0.0875$	Y
Ni^{2+}	0.10 M	-X	0.10 - X		0.0125 + 1/2Y

$K_{sp} = [Ni^{2+}][OH^-]^2 = 1.6 \times 10^{-14} = (0.0125 + 1/2Y)(Y)^2 = 0.0125\ Y^2$

$Y = 1.1 \times 10^{-6} = [OH^-]$ pOH = 5.95 **pH = 8.05**

Mole of $Ni(OH)_2$ precipitated $[0.175\ mol/L\ (OH^-) \div 2](0.250\ L) =$

2.19×10^{-2} moles $Ni(OH)_2$

2.19×10^{-2} moles x 92.7 g/mole = **2.0 g $Ni(OH)_2(s)$**

17.37 K_{sp} $Mn(OH)_2 = 1.2 \times 10^{-11}$ K_{sp} $Fe(OH)_2 = 2 \times 10^{-15}$

	Init. Conc.	Added	Change	Equil. Conc.	Assume	Equil. Conc.
Fe^{2+}	0.100		-Y	0.100 - Y	$Y \approx 0.100$	Fe^{2+}
Mn^{2+}	0	X		X	$X \approx 0.100$	0.100
OH^-	1×10^{-7}	2X	-2Y	2X - 2Y		OH^-

K_{sp} $(Mn(OH)_2) = 1.2 \times 10^{-11} = (0.10)[OH^-]^2$, $[OH^-] = 1.1 \times 10^{-5}$

K_{sp} $(Fe(OH)_2) = 2 \times 10^{-15} = [Fe^{2+}](1.1 \times 10^{-5})^2$

$[Fe^{2+}] \approx 2 \times 10^{-5}$ M

pH = 14 + log 1.1 x 10^{-5} = **9.04**

$[Mn^{2+}] \approx 0.100$ M

17.38 K_{sp} $(BaCrO_4) = 2.4 \times 10^{-10}$ K_{sp} $(PbCrO_4) = 1.8 \times 10^{-14}$

Since the ion concentrations are equal, the **$PbCrO_4$ will precipitate first**, as revealed by its smaller K_{sp} value.

Calculation of $[CrO_4^{2-}]$ when Ba^{2+} begins to precipitate:
$2.4 \times 10^{-10} = [Ba^{2+}][CrO_4^{2-}] = (0.010)[CrO_4^{2-}]$
$[CrO_4^{2-}] = 2.4 \times 10^{-8}$ M

The concentration of Pb^{2+} when the $[CrO_4^{2-}]$ reaches 2.4 x 10^{-8} M will be:
$1.8 \times 10^{-14} = [Pb^{2+}][CrO_4^{2-}] = [Pb^{2+}](2.4 \times 10^{-8})$

$[Pb^{2+}] = 7.5 \times 10^{-7}$ M

In summary, the $PbCrO_4$ will begin to precipitate first. It will continue to precipitate until the CrO_4^{2-} concentration reaches 2.4 x 10^{-8} M. At that point and at higher concentrations of CrO_4^{2-}, both $PbCrO_4$ and $BaCrO_4$ will precipitate. When the concentration of CrO_4^{2-} reaches that 2.4 x 10^{-8} M value, the concentration of Pb^{2+} will have been reduced to 7.5 x 10^{-7} M.

17.39 $K_{a_1} \times K_{a_2} = 4.0 \times 10^{-6} = \frac{[H^+]^2[C_2O_4^{2-}]}{[H_2C_2O_4]}$

For maximum separation: ion product MgC_2O_4 equals K_{sp} (MgC_2O_4)

$8.6 \times 10^{-5} = [Mg^{2+}][C_2O_4^{2-}] = (0.10\ M)[C_2O_4^{2-}]$

$[C_2O_4^{2-}] = 8.6 \times 10^{-4}$ M

$$4.0 \times 10^{-6} = \frac{[H^+]^2[C_2O_4^{2-}]}{[H_2C_2O_4]} = \frac{[H^+]^2(8.6 \times 10^{-4})}{(0.10)}$$

$[H^+] = 2.2 \times 10^{-2}$, pH = 1.66 **pH of less than 1.66**

17.40 Minimum pH for precipitation of Mn^{2+}:

$K_{sp} = 1.2 \times 10^{-11} = (0.10)[OH^-]^2$

$[OH^-] = 1.1 \times 10^{-5}$ pH = 9.04

Minimum pH for precipitation of Cu^{2+}:

$K_{sp} = 4.8 \times 10^{-20} = (0.10)[OH^-]^2$

$[OH^-] = 6.9 \times 10^{-10}$ pH = 4.84

From pH 4.84 to 9.04 Cu^{2+} will precipitate. Above pH 9.04 both will precipitate.

17.41 $H_2CO_3 \rightleftharpoons 2H^+ + CO_3^{2-}$

$K_a = K_{a_1} \times K_{a_2}$ = (from Table 16.2) $4.3 \times 10^{-7} \times 5.6 \times 10^{-11} = 2.4 \times 10^{-17}$

Maximum CO_3^{2-} before precipitation of $CaCO_3$

$K_{sp} = 9 \times 10^{-9} = (0.050)[CO_3^{2-}]$

$[CO_3^{2-}] = 2 \times 10^{-7}$

Maximum CO_3^{2-} before precipitation of $PbCO_3$

$K_{sp} = 7.4 \times 10^{-14} = (0.050)[CO_3^{2-}]$

$[CO_3^{2-}] = 1.5 \times 10^{-12}$

Maximum separation will be when $[CO_3^{2-}]$ = slightly less than 2×10^{-7}

$H_2CO_3 \rightleftharpoons 2H^+ + CO_3^{2-}$

$$K_a = 2.4 \times 10^{-17} = \frac{[H^+]^2(2 \times 10^{-7})}{0.050}$$

$[H^+] = 2.4 \times 10^{-6}$

Less than pH = 5.62

17.42 (a) $Ag^+ + 2Cl^- \rightleftharpoons AgCl_2^-$ $\quad K_{form} = \frac{[AgCl_2^-]}{[Ag^+][Cl^-]^2}$

(b) $K_{form} = \frac{[Ag(S_2O_3)_2^{3-}]}{[Ag^+][S_2O_3^{2-}]^2}$ $\quad$ (c) $K_{form} = \frac{[Zn(NH_3)_4^{2+}]}{[Zn^{2+}][NH_3]^4}$

17.43 (a) $Fe(CN)_6^{4-} \rightleftharpoons Fe^{2+} + 6CN^-$ $\quad K_{inst} = \frac{[Fe^{2+}][CN^-]^6}{[Fe(CN)_6^{4-}]}$

(b) $K_{inst} = \frac{[Cu^{2+}][Cl^-]^4}{[CuCl_4^{2-}]}$ $\quad$ (c) $K_{inst} = \frac{[Ni^{2+}][NH_3]^6}{[Ni(NH_3)_6^{2+}]}$

17.44 $Ag^+ + 2CN^- \rightleftharpoons Ag(CN)_2^-$ $\quad K_{form} = 5.5 \times 10^{18}$ (5.3×10^{18} in the tables)

$AgCN(s) \rightleftharpoons Ag^+ + CN^-$ $\quad K_{sp} = 1.6 \times 10^{-14}$

$AgCN(s) + CN^- \rightleftharpoons Ag(CN)_2^-$ $\quad K_c = ?$

$K_c = \frac{[Ag(CN)_2^-]}{[CN^-]}$ $\quad K_{form} \times K_{sp} = \frac{[Ag(CN)_2^-]}{[Ag^+][CN^-]^2} \times \frac{[Ag^+][CN^-]}{1}$

$$K_{form} \times K_{sp} = \frac{[Ag(CN)_2^-]}{[CN^-]} = K_c$$

$K_c = K_{form} \times K_{sp} = 5.5 \times 10^{18} \times 1.6 \times 10^{-14} = 8.8 \times 10^4$

17.45 $AgI_2^-(aq) \rightleftharpoons Ag^+(aq) + 2I^-(aq)$

A more dilute solution would favor the right side of this equilibrium. The presence of the ions, Ag^+ and I^-, would in turn favor the formation of solid AgI via $Ag^+(aq) + I^-(aq) \rightarrow AgI(s)$.

17.46 K_{sp} (AgI) = 8.5×10^{-17} K_{form} ($Ag(CN)_2^-$) = 5.3×10^{18}

$AgI(s) \rightleftharpoons Ag^+(aq) + I^-(aq)$

$Ag^+(aq) + 2CN^-(aq) \rightleftharpoons Ag(CN)_2^-(aq)$

Combined reactions:

$AgI + 2CN^- \rightleftharpoons Ag(CN)_2^- + I^-$

	Init. Conc.	Change	Equil. Conc.
I^-	0	+X	X
CN^-	0.010	- 2X	0.010 - 2X
$Ag(CN)_2^-$	0	X	X

$$K_{sp} \times K_{form} = \frac{[Ag(CN)_2^-][I^-]}{[CN^-]^2} = 4.5 \times 10^2 = \frac{X^2}{(0.010 - 2X)^2}; \frac{X}{(0.010 - 2X)} = 21$$

$X = 4.9 \times 10^{-3}$ M

Molar solubility of AgI in 0.010 M KCN = 4.9×10^{-3} mol/L

17.47 K_{sp} ($Zn(OH)_2$) = 1.2×10^{-23} $Zn(NH_3)_4^{2+} \rightleftharpoons Zn^{2+} + 4NH_3$ K_{inst} = ?

	Init. Conc.	Change	Equil. Conc.	Assume	Equil. Conc.
Zn^{2+}	5.7×10^{-3}	-X	5.7×10^{-3} - X	$X \approx 5.7 \times 10^{-3}$	$[Zn^{2+}]$
OH^-	$2 \times 5.7 \times 10^{-3}$	0	0.0114		0.0114
NH_3	1.0	-4X	1.0 - 4X	≈ 1.0	1.0
$Zn(NH_3)_4^{2+}$	0	+X	X		5.7×10^{-3}

Using K_{sp}: $1.2 \times 10^{-23} = [Zn^{2+}](0.0114)^2$ $[Zn^{2+}] = 9.2 \times 10^{-20}$

$$K_{inst} = \frac{(9.2 \times 10^{-20})(1.0)^4}{5.7 \times 10^{-3}} = \mathbf{1.6 \times 10^{-17}}$$

[If one uses K_{sp} ($Zn(OH)_2$) = 4.5×10^{-17}, a value of $K_{inst} = 6.1 \times 10^{-11}$ will be obtained.]

17.48 $Cu(OH)_2 + 4NH_3 \rightleftharpoons Cu(NH_3)_4^{2+} + 2OH^-$

$K_c = K_{sp} \times K_{form} = 4.8 \times 10^{-20} \times 4.8 \times 10^{12} = 2.3 \times 10^{-7}$

	Init. Conc.	Change	Equil.
NH_3	2.0	-4X	2.0 - 4X
$Cu(NH_3)_4^{2+}$	0	+X	X
OH^-	≈0	+2X	2X

$$K_c = \frac{(X)(2X)^2}{(2.0 - 4X)^4} = \frac{4X^3}{(2.0)^4} = 2.3 \times 10^{-7}$$

$X = 9.7 \times 10^{-3}$

9.7×10^{-3} mol/L of $Cu(OH)_2$ will dissolve in 2.0 M NH_3

17.49 $AgI(s) + I^-(aq) \rightleftharpoons AgI_2^-(aq)$
$K_c = K_{sp} \times K_{form} = 8.5 \times 10^{-17} \times 1.0 \times 10^{11} = 8.5 \times 10^{-6}$

	Init. Conc.	Change	Equil.
I^-	1.0 M	-X	1.0 -X
AgI_2^-	0	+X	X

$$K_c = 8.5 \times 10^{-6} = \frac{[AgI_2^-]}{[I^-]} = \frac{X}{1.0 - X}$$

$X = 8.5 \times 10^{-6}$

8.5×10^{-6} mol/L of AgI will dissolve in 1.0 M NaI.

17.50 When NH_4Cl dissolves, it forms NH_4^+ and Cl^-. The NH_4^+ is an acid and produces H^+ upon hydrolysis: $NH_4^+ + H_2O \rightarrow NH_3 + H^+(aq)$. The H^+ will react with the OH^- released as $Mg(OH)_2$ dissolves. According to LeChatelier's Principle, as a stress is created the equilibrium between solid $Mg(OH)_2$ and its ions will shift toward the side that has lost the OH^-.

17.51 $AgC_2H_3O_2(s) \rightleftharpoons Ag^+ + C_2H_3O_2^-$ $K_{sp} = 2.3 \times 10^{-3}$

$HC_2H_3O_2 \rightleftharpoons H^+ + C_2H_3O_2^-$ $K_a = 1.8 \times 10^{-5}$

$$1.8 \times 10^{-5} = \frac{[H^+][[C_2H_3O_2^-]}{[HC_2H_3O_2]} = \frac{[C_2H_3O_2^-]^2}{1.0}$$

$[C_2H_3O_2^-] = 4.2 \times 10^{-3}$ Ion Product for the possible precipitate $AgC_2H_3O_2$ is: $1.0 \times 4.2 \times 10^{-3} = 4.2 \times 10^{-3}$. The Ion Product exceeds the value of K_{sp}; therefore, **a precipitate will form.**

17.52 K_{sp} $(Mg(OH)_2) = 7.1 \times 10^{-12} = [Mg^{2+}][OH^-]^2$

$$K_b\ (NH_3) = 1.8 \times 10^{-5} = \frac{[NH_4^+][OH^-]}{[NH_3]}$$

$[Mg^{2+}] = 0.10$ Therefore, $[OH^-]^2 = 7.1 \times 10^{-12}/0.10$

$[OH^-] = 8.4 \times 10^{-6}$ $NH_4^+ + OH^- \rightarrow NH_3 + H_2O$

From the dissolving of $Mg(OH)_2$ the amount of OH^- should equal twice the concentration of dissolved $Mg(OH)_2$. However, some OH^- will be consumed in the above reaction. Final concentration of $OH^- = 0.20 - X = 8.4 \times 10^{-6}$. The value of X will be the amount of NH_3 formed. $[NH_3] = 0.20$ M

$$K_b = 1.8 \times 10^{-5} = \frac{[NH_4^+][OH^-]}{[NH_3]} = \frac{[NH_4^+](8.4 \times 10^{-6})}{0.20}$$

$[NH_4^+] = 0.43$ M Moles of NH_4Cl that must be added to the 1 L are = $0.20 + 0.43 =$ **0.63 moles of NH_4Cl added.**

17.53 $K_{sp} = [Ag^+][C_2H_3O_2^-] = 2.3 \times 10^{-3} = (0.200)\ [C_2H_3O_2^-]$

$[C_2H_3O_2^-] = 1.2 \times 10^{-2}$ $HC_2H_3O_2 \rightleftharpoons H^+ + C_2H_3O_2^-$

$[H^+] = 0.10 - X$ $[HC_2H_3O_2] = X$

$$1.8 \times 10^{-5} = \frac{(0.10 - X)(1.2 \times 10^{-2})}{X}$$

$X = 9.98 \times 10^{-2}$ or 0.10 M

Amount of $NaC_2H_3O_2$ required would be
> [0.10 mol/L + 0.012 mol/L] x 0.200 L x 82 g/mol **> 1.8 g of $NaC_2H_3O_2$**

17.54 $[Ag^+] = 0.20$ M

$K_{sp} = [Ag^+][C_2H_3O_2^-] = 2.3 \times 10^{-3} = (0.20)[C_2H_3O_2^-]$

$[C_2H_3O_2^-] = 1.2 \times 10^{-2}$ M

	$HC_2H_3O_2$	$\rightleftharpoons$	H^+	+	$C_2H_3O_2^-$
Init.	$0.10 - 1.2 \times 10^{-2}$		~X		1.2×10^{-2}

$$K_a = \frac{[H^+][[C_2H_3O_2^-]}{[HC_2H_3O_2]} = 1.8 \times 10^{-5} = \frac{[H^+][1.2 \times 10^{-2}]}{0.10 - 1.2 \times 10^{-2}}$$

$[H^+] = 1.3 \times 10^{-4}$ $\qquad [HF] = [C_2H_3O_2^-] - [H^+]$

F^-	+	H^+	$\rightarrow$	HF
?		1.3×10^{-4}		1.2×10^{-2}

$HF = (1.2 \times 10^{-2}) - (1.3 \times 10^{-4})$
$= 1.2 \times 10^{-2}$

$HF \rightleftharpoons H^+ + F^-$

$$K_a = 6.5 \times 10^{-4} = \frac{[H^+][F^-]}{[HF]} = \frac{1.3 \times 10^{-4}\,[F^-]}{1.2 \times 10^{-2}}$$

$[F^-] = 6.0 \times 10^{-2}$ M

Moles F^- added $= 6.0 \times 10^{-2} + 1.2 \times 10^{-2} = 7.2 \times 10^{-2}$

7.2×10^{-2} mol L^{-1} x 0.200 L x 58.1 g/mol = **0.84 g KF or more must be added to precipitate any $AgC_2H_3O_2$.**

17.55 $NH_3 + H_2O \rightleftharpoons NH_4^+ + OH^-$ $\quad K_b = 1.8 \times 10^{-5} = \frac{[OH^-]^2}{0.10}$

$[OH^-]$ from $NH_3 = 1.3 \times 10^{-3}$ M

	$Mg(OH)_2(s)$ $\rightarrow$	$Mg^{2+}(aq)$ +	$2OH^-(aq)$
Initial		0	1.3×10^{-3}
Change		+X	+2X
Equil.		X	$2X + 1.3 \times 10^{-3}$

$K_{sp} = [Mg^{2+}][OH^-]^2 = 7.1 \times 10^{-12} = (X)(2X + 1.3 \times 10^{-3})^2$

$\approx (X)(1.3 \times 10^{-3})^2$

X = molar solubility of $Mg(OH)_2$ = 4.2×10^{-6} mol/L

17.56 Initial (after dilution): 0.40 M NH_3, 0.080 M Mn^{2+} and 0.080 M Sn^{2+}

$Mn(OH)_2 \rightleftharpoons Mn^{2+} + 2OH^-$

$K_{sp} = 1.2 \times 10^{-11} = (0.080)[OH^-]^2$

$[OH^-] = 1.2 \times 10^{-5}$

$NH_3 + H_2O \rightleftharpoons NH_4^+ + OH^-$

$[NH_3] = 0.40 - 2(0.080) = 0.24$ (after ppt. of $Sn(OH)_2$)

$$K_b = 1.8 \times 10^{-5} = \frac{[NH_4^+][1.2 \times 10^{-5}]}{0.24}$$

$[NH_4^+] = 0.36$ M

$[NH_4^+]$ added $= 0.36 - 2(0.080) = 0.20$ M

$$\frac{0.20 \text{ mol } NH_4^+}{L} \times 0.500 \text{ L} \times \frac{1 \text{ mol } NH_4Cl}{1 \text{ mol } NH_4^+} \times \frac{53.5 \text{ g}}{\text{mol } NH_4Cl} = \mathbf{5.4\ g}$$

17.57 $CuSO_4 + 4NH_3 \rightleftharpoons Cu(NH_3)_4^{2+} + SO_4^{2-}$

$CuSO_4 + 2NH_3 + 2H_2O \rightleftharpoons Cu(OH)_2 + 2NH_4^+ + SO_4^{2-}$

Assume that the maximum amount of $Cu(NH_3)_4^{2+}$ will be formed.

	Cu^{2+} +	$4NH_3$ $\rightleftharpoons$	$Cu(NH_3)_4^{2+}$
Initial	0.050	0.50	0
Change	-X	-4X	+X
Equil.	0.050 - X	0.50 - 4X	X (Assume X = 0.050)
Assume	Y	0.50 - 0.20 + 4Y	0.050 - Y ≈ 0.050

$$K_{form} = 4.8 \times 10^{12} = \frac{0.050}{Y(.30)^4}$$

$Y = [Cu^{2+}] = 1.3 \times 10^{-12}$

For formation of OH^-, look at K_b of NH_3

$$NH_3 + H_2O \rightleftharpoons NH_4^+ + OH^-$$

	NH_3	NH_4^+	OH^-
Initial (after complex)	.30	0	≈0
Change	-X	+X	+X
Equil.	.30 - X	X	X

$$K_b = 1.8 \times 10^{-5} = \frac{X^2}{.30 - X} = \frac{X^2}{.30} \qquad X = [OH^-] = 2.3 \times 10^{-3}$$

For $Cu(OH)_2$

$Cu(OH)_{2(s)}$ $\rightleftharpoons$	Cu^{2+} +	$2OH^-$
Initial	1.3×10^{-12}	2.3×10^{-3}
Change	$-1.3 \times 10^{-12} + X$	$-2(1.3 \times 10^{-12}) + 2X$
Equil.	X	$\approx 2.3 \times 10^{-3}$

$4.8 \times 10^{-20} = (X)(2.3 \times 10^{-3})^2$

$X = [Cu^{2+}] =$ **9.1×10^{-15} M**

18 ELECTROCHEMISTRY

18.1 Electrochemistry is the study of the relationships that exist between chemical reactions and the flow of electricity. An electrochemical reaction is either the nonspontaneous redox reaction caused by the passage of electricity through a system or the spontaneous redox reaction that is able to supply electricity.

18.2 Electric charge in metallic conduction involves the relatively easy movement of electrons through the metallic lattice. Electrons injected into one end of the metallic solid are able to move through the solid and eventually emerge at the other end. In electrolytic conduction, positive ions are pulled toward the negative electrode and negative ions are pulled toward the positive electrode. Conduction takes place as electrons are pulled from the negative ions at the one electrode and given to the positive ions at the other electode.

18.3 In order for electrolytic conduction to occur in an aqueous solution of an electrolyte there must be oxidation at one electrode and reduction at the other electrode.

18.4 (a) In an electrolytic cell, oxidation will take place at the anode and reduction at the cathode.
(b) In a galvanic cell oxidation will take place at the anode just as it does in the electrolytic cell and reduction takes place at the cathode.

18.5 Prepare a sketch similar to that shown in Figure 18.3, but make the following changes:
(a) Replace the Na with Mg.
(b) Replace the Na^+ with Mg^{2+}.
(c) The half-reaction at the cathode will involve two electrons.
(d) Show electrons flowing from the anode to the D.C. source and from the D.C. source to the cathode.

18.6 $2H_2O \rightarrow O_2 + 4H^+ + 4e^-$ (oxidation)
$2H_2O + 2e^- \rightarrow H_2 + 2OH^-$ (reduction)

18.7 The products would be O_2 at the anode due to the oxidation of water and H_2 at the cathode due to the reduction of the hydrogen ion.

18.8 Na_2SO_4 and H_2SO_4 are needed to maintain electrical neutrality by carrying the charge through the solution (electrolytic conduction). H^+ and OH^- ions are produced during the oxidation and reduction of water. Ions of opposite charge are required in the vicinity of these ions to "neutralize" their charge. Pure H_2O does not conduct charge efficiently enough.

18.9 anode: $2I^- \rightarrow I_2 + 2e^-$
cathode: $Ni^{2+} + 2e^- \rightarrow Ni$

cell: $Ni^{2+} + 2I^- \rightarrow Ni + I_2$
or $NiI_2(aq) \rightarrow Ni(s) + I_2(aq)$

18.10 anode: $2H_2O \rightarrow O_2 + 4H^+ + 4e^-$
cathode: $Ni^{2+} + 2e^- \rightarrow Ni$

cell: $2Ni^{2+} + 2H_2O \rightarrow 2Ni + O_2 + 4H^+$
or $2NiSO_4(aq) + 2H_2O \rightarrow 2Ni(s) + O_2(g) + 2H_2SO_4(aq)$

18.11 (a) cathode: $2H_2O + 2e^- \rightarrow H_2 + 2OH^-$
anode: $2Cl^- \rightarrow Cl_2 + 2e^-$

net: $2H_2O + 2Cl^- \rightarrow H_2 + Cl_2 + 2OH^-$

(b) In a stirred solution, Cl_2 reacts with OH^-
$Cl_2 + 2OH^- \rightarrow Cl^- + OCl^- + H_2O$
Net reaction is: $Cl^- + H_2O \rightarrow OCl^- + H_2$

18.12 Advantages of the diaphragm cell are that it prevents the formation of hypochlorite and avoids the formation of dangerous mixtures of H_2 and Cl_2. The disadvantage is that the NaOH is contaminated with NaCl

18.13 The advantage of the mercury electrolysis cell for production of NaOH and Cl_2 is that the NaOH obtained is not contaminated by Cl^-. The disadvantage is the possibility for mercury pollution.

18.14 The Downs cell is used for electrolysis of molten sodium chloride. It is designed to keep the Na and Cl_2 apart so they cannot react with each other to reform NaCl.

18.15 Cryolite reduces the melting point of Al_2O_3 from 2000°C to about 1000°C which is low enough to make the electrolysis of molten Al_2O_3 feasible.

18.16 Al cannot be produced by the electrolysis of an aqueous solution containing a salt such as $Al_2(SO_4)_3$ because H_2O is more easily reduced than Al^{3+}.

18.17 $Mg^{2+}(aq)$ (from sea water) + $2OH^-(aq) \rightarrow Mg(OH)_2(s)$

$Mg(OH)_2(s) + 2HCl \rightarrow MgCl_2(s) + 2H_2O$

$MgCl_2(\ell) \xrightarrow[\text{energy}]{\text{electrical}} Mg(\ell) + Cl_2(g)$

18.18 For the electrolytic purification of copper see discussion on copper in Section 18.3, Figure 18.13, and Margin Figure. The process is economically feasible because the impurities, silver, gold and platinum, can be sold for enough money to nearly pay for the electricity required for the electrolysis.

18.19 Electroplating is the process by which a metal is caused to be deposited ("plated out") on an electrode in an electrolysis cell. To have nickel plated onto an object in a $NiSO_4$ solution, the object must be part of the cathode because the Ni^{2+} must be reduced to Ni metal.

18.20 One faraday equals 96,500 amperes x seconds (coulombs) and is the charge of one mole of electrons.

1 mol $e^- \Leftrightarrow$ 96,500 C

1 C = 1 amp x 1 sec

18.21 A faraday is the amount of electricity equal to the charge of 1 mole of electrons (96,500 coulombs).

18.22 (a) **2** (b) **1** (c) **5** (d) **2** (e) **8**

18.23 $1\text{ C} \times \frac{1\text{ F}}{96{,}500\text{ C}} \times \frac{6.022 \times 10^{23}\ e^-}{1\text{ F}} = \mathbf{6.24 \times 10^{18}\ e^-}$

18.24 (a) **1** (b) **4** (c) **10** (d) **2** (e) **8**

18.25 (a) $8950 \text{ C} \times \frac{1 \text{ mol e}^-}{96{,}500 \text{ C}} \Leftrightarrow$ **0.0927 mol e⁻**

(b) $1.5 \text{ A} \times 30 \text{ s} \times \frac{1 \text{ C}}{\text{A} \times \text{s}} \times \frac{1 \text{ mol e}^-}{96{,}500 \text{ C}} \Leftrightarrow \mathbf{4.7 \times 10^{-4} \text{ mol e}^-}$

(c) $14.7 \text{ A} \times 10 \text{ min} \times \frac{60 \text{ s}}{\text{min}} \times \frac{1 \text{ C}}{\text{A} \times \text{s}} \times \frac{1 \text{ mol e}^-}{96{,}500 \text{ C}} \Leftrightarrow \mathbf{9.1 \times 10^{-2} \text{ mol e}^-}$

18.26 (a) $10{,}500 \text{ C} \times \frac{\text{A} \times \text{s}}{\text{C}} \times \frac{1}{25.0 \text{ A}} \times \frac{1 \text{ min}}{60.0 \text{ s}} \Leftrightarrow$ **7.00 min**

(b) $0.65 \text{ mol e}^- \times \frac{96{,}500 \text{ C}}{1 \text{ mol e}^-} \times \frac{\text{A} \times \text{s}}{\text{C}} \times \frac{1}{15 \text{ A}} \times \frac{1 \text{ min}}{60 \text{ s}} \Leftrightarrow$ **70 min**

(c) $0.20 \text{ mol } (\text{Cu}^{2+}/\text{Cu}) \times \frac{2 \text{ mol e}^-}{1 \text{ mol } (\text{Cu}^{2+}/\text{Cu})} \times \frac{96{,}500 \text{ C}}{1 \text{ mol e}^-} \times \frac{\text{A} \times \text{s}}{\text{C}} \times \frac{1}{12 \text{ A}}$

$\times \frac{1 \text{ min}}{60 \text{ s}} \Leftrightarrow$ **54 min**

18.27 (a) $\frac{84{,}200 \text{ C}}{6.30 \text{ A}} \times \frac{\text{A} \times \text{s}}{1 \text{ C}} \times \frac{1 \text{ min}}{60.0 \text{ s}} \Leftrightarrow$ **223 min**

(b) $\frac{1.25 \text{ mol e}^-}{8.40 \text{ A}} \times \frac{96{,}500 \text{ C}}{1 \text{ mol e}^-} \times \frac{\text{A} \times \text{s}}{\text{C}} \times \frac{1 \text{ min}}{60.0 \text{ s}} \Leftrightarrow$ **239 min**

(c) $\frac{0.500 \text{ mol Al}}{18.3 \text{ A}} \times \frac{3 \times 96{,}500 \text{ C}}{1 \text{ mol Al}} \times \frac{\text{A} \times \text{s}}{1 \text{ C}} \times \frac{1 \text{ min}}{60.0 \text{ s}} \Leftrightarrow$ **132 min**

18.28 (a) $10.0 \text{ mL } O_2 \text{ (STP)} \times \frac{1 \text{ mol } O_2}{22{,}400 \text{ mL } O_2 \text{ (STP)}} \times \frac{4 \text{ mol e}^-}{1 \text{ mol } O_2 \text{ (STP)}}$

$\Leftrightarrow \mathbf{1.79 \times 10^{-3} \text{ mol e}^-}$

(b) $10.0 \text{ g Al} \times \frac{1 \text{ mol Al}}{26.98 \text{ g Al}} \times \frac{3 \text{ mol e}^-}{1 \text{ mol Al}} \Leftrightarrow$ **1.11 mol e⁻**

(continued)

18.28 (continued)

(c) $5.00 \text{ g Na} \times \frac{1 \text{ mol Na}}{22.99 \text{ g Na}} \times \frac{1 \text{ mol e}^-}{1 \text{ mol Na}} \Leftrightarrow \mathbf{0.217 \text{ mol e}^-}$

(d) $5.00 \text{ g Mg} \times \frac{1 \text{ mol Mg}}{24.3 \text{ g Mg}} \times \frac{2 \text{ mol e}^-}{1 \text{ mol Mg}} \Leftrightarrow \mathbf{0.412 \text{ mol e}^-}$

18.29 $25 \text{ A} \times 8.0 \text{ hr} \times \frac{3600 \text{ s}}{1 \text{ hr}} \times \frac{1 \text{ C}}{\text{A} \times \text{s}} \times \frac{F}{96{,}500 \text{ C}} \times \frac{1 \text{ mol Na}}{F} \times \frac{23.0 \text{ g Na}}{\text{mol Na}} \Leftrightarrow \mathbf{170 \text{ g Na}}$

Similar calculation yields: **260 g Cl_2** (only 2 significant figures)

18.30 $0.50 \text{ A} \times 1.0 \text{ hr} \times \frac{1 \text{ C}}{\text{A} \times \text{s}} \times \frac{3600 \text{ s}}{1 \text{ hr}} \times \frac{F}{96{,}500 \text{ C}} \times \frac{1 \text{ mol } O_2}{4F}$

$\times \frac{32.0 \text{ g } O_2}{\text{mol}} \Leftrightarrow \mathbf{0.15 \text{ g } O_2}$

Similar calculation yields: **0.019 g H_2**

$0.15 \text{ g } O_2 \times \frac{22.4 \text{ L } O_2 \text{ (STP)}}{32.00 \text{ g } O_2} = \mathbf{0.10 \text{ L } O_2 \text{ (STP)}}$

Similar calculation yields: **0.21 L H_2 (STP)**

18.31 $115 \text{ A} \times 8.00 \text{ hr} \times \frac{1 \text{ C}}{\text{A} \times \text{s}} \times \frac{3600 \text{ s}}{1 \text{ hr}} \times \frac{F}{96{,}500 \text{ C}} \times \frac{1 \text{ mol Cu}}{2F} \times \frac{63.55 \text{ g Cu}}{\text{mol}}$

$\Leftrightarrow$ **1,090 g Cu** (3 significant figures)

18.32 (a) $8.00 \text{ hr} \times 8.46 \text{ A} \times \frac{1 \text{ C}}{\text{A} \times \text{s}} \times \frac{3600 \text{ s}}{1 \text{ hr}} \times \frac{F}{96{,}500 \text{ C}} \times \frac{1 \text{ mol Ag}}{F}$

$\times \frac{107.9 \text{ g Ag}}{\text{mol}} \Leftrightarrow \mathbf{272 \text{ g Ag}}$

(b) $\frac{272 \text{ g Ag}}{0.00254 \text{ cm Ag}} \times \frac{1 \text{ cm}^3 \text{ Ag}}{10.5 \text{ g Ag}} \Leftrightarrow \mathbf{1.02 \times 10^4 \text{ cm}^2}$

18.33 $\frac{21.4 \text{ g Ag}}{10.0 \text{ A}} \times \frac{\text{A} \times \text{s}}{1 \text{ C}} \times \frac{96{,}500 \text{ C}}{F} \times \frac{1 F}{\text{mol Ag}} \times \frac{1 \text{ mol Ag}}{107.9 \text{ g Ag}} \Leftrightarrow \mathbf{1910 \text{ s}}$

(3 significant figures)

18.34 $\frac{35.3 \text{ g Cr}}{6.00 \text{ A}} \times \frac{\text{A}\bullet\text{s}}{1 \text{ C}} \times \frac{3\,F}{1 \text{ mol Cr}^{3+}} \times \frac{96{,}500 \text{ C}}{F} \times \frac{1 \text{ mol Cr}}{52.00 \text{ g Cr}} \times \frac{1 \text{ hr}}{3600 \text{ s}} \Leftrightarrow \mathbf{9.10\ hr}$

18.35 $\frac{5.00 \text{ g Cu}}{5.00 \text{ A}} \times \frac{\text{A} \times \text{s}}{1 \text{ C}} \times \frac{2 \text{ mol e}^-}{63.55 \text{ g Cu}} \times \frac{96{,}500 \text{ C}}{\text{mol e}^-} \times \frac{1 \text{ min}}{60 \text{ s}} \Leftrightarrow \mathbf{50.6\ min}$

18.36 $\frac{0.225 \text{ g Ni}}{10.0 \text{ min}} \times \frac{2 \text{ mol e}^-}{58.71 \text{ g Ni}} \times \frac{96{,}500 \text{ C}}{\text{mol e}^-} \times \frac{1 \text{ min}}{60 \text{ s}} \times \frac{\text{A} \times \text{s}}{1 \text{ C}} \Leftrightarrow \mathbf{1.23\ A}$

18.37 $\frac{1.33 \text{ g Cl}_2}{45.0 \text{ min}} \times \frac{2 \text{ mol e}^-}{70.9 \text{ g Cl}_2} \times \frac{96{,}500 \text{ C}}{\text{mol e}^-} \times \frac{1 \text{ min}}{60 \text{ s}} \times \frac{\text{A} \times \text{s}}{1 \text{ C}} \Leftrightarrow \mathbf{1.34\ A}$

18.38 $\frac{50.0 \text{ mL O}_2 \text{ (STP)}}{3.00 \text{ hr}} \times \frac{\text{mol O}_2}{22{,}400 \text{ mL O}_2 \text{ (STP)}} \times \frac{4 \text{ mol e}^-}{\text{mol O}_2} \times \frac{96{,}500 \text{ C}}{\text{mol e}^-}$

$\times \frac{1 \text{ hr}}{3600 \text{ s}} \times \frac{\text{A} \times \text{s}}{1 \text{ C}} \Leftrightarrow \mathbf{0.0798\ A}$

18.39 $0.500 \text{ L} \times \frac{0.270 \text{ mol Cr}_2(\text{SO}_4)_3}{1 \text{ L}} \times \frac{6 \text{ mol e}^-}{1 \text{ mol Cr}_2(\text{SO}_4)_3} \times \frac{96{,}500 \text{ C}}{\text{mol e}^-}$

$\times \frac{\text{A} \times \text{s}}{\text{C}} \times \frac{1}{3.00 \text{ A}} \times \frac{1 \text{ min}}{60 \text{ s}} \Leftrightarrow \mathbf{434\ min}$

18.40 (a) $1.25 \text{ g Cu} \times \frac{1 \text{ mol Cu}}{63.55 \text{ g Cu}} \times \frac{2 \text{ mol e}^-}{\text{mol Cu}} \Leftrightarrow \mathbf{0.0393\ mol\ e^-}$

(b) $\frac{3.42 \text{ g X}}{0.0393 \text{ mol e}^-} \times \frac{2 \text{ mol e}^-}{1 \text{ mol X}} \Leftrightarrow \mathbf{174\ g\ X/mol\ X}$

18.41 $0.125 \text{ mol Cu} \times \frac{2 \text{ mol e}^-}{1 \text{ mol Cu}} \times \frac{1 \text{ mol Cr}}{3 \text{ mol e}^-} \Leftrightarrow \mathbf{0.0833\ mol\ Cr}$

18.42 $$\frac{0.250\text{ A}}{0.400\text{ L}} \times 35.0\text{ min} \times \frac{\text{C}}{\text{A x s}} \times \frac{60\text{ s}}{1\text{ min}} \times \frac{1\text{ mol e}^-}{96{,}500\text{ C}}$$

$$\times \frac{1\text{ mol OH}^-}{\text{mol e}^-} \Leftrightarrow \frac{0.0136\text{ mol (OH}^-)}{\text{L}} = 0.0136\text{ M OH}^-$$

pOH = 1.866 **pH = 12.134**

18.43 The electron flow takes place on the surface of the zinc. Electrons are removed from the zinc and are picked up by the Cu^{2+} ions that are in the vicinity. The energy change appears as heat.

18.44 The salt bridge is needed in order to maintain electrical neutrality. Cations from the salt bridge move into one compartment to compensate for the excess negative charge while anions move into the other compartment to compensate for the excess positive charge there.

18.45 In galvanic cells the anode is negative and the cathode is positive. In electrolytic cells the anode is positive and the cathode is negative.

18.46 Draw a sketch like that in Figure 18.14 (a) and make the following changes:
(a) Replace the Zn and Zn^{2+} with Fe and Fe^{2+}.
(b) Replace the Cu and Cu^{2+} with Ni and Ni^{2+}.
(c) Label the iron electrode as the anode and the nickel as the cathode.
(d) Indicate the iron electrode as being negative and the nickel electrode as being positive.
(e) Show the flow of electrons as being from the iron electrode to the nickel electrode.
(f) Show cations flowing from the iron compartment through the salt bridge to the nickel compartment and anions flowing in the reverse direction.
(g) The cell reaction is $Ni^{2+} + Fe \rightarrow Ni + Fe^{2+}$ and if the ions are all 1 M, the cell potential is $E°_{cell} = 0.19$ V.

18.47 In a galvanic cell, the negative electrode is the anode; the positive electrode is the cathode. These can be determined with a voltmeter. Another method would be to do a chemical analysis of the products formed at each electrode. Oxidation occurs at the anode, reduction at the cathode.

18.48 (a) Fe(s) | Fe^{2+} (1M) || Ni^{2+} (1 M) | Ni(s)
(b) Cr(s) | Cr^{3+} (1 M) || Sn^{2+} (1 M) | Sn(s)
(c) Cu(s) | Cu^{2+} (1 M) || Ag^+ (1 M) | Ag(s)

18.49 (a) $Zn(s) + Pb^{2+} (1M) \rightarrow Pb(s) + Zn^{2+} (1\ M)$
Zn is the anode; Pb is the cathode.
(b) $Al(s) + 3Ag^{+} (1\ M) \rightarrow Al^{3+} (1\ M) + 3Ag(s)$
Al is the anode; Ag is the cathode.
(c) $3Mn(s) + 2Fe^{3+} (1\ M) \rightarrow 3Mn^{2+} (1\ M) + 2Fe(s)$
Mn is the anode; Fe is the cathode.

18.50 See Figure 18.15 (a) for a sketch of a hydrogen electrode. When the hydrogen gas is at 1.00 atm and hydrogen ion concentration is 1.00 M, the hydrogen electrode will have a potential of exactly 0.000V.

18.51 (a) (anode) $Zn(s) \rightarrow Zn^{2+} + 2e^-$ (cathode) $Ga^{3+} + 3e^- \rightarrow Ga(s)$
(b) $3Zn(s) + 2Ga^{3+} \rightarrow 3Zn^{2+} + 2Ga(s)$
(c) $E_{cell} = E^\circ_{Ga^{3+}|Ga} - E^\circ_{Zn^{2+}|Zn} = E^\circ_{Ga^{3+}|Ga} - (-0.76\ V) = 0.23\ V$
$E^\circ_{Ga^{3+}|Ga} = \mathbf{-0.53\ V}$
(d) $E^\circ_{cell} = 0.34\ V - (-0.53\ V) = \mathbf{0.87\ V}$
(One must assume $Cu^{2+} + 2e^- \rightarrow Cu$)

18.52 (a) Ca^{2+} (b) F_2 (c) H_2O (d) $S_2O_8^{2-}$ (e) Br_2

18.53 (a) ClO_3^- (b) $Cr_2O_7^{2-}$ (c) MnO_4^- (d) PbO_2

18.54 (a) Fe ($Fe \rightarrow Fe^{2+}$ not Fe^{3+}) (b) Mg (c) I^- (d) SO_4^{2-} (e) Mn

18.55 (a) Na (b) Cl_2 (c) Cu (d) Sn (e) H_2

18.56 (a) $E_{cell} = -0.25\ V - (-0.44\ V) = \mathbf{0.19\ V}$
(b) $E_{cell} = -0.14\ V - (-0.74\ V) = \mathbf{0.60\ V}$
(c) $E_{cell} = 0.80\ V - (+0.34\ V) = \mathbf{0.46\ V}$

18.57 (a) $E_{cell} = -0.13\ V - (-0.76\ V) = \mathbf{0.63\ V}$
(b) $E_{cell} = 0.80\ V - (-1.67\ V) = \mathbf{2.47\ V}$
(c) $E_{cell} = -0.04\ V - (-1.03\ V) = \mathbf{0.99\ V}$

18.58 (a) $2Al(s) + 3NiSO_4(aq) \rightarrow 3Ni(s) + Al_2(SO_4)_3(aq)$
(b) $3PbO_2(s) + H_2O + Cr_2(SO_4)_3(aq) + K_2SO_4(aq) \rightarrow$
$K_2Cr_2O_7(aq) + 3PbSO_4(s) + H_2SO_4(aq)$
(c) $2AgNO_3(aq) + Pb(s) \rightarrow 2Ag(s) + Pb(NO_3)_2(aq)$
(d) $Cl_2(g) + MnCl_2(aq) + 2H_2O \rightarrow 4HCl(aq) + MnO_2(s)$
(e) $2HCl(aq) + Mn(s) \rightarrow MnCl_2(aq) + H_2(g)$

18.59 Spontaneous reactions are (a), (d), (e). The other reactions are spontaneous in the reverse direction.

18.60 (a) E° = -2.76 - (-2.38) = **-0.38 V** non-spontaneous as written
(b) E° = -0.13 - (1.36) = **-1.49 V** non-spontaneous as written
(c) E° = 2.00 - (1.36) = **+0.64 V** spontaneous
(d) E° = 1.33 - (1.49) = **-0.16 V** non-spontaneous as written
(e) E° = 1.23 - (1.36) = **-0.13 V** non-spontaneous as written

18.61 (a) $Pb(s) + SO_4^{2-} + Hg_2Cl_2(s) \rightarrow 2Hg(\ell) + 2Cl^- + PbSO_4(s)$
E° = 0.27 - (-0.36) = **0.63 V**
(b) $2Ag(s) + 2Cl^- + Cu^{2+} \rightarrow 2AgCl(s) + Cu(s)$ **E° = 0.12 V**
(c) $Mn(s) + Cl_2(g) \rightarrow Mn^{2+} + 2Cl^-$ **E° = 2.39 V**
(d) $2Al(s) + 3Br_2(\ell) \rightarrow 2Al^{3+} + 6Br^-$ **E° = 2.76 V**

18.62 (a) E° = -0.25 V - (-1.67 V) = **1.42 V**
(b) E° = 1.69 V - (1.33 V) = **0.36 V**
(c) E° = 0.80 V - (-0.13 V) = **0.93 V**
(d) E° = 1.36 V - (1.28 V) = **0.08 V**
(e) E° = 0.00 V - (-1.03 V) = **1.03 V**

18.63 (a) E° = 0.77 V - (-0.14 V) = **0.91 V**
(b) E° = 0.00 V - (0.34 V) = **-0.34 V**
(c) E° = -2.38 V - (-1.67 V) = **-0.71 V**
(d) E° = -0.76 V - (-1.03 V) = **0.27 V**
(e) E° = 1.69 V - (0.27 V) = **1.42 V**

18.64 (a) $K_c = [Ni^{2+}]/[Sn^{2+}]$ At equilibrium
$\log K_c = n E°/0.0592$ (at 25° C) n = 2 E° = -0.14 V - (-0.25 V) = 0.11 V
K_c = **5 x 10^3**
(b) n = 2 E° = 1.36 V - (1.09 V) = 0.27 V K_c = **1 x 10^9**
(c) n = 1 E° = 0.80 V - (0.77 V) = 0.03 V K_c = **3**

18.65 (a) E° = 0.77 V - (-0.14 V) = 0.91 V
$\log K_c = nE°/0.0592$ (at 25° C) n = 2
$\log K_c = 2 \times 0.91/0.0592 = 30.74$
K_c = **5 x 10^{30}**
(b) K_c = **3 x 10^{-12}** (c) K_c = **1 x 10^{-72}**
(d) K_c = **1 x 10^9** (e) K_c = **1 x 10^{48}**

18.66 $\log K_c = n E°/0.0592$ (assume 25° C)

(a) n = 2 E° = -2.76 V - (-2.38 V) = -0.38 V $\log K_c = -12.84 \approx -13.$
K_c = **1 x 10^{-13}**

(b) E° = -1.49 V n = 2 K_c = **5 x 10^{-51}**

(c) E° = 0.64 V n = 2 K_c = **1 x 10^{22}**

(d) E° = -0.16 V n = 30 K_c = **1 x 10^{-81}**

(e) E° = -0.13 V n = 4 K_c = **2 x 10^{-9}**

18.67 ΔG° = -nF E°

(a) $\Delta G° = -2 \text{ mol x } \frac{96{,}500 \text{ C}}{1 \text{ mol e}^-} \text{ x } 0.91 \text{ V x } \frac{10^{-3} \text{ kJ}}{1 \text{ V x C}} = \mathbf{-180 \text{ kJ}}$ (2 sign. fig.)

(b) ΔG° = -2 x 96,500 x (-0.34) x 10^{-3} = **66 kJ**

(c) ΔG° = -6 x 96,500 x (-0.71) x 10^{-3} = **410 kJ**

(d) ΔG° = -2 x 96,500 x 0.27 x 10^{-3} = **-52 kJ**

(e) ΔG° = -2 x 96,500 x 1.42 x 10^{-3} = **-274 kJ**

18.68 ΔG° = -nF E° (a) E° = -2.76 V - (-2.38 V) = -0.38 V n = 2

$\Delta G° = -2 \text{ mol e}^- \text{ x } \frac{96{,}500 \text{ C}}{1 \text{ mol e}^-} \text{ x } (-0.38 \text{ V}) \text{ x } \frac{10^{-3} \text{ kJ}}{1 \text{ V x C}} = \mathbf{73 \text{ kJ}}$

(b) E° = -1.49 V n = 2 ΔG° = -2 x 96,500 x (-1.49) x 10^{-3} = **288 kJ**

(c) E° = 0.64 V n = 2 ΔG° = -2 x 96,500 x 0.64 x 10^{-3} = **-1.2 x 10^2 kJ**

(d) E° = -0.16 V n = 30 ΔG° = -30 x 96,500 x (-0.16) x 10^{-3} = **4.6 x 10^2 kJ**

(e) E° = -0.13 V n = 4 ΔG° = -4 x 96,500 x (-0.13) x 10^{-3} = **50 kJ**

18.69 Nernst Equation: $\mathbf{E = E° - \frac{0.0592}{n} \text{ x log (mass action expression)}}$

(a) E° = 0.34 V - (-0.76 V) = **1.10 V**
E = 1.10 V - (0.0592/2) log {$[Zn^{2+}]/[Cu^{2+}]$} = **1.07 V**

(b) E° = - 0.14 V - (-0.25 V) = **0.11V**
E = 0.11 V - (0.0592/2) log {$[Ni^{2+}]/[Sn^{2+}]$} = **0.16 V**

(c) E° = 2.87 V - (-3.05 V) = **5.92 V**
E = 5.92 V - (0.0592/2) log $\left([Li^+]^2 [F^-]^2/p_{F_2}\right)$ = **5.94 V**

(d) E° = 0.00 V - (-0.76 V) = **0.76 V**
E = 0.76 V - (0.0592/2) log $\left([Zn^{2+}] p_{H_2}/[H^+]^2\right)$ = **0.64 V**

(e) E° = 0.00 V - (-0.44 V) = **0.44 V**
E = 0.44 V -(0.0592/2) log $\left(p_{H_2}[Fe^{2+}]/[H^+]^2\right)$ = **0.46 V**

18.70 (a) $E^\circ = -0.25\ V - (-1.67\ V) = \mathbf{1.42\ V}$

$2Al + 3Ni^{2+} \rightarrow 2Al^{3+} + 3Ni$

$E = 1.42\ V - (0.0592/6)\ \log[(0.020)^2/(0.80)^3] = \mathbf{1.45\ V}$

$\Delta G = -nF\,E = -6 \times 96{,}500 \times 1.45 \times 10^{-3} = \mathbf{-840\ kJ}$

(b) $E^\circ = -0.14\ V - (-0.25\ V) = \mathbf{0.11 V}$

The equation is balanced as given.

$E = 0.11\ V - (0.0592/2)\ \log[(0.010)/(1.10)] = \mathbf{0.17\ V}$

$\Delta G = -2 \times 96{,}500 \times 0.17 \times 10^{-3} = \mathbf{-33\ kJ}$

(c) $E^\circ = +0.80\ V - (-0.76\ V) = 1.56\ V$

$2Ag^+ + Zn \rightarrow 2Ag + Zn^{2+}$

$E = 1.56\ V - (0.0592/2)\ \log\ [(0.010)/(0.050)^2] = \mathbf{1.54\ V}$

$\Delta G = -2 \times 96{,}500 \times 1.54 \times 10^{-3} = \mathbf{-297\ kJ}$

18.71 (a) $E = [-0.13\ V - (-0.14\ V)] - (0.0592/2)\ \log\ [(1.50)/(0.050)] = \mathbf{-0.03\ V}$

(b) $E = [-0.74\ V - (-0.76\ V)] - (0.0592/6)\ \log\ [(0.020)^3/(0.010)^2] = \mathbf{0.03\ V}$

(c) $E = (1.69\ V - 0.34\ V) - (0.0592/2)\ \log\ [(0.0010)/(0.010)\ (0.10)^4] = \mathbf{1.26\ V}$

18.72 $E = 0.34\ V - (0.0592/2)\ \log\ [1/(2 \times 10^{-4})] = \mathbf{0.23\ V}$

18.73 $E^\circ = 1.33\ V - (1.49\ V) = -0.16\ V$

$\Delta G = \Delta G^\circ + 2.303\ RT\ \ \log\ ([MnO_4^-]^6\ [Cr^{3+}]^{10}/\{[Mn^{2+}]^6\ [Cr_2O_7^{2-}]^5\ [H^+]^{22}\})$

$$\Delta G^\circ = -nF\,E^\circ = -30\ \text{mol e}^- \times \frac{96{,}500\ C}{1\ \text{mol e}^-} \times (-0.16\ V) \times \frac{10^{-3}\ kJ}{1\ V \times C} = 460\ kJ$$

$$\Delta G = 460\ kJ + 2.303 \times 8.314\ J \times K^{-1} \times 298\ K \times \frac{1\ kJ}{10^3\ J}$$

$$\times \log \frac{(0.0010)^6 (0.0010)^{10}}{(0.10)^6\ (0.010)^5\ (1.0 \times 10^{-6})^{22}} = 460\ kJ + 5.71\ kJ \times \log 10^{100} = \mathbf{1.0 \times 10^3\ kJ}$$

Since the reaction as written involves free energy increase, the spontaneous reaction would be the reverse reaction.

18.74 $E = E^\circ - (0.0592/n)\log(\text{mass act. exp.})$ $\quad 1/2H_2 + Ag^+ \rightarrow H^+ + Ag$

$E = (0.80\ V - 0\ V) - (0.0592/1)\ \log\ \{[H^+]/[Ag^+]\ p_{H_2}^{1/2}\}$

$K_{sp} = 5 \times 10^{-15} = [Ag^+]\ [Br^-]$ $\quad [Ag^+] = 5 \times 10^{-15}/0.010 = 5 \times 10^{-13}$

$E = 0.80\ V - 0.0592 \times \log\ (1/(5 \times 10^{-13}) = \mathbf{0.07\ V}$

18.75 Anode: Cl^- (?M) + Ag → AgCl + 1 e^-

Cathode: AgCl + 1 e^- → Ag + Cl^- (1 M)

Cl^- (? M) → Cl^- (1 M) A Concentration Cell

E = 0.0435 V E° = 0 E = E° - (0.0592/1) log {(1)/$[Cl^-]_A$}

0.0435 V/(0.0592 V) = - log (1/$[Cl^-]_A$) $[Cl^-]_A$ = **5.43 M**

18.76 Mg(s) + $2Ag^+$ → Mg^{2+} + 2Ag(s) E = 0.80 V - (-2.38 V) = 3.18 V

(a) E = 3.18 V - 0.0296 log [(0.100)/$(0.100)^2$] = **3.15 V**

(b) $? \text{ mol } Ag^+ = 1.00 \text{ g Ag} \times \frac{1 \text{ mol } Ag^+}{107.9 \text{ g } Ag^+} = 9.27 \times 10^{-3} \text{ mol } Ag^+$

$? \text{ mol } Mg^{2+} \Leftrightarrow 9.27 \times 10^{-3} \text{ mol Ag} \times \frac{1 \text{ mol } Mg^{2+}}{2 \text{ mol } Ag^+} \Leftrightarrow 4.64 \times 10^{-3} \text{ mol } Mg^{2+}$

$M_f = M_i + \Delta M \quad 0.100 \text{ M } (Ag^+) - \frac{9.27 \times 10^{-3} \text{ mol } Ag^+}{0.200 \text{ L } Ag^+ \text{ soln}} = 0.054 \text{ M } Ag^+$

$0.100 \text{ M } (Mg^{2+}) + \frac{4.64 \times 10^{-3} \text{ mol } Mg^{2+}}{0.250 \text{ L } Mg^{2+} \text{ soln}} = 0.119 \text{ M } (Mg^{2+})$

$E = 3.18 \text{ V} - 0.0296 \log \frac{(0.119)}{(0.054)^2} = \mathbf{3.13 \text{ V}}$

(c) $? \text{ mol } Mg^{2+} = 0.080 \text{ g Mg} \times \frac{1 \text{ mol } Mg^{2+}}{24.3 \text{ g Mg}} = 3.3 \times 10^{-3} \text{ mol } Mg^{2+}$

$3.3 \times 10^{-3} \text{ mol } Mg^{2+} \times \frac{2 \text{ mol } Ag^+}{1 \text{ mol } Mg^{2+}} \Leftrightarrow 6.6 \times 10^{-3} \text{ mol } Ag^+$

$0.100 \text{ M } Ag^+ - \frac{6.6 \times 10^{-3} \text{ mol } Ag^+}{0.200 \text{ L } Ag^+ \text{ soln}} = 0.067 \text{ M } Ag^+ \text{(final)}$

$0.100 \text{ M } (Mg^{2+}) - \frac{3.3 \times 10^{-3} \text{ mol } Mg^{2+}}{0.250 \text{ L } Mg^{2+} \text{ soln}} = 0.113 \text{ M } Mg^{2+} \text{(final)}$

$E = 3.18 \text{ V} - 0.0296 \log \frac{(0.113)}{(0.067)^2} = \mathbf{3.14 \text{ V}}$

18.77 $25.0 \text{ g Pb} \times \frac{2 \times 96{,}500 \text{ C}}{207.2 \text{ g Pb}} \times \frac{1.5 \text{ V}}{25 \text{ W}} \times \frac{1 \text{ W}}{\text{V} \times \text{C} \times \text{s}^{-1}} \times \frac{1 \text{ hr}}{3600 \text{ s}} = \mathbf{0.39 \text{ hr}}$

18.78 $\Delta G° = \Delta H° - T\Delta S° = -242\ kJ - 383\ K \times (188.7 - 130.6 - 205.0/2) \times 10^{-3}\ kJ/K = -225\ kJ$

$$1.0 \times 10^3\ W \times \frac{1\ J}{W \times s} \times \frac{100\%}{70\%} \times \frac{2.016\ g\ H_2}{mol\ H_2} \times \frac{1\ mol\ H_2}{225 \times 10^3\ J} = \mathbf{0.013\ g\ H_2/s\ (reacting)}$$

0.10 g O_2/s

18.79 $5.00\ min \times 110\ V \times 1.00\ A \times \frac{1\ W}{V \times A} \times \frac{1\ kJ}{10^3\ W \times s} \times \frac{60\ s}{1\ min} = \mathbf{33.0\ kJ}$

18.80 $C_{12}H_{26}(\ell) + 37/2 O_2(g) \rightarrow 12CO_2(g) + 13H_2O(g)$

$\Delta H° = [12 \times \Delta H°_f(CO_2) + 13 \times \Delta H°_f(H_2O(g))] - [\Delta H°_f(C_{12}H_{26}(\ell))] =$
$[12 \times (-394) + 13 \times (-242) - (291)]\ kJ = -8165\ kJ$

$$1.0\ kW\ hr \times \frac{1\ kJ}{1\ kW \times s} \times \frac{3600\ s}{1\ hr} \times \frac{1\ mol\ (C_{12}H_{26})}{8165\ kJ} \times \frac{170\ g\ (C_{12}H_{26})}{1\ mol\ C_{12}H_{26}}$$

$$\times \frac{10^{-3}\ L}{0.74\ g} \times \frac{100\%}{30\%} = \mathbf{0.34\ L\ (C_{12}H_{26})}$$

18.81 $E = E° - \frac{0.0592}{2} \log \frac{[Zn^{2+}]}{[Pb^{2+}]}$

$$0.4438\ V = 0.6365\ V - \frac{0.0592}{2} \log \frac{.500}{[Pb^{2+}]}$$

$[Pb^{2+}] = \mathbf{1.6 \times 10^{-7}\ M}$

18.82 $E = E° - \frac{0.0592}{3} \log \frac{[Ga^{3+}]}{[Ag^+]^3}$

$$1.122\ V = 1.360\ V - \frac{0.0592}{3} \log \frac{.800}{[Ag^+]^3}$$

$[Ag^+] = \mathbf{9 \times 10^{-5}}$

18.83 (anode) $Zn(s) \rightarrow Zn^{2+} + 2e^-$

(cathode) $2MnO_2(s) + 2NH_4^+ + 2e^- \rightarrow Mn_2O_3(s) + 2NH_3 + H_2O$

18.84 The products formed at the electrodes diffuse away and the cell is rejuvenated.

18.85 (anode) $Zn(s) + 2OH^- \rightarrow Zn(OH)_2(s) + 2e^-$

(cathode) $2MnO_2(s) + 2H_2O + 2e^- \rightarrow 2MnO(OH)(s) + 2OH^-$

---

net: $Zn(s) + 2MnO_2(s) + 2H_2O \rightarrow Zn(OH)_2(s) + 2MnO(OH)(s)$

Electrolyte is aqueous KOH.

18.86 (anode) $Zn(s) + 2OH^-(aq) \rightarrow ZnO(s) + H_2O + 2e^-$

(cathode) $HgO(s) + H_2O + 2e^- \rightarrow Hg(\ell) + 2OH^-(aq)$

---

net: $Zn(s) + HgO(s) \rightarrow ZnO(s) + Hg(\ell)$

18.87 (anode) $Zn(s) + 2OH^-(aq) \rightarrow Zn(OH)_2(s) + 2e^-$

(cathode) $Ag_2O(s) + H_2O + 2e^- \rightarrow 2Ag(s) + 2OH^-(aq)$

---

net: $Zn(s) + Ag_2O(s) + H_2O \rightarrow Zn(OH)_2(s) + 2Ag(s)$

18.88 During discharge:

(anode) $Pb(s) + SO_4^{2-}(aq) \rightarrow PbSO_4(s) + 2e^-$

(cathode) $PbO_2(s) + 4H^+(aq) + SO_4^{2-}(aq) + 2e^- \rightarrow PbSO_4(s) + 2H_2O(\ell)$

When being charged, the reactions are reversed, i. e., forced to the left.

18.89 (anode) $Cd(s) + 2OH^-(aq) \rightarrow Cd(OH)_2(s) + 2e^-$

(cathode) $NiO_2(s) + 2H_2O + 2e^- \rightarrow Ni(OH)_2(s) + 2OH^-(aq)$

---

net: $Cd(s) + NiO_2(s) + 2H_2O \rightarrow Cd(OH)_2(s) + Ni(OH)_2(s)$

18.90 Fuel cells produce electrical energy from fuels undergoing reaction in a carefully designed environment. The reactants in a fuel cell may be continually supplied so that energy can be withdrawn as long as the outside fuel supply is maintained.

18.91 Fuel cells operate under more nearly reversible conditions. Therefore, their thermodynamic efficiency is higher; i.e., more of the available energy can be used to do work.

18.92 $\frac{15.5 \text{ mL HCl}}{25.0 \text{ min}} \times \frac{1 \text{ min}}{60 \text{ s}} \times \frac{0.250 \text{ mol HCl}}{10^3 \text{ mL HCl}} \times \frac{1 \text{ mol OH}^-}{1 \text{ mol HCl}} \times \frac{2 \text{ mol e}^-}{2 \text{ mol OH}^-}$

$$\times \frac{96{,}500 \text{ C}}{\text{mol e}^-} \times \frac{\text{A x s}}{\text{C}} \Leftrightarrow \mathbf{0.249\ A}$$

18.93

	Init	STP
P	767 -27 torr	760 torr
V	288 mL	? mL
T	300 K	273 K

$288 \text{ mL} \times \frac{740 \text{ torr} \times 273 \text{ K}}{760 \text{ torr} \times 300 \text{ K}} = \mathbf{255\ mL\ (STP)}$

$$\frac{1.22 \text{ C x s}^{-1} \text{ x } 30.0 \text{ min}}{255 \text{ mL } H_2 \text{ (STP)}} \times \frac{60 \text{ s}}{1 \text{ min}} \times \frac{11{,}200 \text{ mL } H_2}{6{,}022 \times 10^{23} \text{ e}^-} \Leftrightarrow 1.60 \times 10^{-19} \text{ C/e}^-$$

18.94 $1 \text{ m}^2 \text{ Cr (plate)} \times 0.050 \text{ mm} \times \frac{1 \text{ m}}{1000 \text{ mm}} \times \frac{(100 \text{ cm})^3}{\text{m}^3} \times \frac{7.19 \text{ g Cr}}{1 \text{ cm}^3 \text{ Cr}} \times \frac{6 \times 96{,}500 \text{ A s}}{52.00 \text{ g Cr (plate)}}$

$$\times \frac{1}{25 \text{ min}} \times \frac{1 \text{ min}}{60 \text{ s}} \Leftrightarrow \mathbf{3 \times 10^3\ A}$$

(1 significant figure)

18.95 $\frac{[H^+][C_2H_3O_2^-]}{[HC_2H_3O_2]} = 1.8 \times 10^{-5} = \frac{[H^+]^2}{0.10}$ $[H^+]^2 = 1.8 \times 10^{-6}$

$2H^+ + Fe \rightarrow H_2 + Fe^{2+}$ $E = E° - (0.0592/n) \log \{[Fe^{2+}]/[H^+]^2\}$

$E = [(0.00 \text{ V} - (-0.44 \text{ V})] - (0.0296) \log \{(0.10)/(1.8 \times 10^{-6})\} = \mathbf{0.30\ V}$

18.96 $E = E° - (0.0592/n) \log[Pb^{2+}]$; $0.51 \text{ V} = [0.00 - (-0.13)] - (0.0592/2) \log [Pb^{2+}]$

$[Pb^{2+}] = 1 \times 10^{-13}$ $K_{sp} = [Pb^{2+}][CrO_4^{2-}] = (1 \times 10^{-13})(0.10) = \mathbf{1 \times 10^{-14}}$

18.97 $Zn(s) + 2Ag^+ (0.0500\ M) \rightarrow Zn^{2+} (0.100\ M) + 2Ag(s) \qquad 45°C$
(from Appendix C)

$\Delta H° = -153.9\ kJ + 0\ kJ - (0\ kJ + 105.58\ kJ) = -259.48 \approx -259.5\ kJ$

$\Delta S° = -112.1\ J/K + 42.55\ J/K - (41.6\ J/K + 72.68\ J/K) = -183.83 \approx -183.8\ J/K$

$$\Delta G°_{318} = -259.5\ kJ - (318\ K)(-183.8\ J/K)\left(\frac{kJ}{10^3\ J}\right) = -259.5\ kJ + 58.4\ kJ = -201.1\ kJ$$

$$E°_{318} = \frac{\Delta G°_{318}}{-nF} = \frac{(-201.1\ kJ)\left(\frac{10^3 J}{kJ}\right)}{-(2)\ (96{,}500\ C)\left(\frac{J}{C \times V}\right)} = 1.04\ V$$

$$E_{318} = E°_{318} - \frac{(2.303)(8.314\ J/mol{\bullet}K)(318\ K)}{(2)(96{,}500\ J/V{\bullet}mol)} \log \frac{(0.100)}{(0.0500)^2}$$

$E = 1.04\ V - 0.05\ V =$ **0.99 V**

18.98 $Sn^{2+}(aq) + Pb(s) \rightleftharpoons Pb^{2+}(aq) + Sn(s) \qquad 50°C = 323\ K$
(from Appendix C)

$\Delta H° = -1.7\ kJ + 0\ kJ - (-8.8\ kJ + 0\ kJ) = +\ 7.1\ kJ$

$\Delta S° = 10.5\ J/K + 51.6\ J/K - (-17\ J/K + 64.8\ J/K) = 14.3 \approx 14\ J/K$

$$\Delta G°_{323} = 7.1\ kJ - (323\ K)(14\ J/K)\left(\frac{kJ}{10^3\ J}\right) = 7.1\ kJ - 4.5\ kJ = 2.6\ kJ$$

$\Delta G°_{323} = -2.303\ RT \log K_c$

$$\log K_c = \frac{\Delta G°_{323}}{-2.303\ RT} = \frac{(2.6\ kJ)\left(\frac{10^3 J}{kJ}\right)}{-2.303\ (8.314\ J/K)\ (323\ K)} = -\ 0.42$$

$K_c =$ **0.38**

19 CHEMICAL KINETICS: THE STUDY OF THE RATES OF CHEMICAL REACTIONS

19.1 (a) the nature of reactants and products (b) concentration of reacting species (c) temperature (d) influence of outside agents (catalysts)

19.2 The smaller the particle size (i.e., the larger the total surface area), the faster the reaction.

19.3 Methods used to study the rate of a reaction must be fast, accurate and not interfere with the normal course of the reaction.

19.4 Some common examples:
(a) combustion of gas (rapid)
(b) explosion of gasoline vapor in auto engine (very fast)
(c) digestion of food (moderately slow)
(d) iron rusting (slow)
(e) decay of leaves (slow)

19.5 We say that reaction rate depends upon the nature of the reactants because some substances react rapidly with one another by the nature of the reaction taking place while others react more slowly. In other words, even under conditions of equal concentrations and temperature, different chemical reactions progress at different rates.

19.6 (a) magnesium (b) zinc in 1.00 M HCl
(c) powdered zinc (d) iron nail in 1.00 M HCl at 40°C

19.7 Reaction rate = the speed at which reactants are consumed or the products are formed. It is the ratio of the change in concentration to the change in time units, such as mol liter^{-1} s^{-1}.

19.8 (a) $\text{Rate} = \frac{-\Delta[H_2]}{2\Delta t} = \frac{-\Delta[O_2]}{\Delta t} = \frac{\Delta[H_2O]}{2\Delta t}$

(b) $\text{Rate} = \frac{-\Delta[NOCl]}{2\Delta t} = \frac{\Delta[NO]}{2\Delta t} = \frac{\Delta[Cl_2]}{\Delta t}$

(c) $\text{Rate} = \frac{-\Delta[NO]}{\Delta t} = \frac{-\Delta[O_3]}{\Delta t} = \frac{\Delta[NO_2]}{\Delta t} = \frac{\Delta[O_2]}{\Delta t}$

(d) $\text{Rate} = \frac{-\Delta[H_2O_2]}{\Delta t} = \frac{-\Delta[H_2]}{\Delta t} = \frac{\Delta[H_2O]}{2\Delta t}$

(a) H_2 disappears twice as fast as O_2, and at the same rate that H_2O appears.
(b) NOCl disappears at the same rate as NO appears and twice as fast as Cl_2 appears.
(c) As NO and O_3 disappear, NO_2 and O_2 appear at the same rate.
(d) H_2O_2 and H_2 disappear half as fast as H_2O appears.

19.9 $CH_4 + 2O_2 \rightarrow CO_2 + 2H_2O$

$$\text{rate (for } CO_2) = \frac{0.16 \text{ mol } CH_4}{\text{L x s}} \text{ x } \frac{1 \text{ mol } CO_2}{1 \text{ mol } CH_4} = \mathbf{0.16 \text{ mol } CO_2 \text{ L}^{-1} \text{ s}^{-1}}$$

$$\text{rate (for } H_2O) = \frac{0.16 \text{ mol } CH_4}{\text{L x s}} \text{ x } \frac{2 \text{ mol } H_2O}{1 \text{ mol } CH_4} = \mathbf{0.32 \text{ mol } H_2O \text{ L}^{-1} \text{ s}^{-1}}$$

19.10 $4NH_3 + 3O_2 \rightarrow 2N_2 + 6H_2O$

(a) $\text{rate for water being formed} = \frac{0.68 \text{ mol } N_2}{\text{L x s}} \text{ x } \frac{6 \text{ mol } H_2O}{2 \text{ mol } N_2}$

$$= \mathbf{2.0 \text{ mol } H_2O \text{ L}^{-1} \text{ s}^{-1}}$$

(continued)

19.10 (continued)

(b) rate for NH_3 reacting $= \frac{0.68 \text{ mol } N_2}{\text{L x s}} \text{ x } \frac{(-4 \text{ mol } NH_3)}{2 \text{ mol } N_2}$

$$= \mathbf{-1.4 \text{ mol } NH_3 \text{ } L^{-1} s^{-1}}$$

(c) rate for O_2 being consumed $= \frac{0.68 \text{ mol } N_2}{\text{L x s}} \text{ x } \frac{(-3 \text{ mol } O_2)}{2 \text{ mol } N_2}$

$$= \mathbf{-1.0 \text{ mol } O_2 \text{ } L^{-1} s^{-1}}$$

19.11 $2A \rightarrow 4B + C$

From your graph, slope of the disappearance of A at 25 min =

$$\frac{\Delta[A]}{\Delta t} \approx \mathbf{-9.6 \text{ x } 10^{-3} \text{ mol A } L^{-1} \text{ min}^{-1}}$$

The slope for the rate of formation B at 25 min =

$$\frac{\Delta[B]}{\Delta t} \approx \mathbf{1.9 \text{ x } 10^{-2} \text{ mol B } L^{-1} \text{ min}^{-1}}$$

The slope of the disappearance of A at 40 min =

$$\frac{\Delta[A]}{\Delta t} \approx \mathbf{-6.4 \text{ x } 10^{-3} \text{ mol A } L^{-1} \text{ min}^{-1}}$$

The slope of the formation of B at 40 min =

$$\frac{\Delta[B]}{\Delta t} \approx \mathbf{1.3 \text{ x } 10^{-2} \text{ mol B } L^{-1} \text{ min}^{-1}}$$

Rate of B = -2 x rate of A

$\frac{\Delta[C]}{\Delta t}$ at 25 min $\approx \mathbf{4.8 \text{ x } 10^{-3} \text{ mol C mol}^{-1} \text{ } L^{-1}}$

$\frac{\Delta[C]}{\Delta t}$ at 40 min $\approx \mathbf{3.2 \text{ x } 10^{-3} \text{ mol C } L^{-1} \text{ min}^{-1}}$

19.12 A rate law is an experimentally determined relationship between the rate of reaction and the concentrations of the reactants. Temperature and catalysts affect the value of the rate constant.

19.13 The sum of exponents on the concentrations in the rate law is the order of a reaction.

19.14 (a) s^{-1} (b) liter mol^{-1} s^{-1} (c) $liter^2$ mol^{-2} s^{-1}

19.15 As the concentration of CO varies, the rate at which it is removed from the earth's atmosphere by fungi remains constant (no concentration dependence); therefore, it appears to be a zero-order reaction.

19.16 (a) First order with respect to A and B, overall order is two.
(b) Second order with respect to E, overall order is two.
(c) Second order with respect to G and H, overall order is four.

19.17 (a) liter/mol s (b) liter/mol s (c) $liter^3/mol^3$ s

19.18 (a) The rate will double. (b) The rate will increase fourfold.
(c) The rate will increase eightfold. (d) The rate will increase sixteenfold.
(e) The rate will increase by a factor of $2^{1/2}$ or 1.4.
(f) The rate will decrease by a factor of 4 or (2^{-2} = 1/4 or 0.25).

19.19 -1; i.e., Rate = $k[A]^{-1}$

19.20 rate = 1.63×10^{-1} L $mol^{-1}s^{-1}$ x [ICl][H_2]
(a) rate = (1.63×10^{-1} L $mol^{-1}s^{-1}$)(0.25 mol L^{-1})(0.25 mol L^{-1})
= **1.0×10^{-2} mol L^{-1} s^{-1}**

(b) rate = **2.0×10^{-2} mol L^{-1} s^{-1}** (c) rate = **4.1×10^{-2} mol L^{-1} s^{-1}**

19.21 (a) rate = (2.35×10^{-6} L^2 mol^{-2} s^{-1})(1.00 mol L^{-1})^2(1.00 mol L^{-1})

= **2.35×10^{-6} mol L^{-1} s^{-1}**

(b) rate = (2.35×10^{-6} L^2 mol^{-2} s^{-1})(0.250 mol L^{-1})^2(1.30 mol L^{-1})

= **1.91×10^{-7} mol L^{-1} s^{-1}**

19.22 (a) Prepare the requested graph. From the graph one can obtain:
$Rate_{(t = 500)}$ = **~2.5×10^{-3} mol L^{-1} s^{-1}**
$Rate_{(t = 1000)}$ = **~1.6×10^{-3} mol L^{-1} s^{-1}**
$Rate_{(t = 1500)}$ = **~1.3×10^{-3} mol L^{-1} s^{-1}**

(continued)

19.22 (continued)

(b) rate = k $[N_2O_5]$

$k_{500} = \frac{2.5 \times 10^{-3}}{3.52} = \mathbf{7.1 \times 10^{-4} s^{-1}}$ $k_{1000} = \frac{1.6 \times 10^{-3}}{2.48} = \mathbf{6.5 \times 10^{-4} s^{-1}}$

$k_{1500} = \frac{1.3 \times 10^{-3}}{1.75} = \mathbf{7.4 \times 10^{-4} s^{-1}}$ $k_{average} = \mathbf{7.0 \times 10^{-4} s^{-1}}$

19.23 (a) rate = $k[NO_2]^x[O_3]^y$ In the first and second experiments, the initial NO_2 concentration is constant, the initial concentration of O_3 is doubled, and the rate is doubled. Therefore, the value of y must be 1. In the second and third experiments, the initial concentration of O_3 is constant, the initial concentration of NO_2 is halved, and the rate is halved. Therefore, the value of x must be 1. **Rate = $k[NO_2][O_3]$**

(b) **k** = Rate/$[NO_2][O_3]$ = 0.022 mol L^{-1} s^{-1}/
[(5.0 x 10^{-5} mol L^{-1})(1.0 x 10^{-5} mol L^{-1})] = **4.4 x 10^7 L mol^{-1} s^{-1}**

19.24 (a) Rate = $k[A]^x[B]^y$ From experiments 1 and 2, x = 1
From experiments 3 and 4, y = 2.

Rate of formation of C = $k[A][B]^2$

(b) **k** = Rate/$[A][B]^2$ = 1.20 x 10^{-3} mol $L^{-1}s^{-1}$/
[(0.010 mol L^{-1})(0.010 mol $L^{-1})^2$] = **1.2 x 10^3 L^2 mol^{-2} s^{-1}**

(c) Rate of formation of C = 1.2 x 10^3 L^2 mol^{-2} s^{-1} x
(0.020 mol L^{-1})(0.060 mol $L^{-1})^2$ = 8.6 x 10^{-2} mol L^{-1} s^{-1}

$\frac{8.6 \times 10^{-2} \text{ mol}}{\text{L x s}} \times \frac{2 \text{ mol D}}{\text{mol C}} = \mathbf{1.7 \times 10^{-1} \text{ mol D } L^{-1} s^{-1}}$

19.25 (a) Rate = $k[NOCl]^x$ When the concentration was doubled, the rate changed by a factor of 1.44 x 10^{-8}/3.60 x 10^{-9} or 4.00. Therefore, x must be 2.

Rate = $k[NOCl]^2$

(b) **k** = Rate/$[NOCl]^2$ = 3.60 x 10^{-9} mol L^{-1} s^{-1}/(0.30 mol $L^{-1})^2$
= **4.0 x 10^{-8} L mol^{-1} s^{-1}**

(c) $(0.45/0.30)^2$ = **2.2 times faster**

19.26 (a) Rate = $k[NO]^x[Cl_2]^y$
From experiments 1 and 2, one can see that when [NO] is held constant while the $[Cl_2]$ is doubled, the rate also doubles. Therefore, y = 1. From experiments 1 and 3, one can see that when the $[Cl_2]$ is held constant while the [NO] is doubled, the rate quadruples. Therefore, x = 2. **Rate = $k[NO]^2[Cl_2]$**

(b) k = rate/$[NO]^2[Cl_2]$ = 2.53 x 10^{-6} mol L^{-1} s^{-1}/[(0.10 mol $L^{-1})^2$(0.10 mol L^{-1})]
= **2.5 x 10^{-3} L^2 mol^{-2} s^{-1}**

19.27 The half-life of a reaction is the time required for the concentration of a given reactant to be decreased by a factor of 2 (i.e., to half of its initial value).

19.28 The half-life of a first-order reaction is unaffected by the concentrations of the reactants.

19.29 See Figure 19.4, except each $t_{1/2}$ is twice the length of the preceding $t_{1/2}$.

19.30 (a) $\ln \frac{[A]_0}{[A]_t} = kt$ $\qquad \ln \frac{(4.50 \times 10^{-3})}{[A]_t} = (1.46 \times 10^{-1}\ s^{-1})\ (20.0\ s)$

$\ln \frac{(4.50 \times 10^{-3})}{[A]_t} = 2.92$ $\qquad \frac{(4.50 \times 10^{-3})}{[A]_t} = 18.54 \approx 19.$

$[A]_t$ = **2.4 x 10^{-4} mol/L**

(b) $t_{1/2} = \frac{0.693}{k} = \frac{0.693}{1.46 \times 10^{-1} s^{-1}} = \mathbf{4.75\ s}$

(c) $(4.50 \times 10^{-3}\ M)\ (1/2)^3 = 4.50 \times 10^{-3}\ M \times 1/8 = \mathbf{5.62 \times 10^{-4}\ M}$

19.31 $t_{1/2} = \frac{1}{k[B]_0}$

$2.11\ \text{min} = \frac{1}{k(0.10\ M)}$

$k = 1/[(0.10\ M)\ (2.11\ \text{min})]$

$k = 4.7\ M^{-1}\ \text{min}^{-1}$

$t_{1/2} = \frac{1}{(4.7\ M^{-1}\ \text{min}^{-1})\ (0.010\ M)} = \mathbf{21\ min}$

19.32 (See Question 19.31)

$k = 4.7\ M^{-1}\ \text{min}^{-1} \times \frac{M}{mol\ L^{-1}} \times \frac{\text{min}}{60\ s} = \mathbf{7.8 \times 10^{-2}\ L\ mol^{-1}\ s^{-1}}$

19.33 (a) $t_{1/2} = 0.693/k = 0.693/3.2 \times 10^{-2}\ s^{-1} =$ **22 s**

(b) $\ln \frac{[A]_0}{[A]_t} = kt$ $\qquad \ln \frac{[A]_0}{(0.010\ M)} = (3.2 \times 10^{-2}\ s^{-1})\ (60\ s) = 1.92$

$\frac{[A]_0}{(0.010\ M)} = 6.8$ $\qquad [A]_0 = 6.8 \times 10^{-2}$ M or **0.068 M**

19.34 The rate of a reaction is proportional to the number of collisions per second between reacting molecules. As the number of molecules is increased (increase in concentration), the number of collisions is increased by the same factor.

19.35 An overall chemical reaction represents the net chemical change. This does not mean that all the reactants come together simultaneously. To predict the rate law, a reaction mechanism must be known. From the mechanism, reaction order is known. The reaction order can also be obtained experimentally and from that a rate law can be written.

19.36 A reaction mechanism is a series of elementary processes that lead to the formation of the products.

19.37 A one-step mechanism would involve the simultaneous collision of six molecules, five of which would have to be O_2. This is very improbable.

19.38 (a) Rate = $k[NO][Br_2]$

(b) Through the second step the net reaction is:

$2NO + Br_2 \rightarrow 2NOBr$ $\qquad$ Rate = $k[NO]^2[Br_2]$

(c) Step 2 must be rate determining (i.e., the slow step)

(d) This is a termolecular collision, which is very unlikely for a fairly rapid reaction.

(e) No. A mechanism is only theory, and more information can support it or prove it wrong, but the actual path can never be known with complete certainty.

19.39 The rate law is Rate = $k[NO_2]^2$. CO does not appear in the rate-determining step (slow step), and, therefore, CO does not affect the rate and the reaction is said to be zero-order with respect to CO.

19.40 $(CH_3)_3CBr \rightarrow (CH_3)_3C^+ + Br^-$ $\quad$ slow

$(CH_3)_3C^+ + OH^- \rightarrow (CH_3)_3COH$ $\quad$ fast

for which Rate = $k[(CH_3)_3CBr]$

19.41 (a) $2A + B \rightarrow C + 2D$ (b) Rate = $k[A]^2$
(c) Rate = $k[A]^2[B]$

19.42 $NO_2 + O_3 \rightarrow NO_3 + O_2$ slow
$NO_3 + NO_2 \rightarrow N_2O_5$ fast

19.43 The observed rate of reaction is much smaller (for example, approximately 5×10^{12} for $2HI(g) \rightarrow H_2(g) + I_2(g)$) than what is expected if all the collisions were effective. A minimum amount of energy is required to cause a reaction to occur, and the molecules must collide with the proper orientation.

19.44 When two molecules collide, reaction may or may not ocur. Only properly oriented collisions with sufficient energy may result in effective collisions to produce a product different than the reactants.

19.45

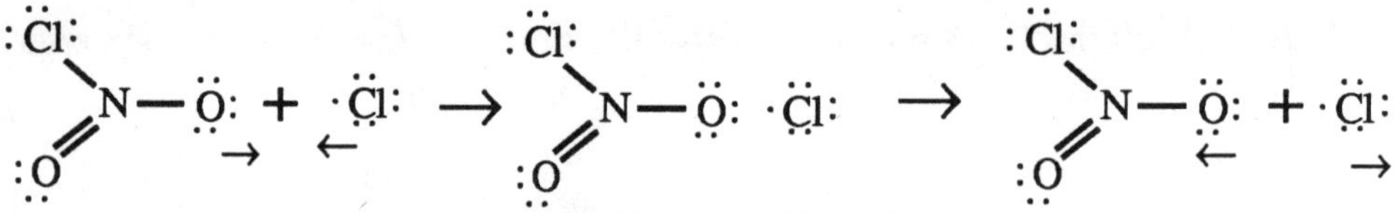

$Cl(O{=})N{-}O + \cdot Cl \rightarrow Cl(O{=})N{-}O \cdots Cl \rightarrow Cl(O{=})N{-}O + \cdot Cl$

(not an effective collision)

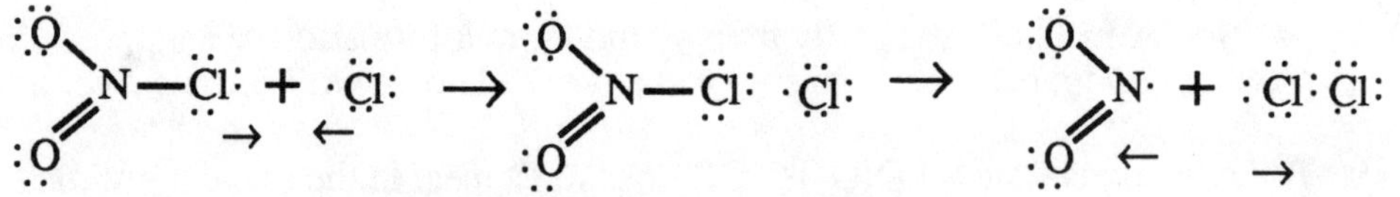

$O(O{=})N{-}Cl + \cdot Cl \rightarrow O(O{=})N{-}Cl \cdots Cl \rightarrow O(O{=})N\cdot + Cl{:}Cl$

(an effective collision)

19.46 $NO + O_2 \rightleftharpoons NO_3$ fast
$NO_3 + NO \rightarrow 2NO_2$ slow

19.47 The **activation energy** represents the kinetic energy required to bring the reactants to the point where they can react to form products. It is the minimum kinetic energy that must be available in a collision.

19.48 Activation energy is the minimum energy required for reaction. As T is increased, more molecules have greater kinetic energy, i.e., a larger fraction of molecules will have the minimum energy required to overcome the activation energy barrier during a collision.

19.49 Reactions, including biochemical ones, slow down at low temperatures primarily because a smaller fraction of molecules possesses the required activation energy.

19.50 Draw a potential energy diagram like that in Figure 19.8. Label the axes, E_a (forward), E_a (reverse) and ΔH reaction.

19.51 The activated complex is the intermediate species that is highly unstable, transient and very reactive. The activated complex exists in a transition state between products and reactants. The transition state is on the barrier (high potential energy) of the potential energy curve.

19.52 $\ln\left(\frac{k_1}{k_2}\right) = \frac{E_a}{R}\left(\frac{1}{T_2} - \frac{1}{T_1}\right)$

$$\ln (0.163/0.348) = \frac{E_a}{8.314 \text{ J mol}^{-1} \text{ K}^{-1}}\left[\frac{1}{513 \text{ K}} - \frac{1}{503 \text{ K}}\right]$$

$$-0.759 = (E_a/8.314 \text{ J mol}^{-1})(0.001949 - 0.001988)$$

$$\mathbf{E_a} = 1.62 \times 10^5 \text{ J mol}^{-1} = \mathbf{163\ kJ/mol}$$

$$k = Ae^{-E_a/RT}$$

$$A = \frac{k}{e^{-E_a/RT}} = ke^{E_a/RT}$$

$$A = 0.163 \text{ L mol}^{-1} \text{ s}^{-1} \times e^{[1.63 \times 10^5 \text{ J mol}^{-1}/(8.314 \text{ J mol}^{-1} \text{ K}^{-1})(503 \text{ K})]}$$

$$= 0.163 \text{ L mol}^{-1} \text{ s}^{-1}\ e^{39.0} = 0.163 \text{ L mol}^{-1} \text{ s}^{-1} \times 8.66 \times 10^{16}$$

$$\mathbf{A = 1.41 \times 10^{16}\ L\ mol^{-1}\ s^{-1}}$$

19.53 (See Problem 19.52)

$$\ln\left(\frac{k_1}{k_2}\right) = \frac{E_a}{R}\left(\frac{1}{T_2} - \frac{1}{T_1}\right)$$

$\ln(1.32 \times 10^{-2}/1.64) = (E_a/8.314\ \text{J mol}^{-1}\text{K}^{-1})[(1/548\ \text{K}) - (1/473\ \text{K})]$

$-4.822 = (E_a/8.314\ \text{J mol}^{-1}\ \text{K}^{-1})(0.001825 - 0.002114)$

$E_a = (-4.822)(8.314\ \text{J mol})/(-0.000289)$

$\mathbf{E_a} = 1.387 \times 10^5\ \text{J mol}^{-1} = \mathbf{139\ kJ\ mol^{-1}}$

$k = Ae^{-E_a/RT}$ $\qquad A = \dfrac{k}{e^{-E_a/RT}} = ke^{E_a/RT}$

$A = (1.32 \times 10^{-2}\ \text{L mol}^{-1}\ \text{s}^{-1})\ e^{1.39 \times 10^5\ \text{J mol}^{-1}/(8.314\ \text{J mol}^{-1}\ \text{K}^{-1})(473\ \text{K})}$

$\mathbf{A} = (1.32 \times 10^{-2}\ \text{L mol}^{-1}\ \text{s}^{-1})(e^{35.3}) = \mathbf{3 \times 10^{13}\ \ L\ mol^{-1}\ s^{-1}}$

19.54 $\ln\left(\dfrac{k_1}{k_2}\right) = \dfrac{E_a}{R}\left(\dfrac{1}{T_2} - \dfrac{1}{T_1}\right)$

$$\ln\left(\frac{1.57 \times 10^{-3}\ \text{L mol}^{-1}\ \text{s}^{-1}}{k_2}\right) = \frac{182 \times 10^3\ \text{J mol}^{-1}}{8.314\ \text{J mol}^{-1}\ \text{K}^{-1}}\left(\frac{1}{873\ \text{K}} - \frac{1}{973\ \text{K}}\right)$$

$\ln(1.57 \times 10^{-3}\ \text{L mol}^{-1}\ \text{s}^{-1}/k_2) = 2.58$

$$\frac{1.57 \times 10^{-3}\ \text{L mol}^{-1}\ \text{s}^{-1}}{k_2} = e^{2.58} = 13$$

$$k_2 = \frac{1.57 \times 10^{-3}\ \text{L mol}^{-1}\ \text{s}^{-1}}{13} = \mathbf{1.2 \times 10^{-4}\ \ L\ mol^{-1}\ s^{-1}}$$

19.55 $\ln\left(\frac{k_1}{k_2}\right) = \frac{E_a}{R}\left(\frac{1}{T_2} - \frac{1}{T_1}\right)$

$$\ln\left(\frac{1.32 \times 10^{-2}\ \text{L mol}^{-1}\ \text{s}^{-1}}{k_2}\right) = \frac{138 \times 10^{3}\ \text{J mol}^{-1}}{8.314\ \text{J mol}^{-1}\ \text{K}^{-1}}\ [(1/573\ \text{K}) - (1/473\ \text{K})]$$

$\ln(1.32 \times 10^{-2}\ \text{L mol}^{-1}\ \text{s}^{-1}/k_2) = -6.12$

$$\frac{1.32 \times 10^{-2}\ \text{L mol}^{-1}\ \text{s}^{-1}}{k_2} = e^{-6.12} = 0.0022$$

$\mathbf{k_2 = 6.0\ L\ mol^{-1}\ s^{-1}}$

19.56 $\ln\left(\frac{k_1}{k_2}\right) = \frac{E_a}{R}\left(\frac{1}{T_2} - \frac{1}{T_1}\right)$ $\qquad \frac{k_1}{k_2} = \frac{1}{4}$

$$\ln(1/4) = \frac{E_a}{8.314\ \text{J mol}^{-1}\ \text{K}^{-1}}\ [(1/373\ \text{K}) - (1/303\ \text{K})]$$

$\mathbf{E_a}$ $= 1.86 \times 10^4\ \text{J mol}^{-1} =$ $\mathbf{18.6\ kJ\ mol^{-1}}$

19.57 $\ln\left(\frac{k_1}{k_2}\right) = \frac{E_a}{R}\left(\frac{1}{T_2} - \frac{1}{T_1}\right)$

$$\ln\frac{(3.2 \times 10^{-2})}{(9.3 \times 10^{-2})} = \frac{E_a}{8.314\ \text{J mol}^{-1}\ \text{K}^{-1}}\ [(1/848\ \text{K}) - (1/823\ \text{K})]$$

$$\frac{(-1.1)\ (8.314\ \text{J mol}^{-1})}{(0.001179 - 0.001215)} = E_a = 2.48 \times 10^{5}\ \text{J mol}^{-1} = \mathbf{250\ kJ\ mol^{-1}}$$

19.58 Recall the equation: $\ln k = \ln A - \frac{E_a}{RT}$

If one plots ln k vs. 1/T, the slope will be equal to $-E_a/R$. Don't forget to use kelvins! The value of E_a was determined to be **1.6×10^2 kJ** by this method.

19.59 (See Problem 19.58)

$$\ln k = \ln A - \frac{E_a}{R}\left(\frac{1}{T}\right)$$

k is proportional to 1/time, therefore:

$$\ln \frac{1}{t} = \ln A - \frac{E_a}{R}\left(\frac{1}{T}\right)$$

Plot ln (1/t) vs. 1/T (T = kelvins)

$$\text{Slope} = -\frac{E_a}{R}$$

E_a = 64 kJ mol^{-1}

Estimated by extrapolation, when 1/T = 1/288 = 3.47 x 10^{-3}, ln 1/t = -2.64 or

ln t = 2.64.

Therefore, it was estimated that **time = ~13 min.**

19.60 A homogeneous catalyst is in the same phase as the reactants while a heterogeneous catalyst is in a different phase (e.g., a catalytic surface in contact with reacting gases).

19.61 (a) A **heterogeneous catalyst** is a substance that provides a low energy pathway (E_a) to the products, but is not in the same phase as the reactants. These catalysts appear to adsorb reactant molecules and certain bonds within the reactants are weakened or broken.
(b) An **inhibitor** interferes with the effectiveness of a catalyst by interfering with adsorption.

19.62 Catalytic converters on automobile mufflers are poisoned by the lead in leaded gasoline. Once poisoned, the converters will not function to reduce automobile pollution.

19.63 The catalyst lowers the activation energy by giving the reactants a different pathway (mechanism) for the chemical reaction. This path has a lower E_a, thereby increasing the number of effective collisions.

19.64 (a) no effect
(b) no effect
(c) lowers activation energy by changing the nature of the transition state.

19.65 Free radicals are extremely reactive intermediate substances consisting of atoms or groups of atoms that possess unpaired electrons. They are formed either thermally or by absorption of photons of appropriate frequencies.

19.66 Recall $E = h\nu$ and $\lambda = \frac{c}{\nu}$

$h = 6.63 \times 10^{-34}$ J s, $c = 3.00 \times 10^{8}$ m s^{-1}, 1nm = 10^{-9} m

$$\nu = \frac{E}{h} = \frac{348{,}000 \text{ J mol}^{-1} \times (1 \text{ mol}/6.022 \times 10^{23} \text{ molecules}) \times (1 \text{ molecule})}{6.63 \times 10^{-34} \text{ J s}}$$

$$= 8.72 \times 10^{14} \text{ s}^{-1} = \mathbf{8.72 \times 10^{14}} \textbf{ Hz}$$

$$\lambda = \frac{c}{\nu} = \frac{(3.00 \times 10^{8} \text{ m s}^{-1})(10^{9} \text{ nm m}^{-1})}{8.72 \times 10^{14} \text{ s}^{-1}} = \mathbf{344 \text{ nm}}$$

19.67 Free radicals are dangerous in living organisms because they are so reactive and may cause aging and cancer.

19.68 In chain reactions a single reactive intermediate (radical) produces many product molecules before termination. Therefore, products are produced at a rate faster than the initiation step alone.

19.69 (a) initiation step 1 (b) propagation steps 3, 5, 6 (c) termination step 2

19.70 rate = $k[B]^{\underline{x}}$
log (rate) = log ($k[B]^{\underline{x}}$)
log (rate) = log k+ log $[B]^{\underline{x}}$
log (rate) = log k+ $\underline{x}$ log [B] Compare the equation to: y = b + mx. A plot of log (rate) or y versus log[B] or x will yield a straight line that has **a slope of $\underline{x}$** or m and an intercept of log k or b.

19.71 $t = \# + 4$ Where # = no. of chirps in 8 seconds

(a) $20 = \# + 4$ no. of chirps = **16** at 20°C
$25 = \# + 4$ no. of chirps = **21** at 25°C
$30 = \# + 4$ no. of chirps = **26** at 30°C
$35 = \# + 4$ no. of chirps = **31** at 35°C

(b) E_a = slope x (-R) = (-3.9 x 10^3) x (-8.314 x 10^{-3} kJ/moles) = **32kJ/mole**

(c) t(°C) = # + 4; 120°C = # + 4
# of chirps = **116** (From the graph, a somewhat smaller value was calculated. However, in reality, the value is probably very small since the cricket is unlikely to survive 8 sec at 120°C.)

19.72 $\ln\left(\frac{k_1}{k_2}\right) = \frac{E_a}{R}\left(\frac{1}{T_2} - \frac{1}{T_1}\right)$

T_1 = boiling point of water at 760 torr = 100°C

T_2 = boiling point of water at 355 torr = 80°C (See Table 11.2)

k is proportional to 1/time

$$\ln [(1/t_1)/(1/t_2)] = \frac{418 \times 10^3 \text{ J mol}^{-1}}{8.314 \text{ J mol}^{-1} \text{ K}^{-1}} [(1/353 \text{ K}) - (1/373 \text{ K})]$$

$\ln (t_2 / 3.0 \text{ min}) = 7.64$

$t_2 = 6.2 \times 10^3$ min or over **100 hours**

19.73 $2O_3 + h\nu \rightarrow 3O_2$
Since this is a chain reaction, small amounts of intermediate can bring about the destruction of very large amounts of ozone.

19.74 A graph of the data in Exercise 19.11 shows that a plot of log[A] versus time is a curved line while a plot of 1/[A] against time gives a straight line. Therefore, the reaction is **second order.**
k = 2.5 x 10^{-2} L $mol^{-1} min^{-1}$

20 METALS AND THEIR COMPOUNDS; THE REPRESENTATIVE METALS

20.1 Most metals are found in the combined state. Sources of metals are the ocean and land-based deposits of metal carbonates, sulfates, oxides, and sulfides.

20.2 An ore is a material that contains a desirable constituent in sufficiently high concentration that its extraction from the ore is economically worthwhile.

20.3 The three steps are: concentration, reduction, and refining.

20.4 Gold is about nine times as dense as sand and mud and,as a result, is not as easily washed away by the swirling action of the pan.

20.5 An amalgam is a solution of a metal in mercury. Gold ore is mixed with mercury, which dissolves the metallic gold. The mercury is then separated from the stone and distilled, leaving the pure gold behind.

20.6 In flotation, the ore is finely ground and added to a mixture of oil and water. A stream of air is then blown through the mixture and the oil-covered mineral is carried to the surface where it can be removed. In roasting, a sulfide ore is heated in air, converting the metal sulfide to an oxide that is more conveniently reduced.

20.7 (a) $Al_2O_3 + 2OH^- \rightarrow 2AlO_2^- + H_2O$

(b) $AlO_2^- + H_2O + H^+ \rightarrow Al(OH)_3$

(c) $2Al(OH)_3 \xrightarrow{heat} Al_2O_3 + 3H_2O$

20.8 In compounds metals exist in positive oxidation states and, therefore, must be reduced to obtain the metal in a pure metallic form.

20.9 It must have a small or negative ΔH°_f so that the reaction proceeds to completion at a reasonable temperature, i.e., that ΔG°_T is negative.

20.10 It has a large, negative $\Delta H°_f$ and, as a result, will not decompose except at very high temperatures.

20.11 $2Ag_2O \rightarrow 4Ag + O_2$

ΔH (reaction) = [(4 x 0) + (1 x 0)] - [2 mol x (-30.5 kJ/mol)] = 61.0 kJ

ΔS (reaction) = -2 mol x ΔS_f = -2 mol x (-66.1 J/mol K) = 132.2 J/K

$\Delta G = \Delta H - T\Delta S$ $\quad\quad$ $\Delta G = -RT \ln K$ $\quad$ (if K = 1, then $\Delta G = 0$)

When K = 1, 0 = $\Delta H - T\Delta S$ $\quad\quad$ 0 = 61.0 kJ - T(132.2 J/K)

T = (61.0 x 10^3/132.2)K = **461 K** $\quad\quad$ **188°C**

20.12 $2Au_2O_3 \rightarrow 4Au + 3O_2$

ΔH = [(4 x 0) + (3 x 0)] - [2 mol x 80.8 kJ/mol] = -161.6 kJ

ΔS = [(4 mol x 47.7 J/mol•K) + (3 mol x 205 J/mol•K)] - [2 mol x 125 J/mol•K]
= 555.8 J/k = 0.5558 kJ/K

$\Delta G = \Delta H - T\Delta S$ $\quad$ ΔG = - at K > 1 $\quad$ ΔG = -161.6 kJ - T(0.5558 kJ/K)

ΔG will equal a negative value at all values of T since T can only have a positive value.

20.13 $ZnO(s) \rightarrow Zn(s) + 1/2\ O_2(g)$ $\quad\quad$ $\Delta H° = +348$ kJ

$\Delta S°$ = (41.8 + 1/2 x 205.0 - 43.5) J K^{-1} = 100.8 J K^{-1}

$\Delta G° = \Delta H° - T\Delta S°$ = 348 kJ - T(100.8 x 10^{-3} kJ K^{-1}) = 0

T = 3450 K (or **3180°C**)

20.14 $\Delta G° = -2.303\ RT\ \log K_p$, $\quad$ $\Delta G° = \Delta H° - T\Delta S°$

When $K_p = 1$, then $\Delta G° = 0$ and $\Delta H° = T\Delta S°$ $\quad\quad$ $\Delta H°$ = 155 kJ

$\Delta S° = S°(Cu(s)) + 1/2\ S°(O_2(g)) - S°(CuO(s))$ = 33.3 J K^{-1} +
1/2 x 205.0 J K^{-1} - 43.5 J K^{-1} = 92.3 J K^{-1}

T = 1.55 x 10^5 J/92.3 J K^{-1} = 1680 K = **1410°C** (3 sign. figs.)

20.15 $\Delta G° = -RT \ln K_p$ and $\Delta G° = \Delta H° - T\Delta S°$; $\Delta G° = 754.4\text{ kJ} - T\left(\frac{257.9\text{ J}}{\text{K}}\right)$

$\Delta G°_{373} = 658.2$ kJ, $\Delta G°_{773} = 555$ kJ, $\Delta G°_{2273} = 168.2$ kJ

$\ln K_p = -\Delta G°/RT$ $K_p = e^{-\Delta G°/RT}$

$K_{p(373)} = 7 \times 10^{-93}$ $K_{p(773)} = 3 \times 10^{-38}$

$K_{p(2273)} = 1.36 \times 10^{-4}$

20.16 (a) $3Fe_2O_3 + CO \rightarrow 2Fe_3O_4 + CO_2$
$Fe_3O_4 + CO \rightarrow 3FeO + CO_2$
$FeO + CO \rightarrow Fe + CO_2$

(b) $CaCO_3 \xrightarrow{\text{heat}} CaO + CO_2$
$CaO + SiO_2 \rightarrow CaSiO_3$

20.17 It is plentiful and least expensive.

20.18 $2PbO + C \rightarrow 2Pb + CO_2$
$PbO + H_2 \rightarrow Pb + H_2O$

20.19 The chemical reducing agent itself would be even more difficult to prepare.

20.20 They tend to have relatively low melting points.

20.21 $2NaCl(\ell) \xrightarrow{\text{electrolysis}} 2Na(\ell) + Cl_2(g)$

$2Al_2O_3(\ell) \xrightarrow[\text{cryolite}]{\text{electrolysis}} 4Al(\ell) + 3O_2(g)$

20.22 To remove impurities and to lower the carbon content. The Bessemer converter, open hearth furnace and the basic oxygen process are described in Section 20.1 (Figures 20.4 and 20.5). The basic oxygen process is the principal steel-making method today.

20.23 The Mond process is used in the refining of nickel. In this process impure nickel is treated with carbon monoxide at moderately low temperatures. The $Ni(CO)_4$ gas that is produced is then separated and heated to 200°C and decomposes to give pure nickel plus CO.

20.24 The lower the electronegativity, the more metallic the element. Metallic character decreases left to right and increases from top to bottom. Lithium is more metallic than Be, which is more metallic than B: the rest are nonmetallic. Carbon is less metallic than Si, which is less metallic than Ge, etc. (In Group IVA, C is a nonmetal and Pb is a metal.)

20.25 (a) Li (b) Al (c) Cs (d) Sn (e) Ga

20.26 Ga_2O_3

20.27 Al_2O_3

20.28 Amphoteric - able to function as either an acid or base.

$2Al + 6H^+ \rightarrow 2Al^{3+} + 3H_2$

$2Al + 2OH^- + 2H_2O \rightarrow 2AlO_2^- + 3H_2$

$Be + 2H^+ \rightarrow Be^{2+} + H_2$

$Be + 2H_2O + 2OH^- \rightarrow Be(OH)_4^{2-} + H_2$

20.29 Many are amphoteric and form compounds with some covalent properties.

20.30 Ionic potential = ϕ = charge/ion radius. The larger the value of ϕ for the cation, the greater the covalent character in a metal-nonmetal bond.

20.31 (a) $GeCl_4$ (b) Bi_2O_5 (c) PbS (d) Li_2S (e) MgS

20.32 (a) SnO (b) $AlCl_3$ (c) BeF_2 (d) PbS (e) SnS

20.33 Charge transfer from anion to cation which absorbs photons in the visible portion of the spectrum is the phenomenon responsible for the color of compounds such as SnS_2 and PbS.

20.34 Based on theory, the answers are: (a) Ag_2S (b) CuBr (c) SnS_2 (d) Al_2S_3

20.35 Red-violet

20.36 Since their oxides are basic and alkali means basic, the Group IA elements are called alkali metals. Oxidation states of zero and 1+ are observed for Group IA metals.

20.37 They are relatively rare and, therefore, more expensive to produce. Compounds of Na and K usually serve just as well.

They are less expensive to produce.

20.39 $Na^+(g) \rightarrow Na^+(aq)$ $\Delta H° = ?$

(1) $Na(s) \rightarrow Na^+(aq) + e^-$ $\Delta H°_1 = -239.7$

(2) $Na(g) \rightarrow Na(s)$ $\Delta H°_2 = -108.7$

(3) $Na^+(g) + e^- \rightarrow Na(g)$ $\Delta H°_3 = -493.7$

$\Delta H° = \Delta H°_1 + \Delta H°_2 + \Delta H°_3 =$ **-842.1 kJ mol^{-1}**

20.40 $Na(s) \rightarrow Na^{2+}(aq) + 2\,e^-$ (overall process) $\Delta H° =$ (negative)

$Na(s) \rightarrow Na(g)$ $\Delta H°_{sublimation} = 108.7$ kJ mol^{-1}

$Na(g) \rightarrow Na^+(g) + 1\,e^-$ $I.E._{(1)} = 495.8$ kJ mol^{-1}

$Na^+(g) \rightarrow Na^{2+}(g) + 1\,e^-$ $I.E._{(2)} = 4{,}565$ kJ mol^{-1}

$Na^{2+}(g) \rightarrow Na^{2+}(aq)$ $\Delta H°_{hydration} = ?$

(ΔH_{hyd} would have to be a larger negative value than **-5170 kJ mol^{-1}**)

[If I. $E._{(1)}$ is 493.7 (as in Question 20.39), the answer is -5167 kJ mol^{-1}]

20.41

K(s)	→	$K^+(aq)$	+ 1e$^-$	$\Delta H° = ?$
K(s)	→	K(g)		$\Delta H° = 90.0$ kJ/mol
K(g)	→	$K^+(g)$	+ 1e$^-$	$\Delta H° = 418$ kJ/mol
$K^+(g)$	→	$K^+(aq)$		$\Delta H° = -759$ kJ/mol
				$\Delta H° = -251$ kJ/mol

$\Delta G° = -nFE° = \Delta H° - T\Delta S°$

$$\Delta S° = \frac{-nFE° - \Delta H°}{-T} = \frac{[1 \times 96{,}500 \text{ C/mol} \times 2.92 \text{ V} \times \text{J/(V} \times \text{C)}]}{298 \text{ K}} + \frac{(-251 \times 10^3 \text{ J/mol})}{298 \text{ K}}$$

$$= \mathbf{103\ J\ K^{-1}\ mol^{-1}}$$

20.42 Na and K are the most abundant alkali metals. Francium is least abundant because it is radioactive with a short half-life.

20.43 Alkali metals occur as compounds in the ocean and in salt deposits.

20.44 $MCl(\ell) + Na(g) \rightleftharpoons NaCl(\ell) + M(g)$ [M = K, Rb, Cs]

20.45 Use of sodium in the cooling of nuclear reactors takes advantage of its low melting point, relatively high boiling point, and good thermal conductivity. Using sodium in vapor lamps takes advantage of sodium's emission spectra.

20.46 (a) yellow (b) red (c) violet

20.47 To detect potassium in the presence of sodium, view their flame through blue "cobalt glass" to filter out the yellow light from sodium.

20.48 $2Rb(s) + 2H_2O(\ell) \rightarrow 2Rb^+(aq) + 2OH^-(aq) + H_2(g)$

20.49 Because of the very large hydration energy of the tiny Li^+ ion.

20.50 Alkali metals dissolve without reaction in liquid ammonia to give blue solutions. These solutions contain solvated electrons that act as very good reducing agents.

20.51 (a) $2Li + Br_2 \rightarrow 2LiBr$

$2Na + Br_2 \rightarrow 2NaBr$

(b) $2Li + S \rightarrow Li_2S$

$2Na + S \rightarrow Na_2S$

(c) $6Li + N_2 \rightarrow 2Li_3N$

$Na + N_2 \rightarrow$ no reaction

20.52 $4Li + O_2 \rightarrow 2Li_2O$

$2Na + O_2 \rightarrow Na_2O_2$

$M + O_2 \rightarrow MO_2$ [M = K, Rb, Cs]

20.53 KO_2 is used in a recirculating breathing apparatus.

$4KO_2(s) + 2CO_2(g) \rightarrow 2K_2CO_3(s) + 3O_2(g)$

$2KO_2 + 2H_2O \rightarrow 2KOH + O_2 + H_2O_2$

20.54 Because of hydrolysis; $Na_2O_2(s) + 2H_2O(\ell) \rightarrow 2Na^+(aq) + 2OH^-(aq) + H_2O_2(aq)$

20.55 A metal that in the production of electronic vacuum tubes reacts with traces of O_2 and H_2O and removes them from an otherwise inert atmosphere is sometimes referred to as a getter.

20.56 Other common names for sodium hydroxide are caustic soda and lye. Some of the uses of sodium hydroxide include making soap;neutralizing acids; in drain cleaners; making detergents, pulp, and paper; and removing sulfur from petroleum.

20.57 Trona ore is: $Na_2CO_3 \cdot NaHCO_3 \cdot 2H_2O$

Solvay process:
$CaCO_3 \xrightarrow{heat} CaO + CO_2$
$CO_2 + H_2O \rightarrow H_2CO_3$
$H_2CO_3 + NH_3 \rightarrow NH_4^+ + HCO_3^-$
$HCO_3^- + Na^+ + Cl^- \rightarrow NaHCO_3 + Cl^-$
$2NaHCO_3 \xrightarrow{heat} Na_2CO_3 + H_2O + CO_2$
$CaO + H_2O \rightarrow Ca(OH)_2$
$2NH_4Cl + Ca(OH)_2 \rightarrow CaCl_2 + 2NH_3 + 2H_2O$
Net reaction: $2NaCl + CaCO_3 \rightarrow Na_2CO_3 + CaCl_2$

20.58 HCO_3^- is a buffer since it can neutralize the effect of either added acids or bases.
$HCO_3^- + H_3O^+ \rightarrow H_2CO_3 + H_2O$
$HCO_3^- + OH^- \rightarrow CO_3^{2-} + H_2O$
Common name: baking soda
Fire extinguisher: $2NaHCO_3(s) \xrightarrow{heat} Na_2CO_3(s) + H_2O(g) + CO_2(g)$

20.59 Potash is K_2CO_3.
It is made in the following way:
$KOH + CO_2 \rightarrow KHCO_3$
$2KHCO_3 \xrightarrow{heat} K_2CO_3 + H_2O + CO_2$

20.60 Li_2CO_3 has been used to treat manic depression.

20.61 Group IIA elements are called alkaline earth metals because their oxides are basic and their ores are found in the earth. Their densities, hardness, melting points and ionization energies are all greater than those of Group IA metals.

20.62 Calcium and magnesium are found in mineral deposits of limestone ($CaCO_3$), gypsum ($CaSO_4 \cdot 2H_2O$), dolomite ($CaCO_3 \cdot MgCO_3$), carnallite ($MgCl_2 \cdot KCl \cdot 6H_2O$) and in the sea.
Radium is found in pitchblende - a uranium ore.

20.63 From dolomite:

$CaCO_3{\cdot}MgCO_3 \xrightarrow{heat} CaO{\cdot}MgO + 2CO_2$

$CaO + H_2O \rightarrow Ca^{2+} + 2OH^-$

$MgO + H_2O \rightarrow Mg(OH)_2(s)$

$Mg(OH)_2(s) + 2HCl \rightarrow MgCl_2 + 2H_2O$

$MgCl_2(\ell) \xrightarrow{electrolysis} Mg(\ell) + Cl_2(g)$

From sea water:

$CaO + H_2O + Mg^{2+} \rightarrow Ca^{2+} + Mg(OH)_2(s)$

$Mg(OH)_2(s) + 2HCl \rightarrow MgCl_2 + 2H_2O$

$MgCl_2(\ell) \xrightarrow{electrolysis} Mg(\ell) + Cl_2(g)$

20.64 The hydration energies for Group IIA elements are larger than those of the alkali metals which offsets their larger ionization energies; as a result, the reduction potentials of the Group IIA metals are nearly as negative as those for the Group IA metals.

20.65 Calcining: heating a substance strongly decomposes carbonate to the oxide and CO_2.
Lime is CaO.

$CaO + H_2O \rightarrow Ca(OH)_2$

Lime is an inexpensive, relatively strong base.

20.66 $2Mg + O_2 \rightarrow 2MgO$ + light and heat

20.67 (a) brick-red (b) crimson (c) yellowish-green

20.68 The oxides of beryllium and magnesium provide a water-insoluble protective film that prevents further oxidation. Ca, Sr, and Ba react with water forming water-soluble oxides and are, therefore, not useful as structural materials.

20.69 $2Mg + O_2 \rightarrow 2MgO$

$Mg + S \rightarrow MgS$

$3Mg + N_2 \rightarrow Mg_3N_2$

20.70 $Be + 2H^+ \rightarrow Be^{2+} + H_2$

$Be + 2H_2O + 2OH^- \rightarrow Be(OH)_4{}^{2-} + H_2(g)$

20.71 Covalently linked chain of $BeCl_2$ units. See Figure 20.18 and the drawing in the text beside Section 20.6. Formation of the additional two Be–Cl coordinate covalent bonds suggests that the Be in $BeCl_2$ seeks additional electrons to complete its octet.

20.72 Beryllium compounds are covalent because the Be atom is so very small and highly charged it draws electron density toward the Be^{2+} ion making bonds more covalent than ionic.
Organomagnesium compounds contain portions of organic molecules covalently bonded to Mg.

20.73 (a) $Ca + 2H_2O \rightarrow Ca(OH)_2 + H_2$

(b) $2K + 2H_2O \rightarrow 2KOH + H_2$

20.74 Solubilities of the alkaline earth hydroxides increase and the sulfates decrease from top to bottom in Group IIA.

20.75 Calcium carbonate is used to make lime, as a mild abrasive, as an antacid, and as the main ingredient in chalk.

20.76 Milk of magnesia is $Mg(OH)_2$ suspended in H_2O.

20.77 Gypsum is: $CaSO_4 \cdot 2H_2O$
Making of plaster of Paris: $CaSO_4 \cdot 2H_2O \xrightarrow{heat} CaSO_4 \cdot 1/2H_2O + 3/2H_2O$
Reaction of plaster of Paris with water: $CaSO_4 \cdot 1/2H_2O + 3/2H_2O \rightarrow CaSO_4 \cdot 2H_2O$

20.78 Barium sulfate can be used in x-ray diagnosis because it is opaque to x-rays and is very insoluble.

20.79 Epsom salts is: $MgSO_4 \cdot 7H_2O$. MgO is used in refractory bricks, the manufacture of paper, and, medicinally, as an antacid.

20.80

Group	Oxidation Numbers
IIIA	1+, 3+ (except Al only 3+)
IVA	2+, 4+
VA	3+, 5+

20.81 The post-transition metals are Ga, In, Tl, Sn, Pb and Bi.
Lower oxidation states become more stable going down a group because the energy needed to remove additional electrons is less likely to be recovered by forming additional bonds since bond strengths decrease as the atoms become larger.

20.82 An ore of aluminum is Bauxite, Al_2O_3.

$Al_2O_3(s)$ + impurities(s) $\xrightarrow{2OH^-}$ $2AlO_2^-(aq)$ + impurities(s) + H_2O

$AlO_2^- + H_3O^+ \rightarrow Al(OH)_3(s)$

$2Al(OH)_3 \xrightarrow{heat} Al_2O_3 + 3H_2O$

20.83 Aluminum is used as a structural metal, in kitchen utensils, automobiles, aircraft, beverage cans, aluminum foil, other consumer products, electrical wiring and in alnico for magnets.

20.84 The Al_2O_3 on the surface of aluminum protects the metal beneath and prevents corrosion. Forming an amalgam with the Al surface prevents Al_2O_3 from adhering and causes rapid corrosion.

20.85 $2Al(s) + 6H^+(aq) \rightarrow 2Al^{3+}(aq) + 3H_2(g)$

$2Al(s) + 2OH^-(aq) + 2H_2O(\ell) \rightarrow 2AlO_2^-(aq) + 3H_2(g)$

20.86 γ-Al_2O_3 is quite reactive; α-Al_2O_3 is quite inert. Gems composed of Al_2O_3 are ruby and sapphire.

20.87 The thermite reaction is: $Fe_2O_3(s) + 2Al(s) \rightarrow Al_2O_3(\ell) + 2Fe(\ell)$ + heat

20.88 See the drawing in the Section on "Chemical Properties and Compounds (Aluminum)" and the figure in the margin.

20.89 Solutions of aluminum salts are acidic because hydrolysis will produce hydrated protons:

$Al(H_2O)_6^{3+} + H_2O \rightleftharpoons [Al(H_2O)_5OH]^{2+} + H_3O^+$

20.90 $Al(H_2O)_6^{3+} + OH^- \rightarrow Al(H_2O)_5OH^{2+}(aq) + H_2O$

$Al(H_2O)_5OH^{2+} + OH^- \rightarrow Al(H_2O)_4(OH)_2^+(aq) + H_2O$

$Al(H_2O)_4(OH)_2^+ + OH^- \rightarrow Al(H_2O)_3(OH)_3(s) + H_2O$

$Al(H_2O)_3(OH)_3(s) + OH^- \rightarrow Al(H_2O)_2(OH)_4^-(aq) + H_2O$

20.91 The aluminate ion may be written as either: AlO_2^- or $Al(H_2O)_2(OH)_4^-$

20.92 Aluminum sulfate is used in water treatment plants because as solutions of $Al_2(SO_4)_3$ are made basic, hydrated aluminum hydroxide precipitates. As the precipitated gelatinous $Al(OH)_3$ settles, it carries fine sediment and bacteria with it.

20.93 An alum is a double salt of the general formula $M^+M^{3+}(SO_4)_2{\bullet}12H_2O$ ($M^+ = Na^+$, NH_4^+, or K^+ and $M^{3+} = Al^{3+}$, Cr^{3+}, or Fe^{3+}). An example is $NaAl(SO_4)_2 \bullet 12H_2O$. This is the alum in baking powders. It evolves CO_2 from $NaHCO_3$ because the aluminum ion hydrolyzes releasing H_3O^+ which reacts with the $NaHCO_3$.

20.94 $Al_2O_3 + 2OH^- \rightarrow 2AlO_2^- + H_2O$

20.95 Tin from its ore, cassiterite: $SnO_2 + C \rightarrow Sn + CO_2$

Lead from its ore, galena: $2PbS + 3O_2 \rightarrow 2PbO + 2SO_2$

$2PbO + C \rightarrow 2Pb + CO_2$

Bismuth from its ore: $2Bi_2S_3 + 9O_2 \rightarrow 2Bi_2O_3 + 6SO_2$

$2Bi_2O_3 + 3C \rightarrow 4Bi + 3CO_2$

20.96 When the tin coating of a tin can is scratched, the iron is more easily oxidized than the tin. At the scratch in the surface of a tin can a galvanic cell is established in which iron is the anode and, thus, oxidation of the iron is preferred to oxidation of the tin..

20.97 Allotropes are different physical forms of the same element. Tin exhibits allotropism.

20.98 Unlike nearly all other substances, bismuth expands slightly when it freezes. Wood's metal is an alloy of 50% Bi, 25% Pb, 12.5% Sn, and 12.5% Cd. Wood's metal has a low melting point (70°C) and is used in fuses and in triggering mechanisms for automatic sprinkler systems.

20.99 $Sn + 4HNO_3 \rightarrow SnO_2 + 4NO_2 + 2H_2O$

$3Pb + 8HNO_3 \rightarrow 3Pb(NO_3)_2 + 2NO + 4H_2O$

$Sn + 2Cl_2 \rightarrow SnCl_4$

$Pb + Cl_2 \rightarrow PbCl_2$

In each case, tin yields the 4+ ion; lead yields the 2+ ion.

20.100 Bi^{5+} is such a powerful oxidizing agent it would oxidize Cl^- to Cl_2.

20.101 $Sn(s) + 2OH^-(aq) + 2H_2O(\ell) \rightarrow Sn(OH)_4^{2-}(aq) + H_2(g)$

$Pb(s) + 2OH^-(aq) + 2H_2O(\ell) \rightarrow Pb(OH)_4^{2-}(aq) + H_2(g)$

20.102 An oxide of lead is PbO (litharge). Litharge is used in pottery glazes and in making lead crystal.
Another oxide of lead is Pb_3O_4 (red lead). It is used in corrosion-inhibiting paint.
Another oxide of lead is PbO_2. It is used as cathode material in the lead storage battery.

20.103 Lead based paints slowly darken due to the reaction of Pb^{2+} with airborne H_2S to give black PbS.

20.104 Some uses of bismuth compounds include their use in cosmetics and pharmaceuticals. The bismuthyl ion is BiO^+.

21 THE TRANSITION METALS

21.1 A transition element is one that possesses a partially filled or filled d subshell and fits between Groups IIA and IIIA.

21.2 The inner transition elements possess a partially filled f sublevel that is two quantum levels below the occupied level with the largest principal quantum value.

21.3 Compounds of elements in the corresponding A and B groups have similar composition, structure, and maximum positive oxidation states.

21.4 None of these nine elements of Group VIII have counterparts among the representative elements. They have greater horizontal similarities within triads.

21.5 Fe, Co, Ni

21.6 This question will be answered differently by different students.

21.7 Four general properties of the transition elements are:
(1) They exhibit multiple oxidation states.
(2) Many transition metal compounds are paramagnetic.
(3) Many of their compounds are colored.
(4) They tend to form complex ions.

21.8 SO_4^{2-}

21.9 $KMnO_4$

21.10 See Table 7.2 for the electron configurations of the elements Sc through Zn.

21.11 Most have a pair of electrons in an s orbital in the shell with the largest value of the principal quantum number. These are the first and easiest electrons to be lost during ionization; therefore, a common oxidation state is the +2 in which only the s electrons have been removed.

21.12 Multiple oxidation states result from the ease with which transition elements can lose not only their outer s electrons, but also their ability to lose underlying d electrons as well.

21.13 Going from left to right in a period, the lower oxidation states become relatively more stable. Going from top to bottom in a d-block group, the higher oxidation states become relatively more stable.

21.14 CrO_4^{2-}

21.15 Ni^{3+}

21.16 Cr^{2+}

21.17 Cu^{3+}

21.18 The lanthanide contraction results in the period 6 d-transition elements following lanthanum having almost the same size as the corresponding period 5 d-transition element above each one. It occurs because the f subshell lying 2 shells below the outer shell is filled with 14 electrons and the charge of the nucleus increased. The added f electrons are not completely effective at shielding the outer electrons from the increased nuclear charge. Therefore, an extra decrease in size occurs resulting in Zr and Hf being about the same size; also Nb and Ta are about the same size; also Mo and W; etc.

21.19 The differences in electron configuration occur in a subshell that is two shells below the outer shell. Therefore, many chemical properties of the lanthanide elements are very similar.

21.20 The minimum in atomic radius that occurs near the center of each transition series may be explained as being a decrease as inner d electrons are added until the d subshell becomes half-filled; additional electrons cause the d orbitals to expand because of increased interelectron repulsion.

21.21 They have the same electron structure in their outer shells and because of the lanthanide contraction, Hf is very nearly the same size as Zr. Thus, their chemical properties are very similar.

21.22 Since their atomic radii are equal (139 pm), one would conclude that the densities of W and Mo would be in the same ratio as their atomic masses: $(10.2\ g/cm^3) \times (183.85/95.94) = 19.5\ g/cm^3$ vs. the experimental value of $19.3\ g/cm^3$.

21.23 The effects of increased nuclear charge by the addition of protons and increased shielding by the electrons added to the d subshell just beneath the outer shell nearly off-set each other in the transition elements.

21.24 Fe, Co and Ni

21.25 Paramagnetic substances are weakly drawn into a magnetic field,whereas, ferromagnetic substances are strongly drawn ($\sim 10^6$ times stronger than paramagnetic substances) into a magnetic field. Both properties are a result of unpaired electrons. A ferromagnetic substance can become permanently magnetized by placing it in a strong magnetic field so that the domains become aligned with the field. The domains remain aligned even after removal of the external field.

21.26 When a substance that possesses ferromagnetism is melted, it becomes simply paramagnetic because the domains become randomly oriented by thermal motion.

21.27 Sc, Y, La and lanthanide elements (atomic numbers 58 to 71)

$$2M + 6H_2O \rightarrow 2M(OH)_3 + 3H_2$$

21.28 $CrO_2(OH)_2$ contains two lone oxygens. Because of their high electronegativity, electron density is drained from the O-H bonds, making it easy for the hydrogen to be removed as H^+. This is not the case, however, for $Cr(OH)_3$.

21.29 Oxides of metals in high oxidation states tend to be acidic anhydrides since when they hydrate they tend to have one or more non-hydrated oxygens that, along with an increased charge of the central atom strengthens the X-O bond and weakens the O-H bond.

21.30 TiO_2 is a better paint pigment than white lead because of its seemingly lower toxicity and because it doesn't darken in the presence of H_2S. Titanium is used in the manufacture of airplanes because it is considerably less dense than steel, strong, corrosion-resistant and does not lose its strength at high temperature.

21.31 The energy for removal of 4 electrons is so high that Ti^{+4} ion does not have a real existence.

21.32 $TiCl_4 + 2H_2O \rightarrow TiO_2 + 4HCl$

21.33 Chromium is used to coat other metals because it is lustrous and very resistant to corrosion. The most important oxidation states of chromium are 0, 3+, and 6+.

21.34 One type of stainless steel is an alloy of chromium, nickel, and iron. Unlike ordinary steel, it is very resistant to corrosion. It is not ferromagnetic.

21.35 Cr(III) in water is acidic because it forms $Cr(H_2O)_6^{3+}$ which is a weak acid and undergoes the following reaction with base: $Cr(H_2O)_6^{3+}(aq) + 3OH^-(aq) \rightarrow Cr(H_2O)_3(OH)_3(s) + 3H_2O(\ell)$

21.36 $Cr(H_2O)_3(OH)_3 + H_3O^+ \rightarrow Cr(H_2O)_4(OH)_2^+ + H_2O$

$Cr(H_2O)_3(OH)_3 + OH^- \rightarrow Cr(H_2O)_2(OH)_4^- + H_2O$

21.37 See the drawing at the end of the section in the text on chromium

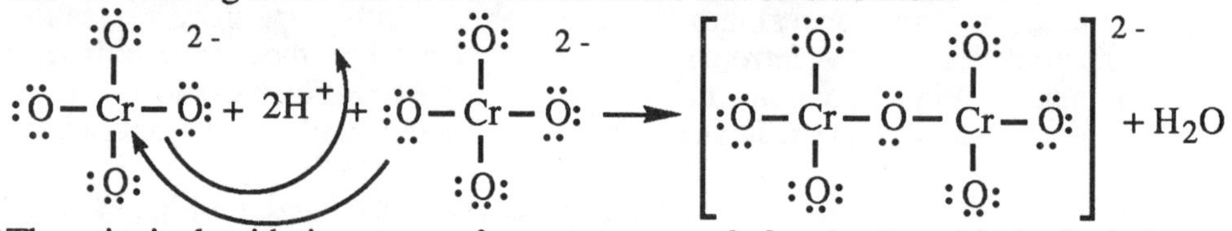

$$CrO_4^{2-} + 2H^+ + CrO_4^{2-} \longrightarrow [O_3Cr{-}O{-}CrO_3]^{2-} + H_2O$$

21.38 The principal oxidation states of manganese are 0, 2+, 4+, 7+ with the 2+ being the most stable oxidation state.

21.39 $KMnO_4$ is a deeply colored strong oxidizing agent and its reduction product in acidic solution, Mn^{2+}, is nearly colorless. In neutral or basic solution, the reduction product is brown, insoluble MnO_2 which obscures the endpoint. Therefore, $KMnO_4$ is a useful titrant only in acidic solutions.

21.40 $2HCl + Mn \rightarrow MnCl_2 + H_2$

21.41 $3MnO_4^{2-} + 4H^+ \rightarrow 2MnO_4^- + MnO_2 + 2H_2O$

21.42 The abundance of iron and the ease with which it is extracted from its ores accounts for its many practical uses.

21.43 FeO, Fe_2O_3, Fe_3O_4. Only Fe_3O_4 is magnetic.

21.44 $Fe \rightarrow Fe^{2+} + 2e^-$

$2e^- + 1/2O_2 + H_2O \rightarrow 2OH^-$

$Fe^{2+} + 2OH^- \rightarrow Fe(OH)_2$

$4Fe(OH)_2 + O_2 + 2H_2O \rightarrow 4Fe(OH)_3$

21.45 When Fe^{3+} and $Fe(CN)_4^{4-}$ are mixed, $Fe_4[Fe(CN)_6]_3 \cdot 16H_2O$ is formed. This product is known as Prussian blue.

21.46 $2H^+(aq) + Fe(s) \rightarrow Fe^{2+}(aq) + H_2(g)$

21.47 Cobalt is used in many alloys with special properties such as in high-temperature alloys that are employed in tools for cutting and machining other metals at high speed and in catalysts. The principal oxidation states of cobalt are +2 and +3.

21.48 Nickel is a very useful metal because it is resistant to corrosion and its alloys have desirable properties such as resistance to impact.

21.49 Solutions of nickel salts are green because the ion usually formed in water, $Ni(H_2O)_6^{2+}$, is green. NiO_2 is used as the cathode in nickel-cadmium batteries.

21.50 Their ease of oxidation decreases from Cu to Ag to Au, and their occurrence in nature as the free metal increases from Cu to Ag to Au.

21.51 Copper is used in electrical wire. Silver is used in photography. Gold is used in the plating of low voltage electrical contacts.

21.52 Observed oxidation states: copper, 0, +1 and +2; silver, 0, +1, +2, +3; gold, 0, +1 and +3.

21.53 H^+ is not a strong enough oxidizing agent to dissolve the coinage metals. Copper and silver react with dilute HNO_3.

$3Cu + 8HNO_3 \rightarrow 3Cu(NO_3)_2 + 2NO + 4H_2O$

$(3Cu(s) + 8H^+(aq) + 2NO_3^-(aq) \rightarrow 3Cu^{2+}(aq) + 2NO(g) + 4H_2O(\ell))$

$3Ag + 4HNO_3 \rightarrow 3AgNO_3 + NO + 2H_2O$

$(3Ag(s) + 4H^+(aq) + NO_3^-(aq) \rightarrow 3Ag^+(aq) + NO(g) + 2H_2O(\ell))$

21.54 $Au + 6H^+ + 3NO_3^- + 4Cl^- \rightarrow AuCl_4^- + 3H_2O + 3NO_2$

21.55 AgCl, AgBr and AgI are used in photographic films and papers.

21.56 To test for the presence of Ag^+, add HCl to the solution suspected to contain Ag^+. If a precipitate forms, separate it from the solution and dissolve it with aqueous ammonia. Then acidify the ammonia solution. A white precipitate of AgCl confirms the presence of Ag^+ in the original solution.

$Ag^+ + Cl^- \rightarrow AgCl(s)$

$AgCl(s) + 2NH_3 \rightarrow Ag(NH_3)_2^+ + Cl^-$

$Ag(NH_3)_2^+ + Cl^- + 2H^+ \rightarrow AgCl(s) + 2NH_4^+$

21.57 The deep blue $Cu(NH_3)_4^{2+}$ ion is formed when ammonia is added to an aqueous solution of pale blue $Cu(H_2O)_4^{2+}$.

21.58 Superconductors are materials that do not have electrical resistance. If ceramic materials could be developed that would act as superconductors at reasonable temperatures, they would enable us to transmit electrical power over long distances with no loss and substantial cost savings.

21.59 A test for superconductivity is to see if the substance is repelled by a magnet after being cooled to the necessary temperature.

21.60 $Zn(s) + H_2SO_4(aq) \rightarrow ZnSO_4(aq) + H_2(g)$

$Cd(s) + H_2SO_4(aq) \rightarrow CdSO_4(aq) + H_2(g)$

$Hg(\ell) + H_2SO_4(aq) \rightarrow$ no reaction

21.61 The protection of a metal from corrosion by being in contact with a metal that is more easily oxidized and which will be preferentially oxidized is known as cathodic protection.

21.62 Galvanizing is the covering of a metal with a coating of zinc. It protects steel by providing a barrier to oxygen and moisture and by cathodic protection.

21.63 Cadmium is used in place of zinc when a basic environment is anticipated since zinc is attacked by base but cadmium is not. Cadmium is not used more often because it is less abundant than zinc (therefore more expensive) and because its salts are very toxic.

21.64 Two common alloys that contain zinc are brass and bronze.

21.65 Common uses of zinc oxide include paint pigment, sun screen, and fast-setting dental cements.

21.66 An aqueous solution of $HgCl_2$ is a poor conductor of electricity because $HgCl_2$ is a weak electrolyte. $HgCl_2 + H_2O \rightleftharpoons Hg(OH)Cl + H^+ + Cl^-$ (~1% reacted)

21.67 To test for Hg_2^{2+} add HCl, which will cause Hg_2Cl_2 to precipitate. Treatment of the Hg_2Cl_2 with aqueous NH_3 will give a black precipitate because of disproportionation to $Hg(\ell)$ and $Hg(NH_2)Cl(s)$.

21.68 (a) cobalt (b) nickel

21.69 $K_{sp} = [Hg^+] [Cl^-] = 2.2 \times 10^{-18}$, 4.4×10^{-18}, 1.1×10^{-17}, and 2.2×10^{-17}

Not Constant

$K_{sp} = [Hg_2^{2+}] [Cl^-]^2 = 1.1 \times 10^{-18}$, 1.1×10^{-18}, 1.1×10^{-18}, and 1.1×10^{-18}

Constant Value **correct formula = Hg_2^{2+}**

21.70
(a) $Ni(H_2O)_6^{2+}$	green
(b) $Ni(NH_3)_6^{2+}$	blue
(c) $Co(H_2O)_6^{2+}$	pink
(d) $Cu(H_2O)_4^{2+}$	pale blue
(e) $Cu(NH_3)_4^{2+}$	deep blue
(f) $Mn(H_2O)_6^{2+}$	pale pink

21.71 To test for cyanide, one could add Fe^{2+}, then Fe^{3+} to form the deep blue precipitate $Fe_4[Fe(CN)_6]_3 \cdot 16H_2O$.

21.72 Ligands are Lewis bases that have one or more unshared pairs of electrons which can be shared with a metal cation. Compounds of the transition elements in which the metal cation is attached to one or more ions or molecules by coordinate covalent bonds are called coordination compounds. Ligands which have one atom that can bond to a metal cation are called monodentate ligands. Ligands which have more than one donor atom that can bond to a metal cation are called polydentate ligands. A chelate is formed by a polydentate ligand that can hold the metal cation in its "claws." The coordination number refers to the total number of ligand atoms that are bonded to a given metal ion in a complex.

21.73 Also see Section 10.7 in the textbook.

(a) Oxalate

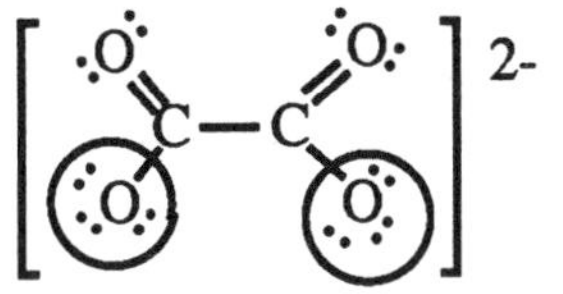

$[C_2O_4]^{2-}$

(b) Ethylenediamine

H_2C—CH_2

H_2N: :NH_2

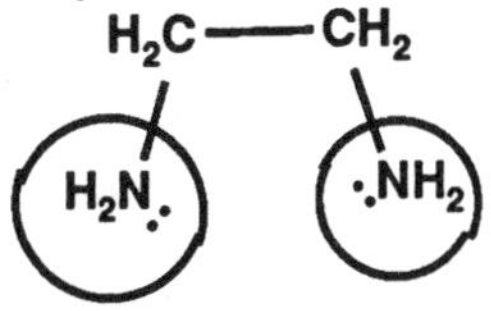

(continued)

21.73 (continued)

(c) Ethylenediaminetetraacetate ion

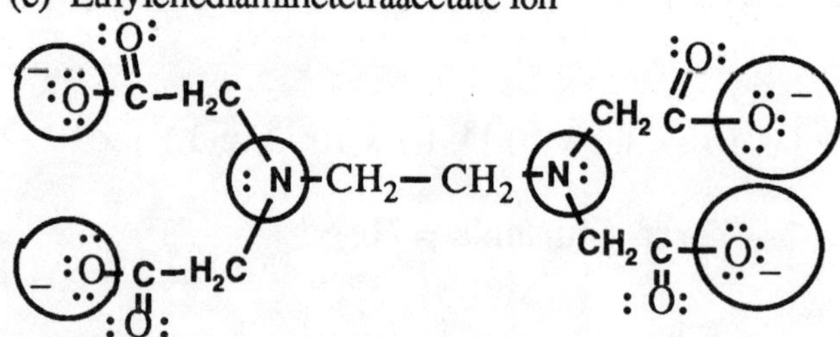

21.74 (a) en stands for ethylenediamine

(b) EDTA stands for ethylenediaminetetraacetic acid.

21.75 See Figure 21.16 for assistance in drawing the linear, tetrahedral, square planar, and octahedral structures requested.

21.76

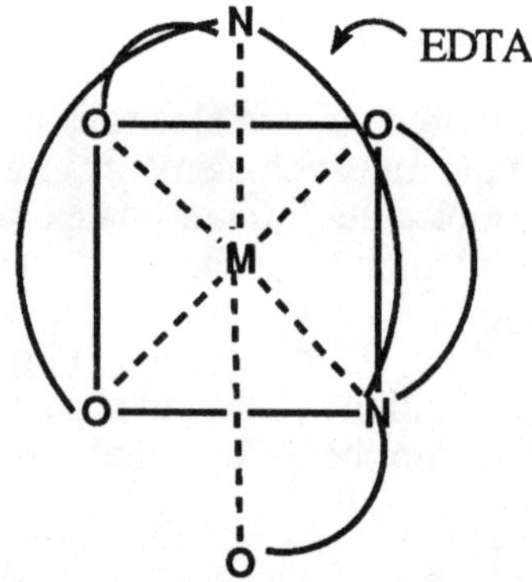

21.77 NTA can coordinate to four sites in an octahedral complex in much the same manner as EDTA.

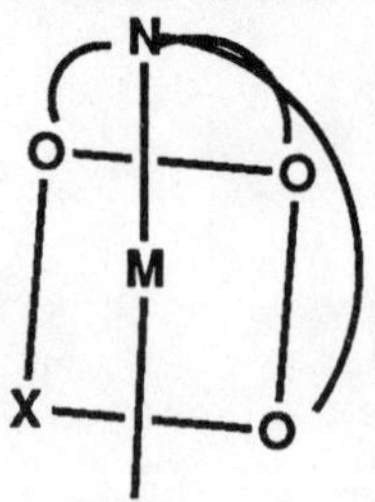

The remaining two sites (X) can be occupied by other ligands or by H_2O molecules. The NTA would increase the solubility of metal salts by shifting to the right equilibria such as: $MX_n(s) + NTA \rightleftharpoons M(NTA)X_{n-4}\,(aq) + 4X(aq)$

21.78 Isomers are two different compounds that have the same molecular formula but differ in the way their atoms are arranged. Stereoisomers result when a given molecule or ion can exist in more than one structural form in which the same atoms are bonded to one another but find themselves oriented differently in space.

21.79 Isomers of $[Co(NH_3)_2Cl_4]^-$

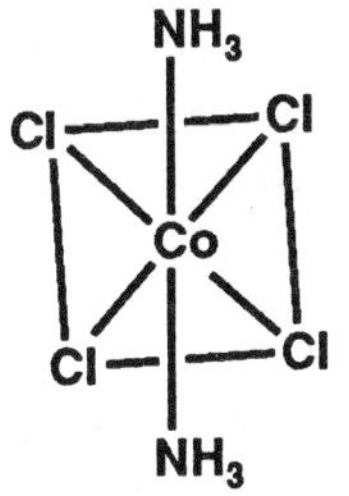

trans

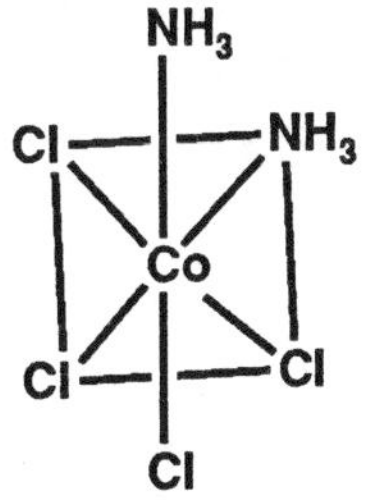

cis

Isomers of $[Co(NH_3)_3Cl_3]$

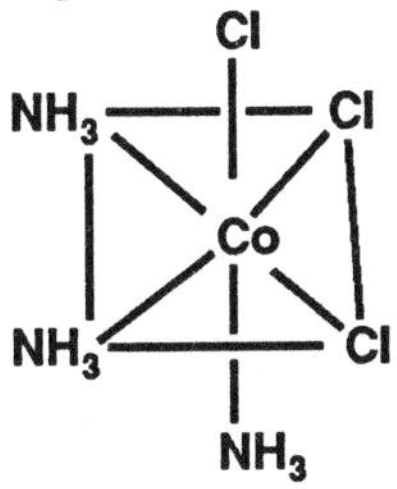

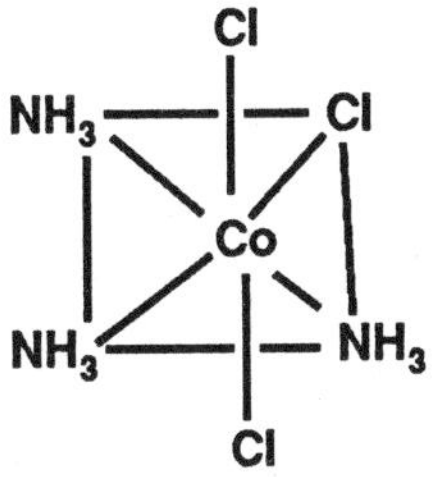

21.80 One isomer cannot be superimposable on its mirror image if a molecule or ion is chiral.

21.81 Chiral substances are said to be optically active because they or their solutions rotate the plane of plane-polarized light when plane-polarized light is passed through them or their solutions.

21.82 Isomers of $[Cr(en)_2Cl_2]^+$

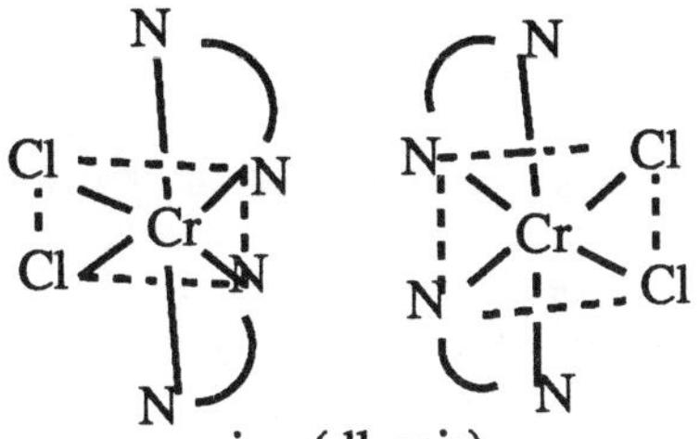

cis - (dl pair)

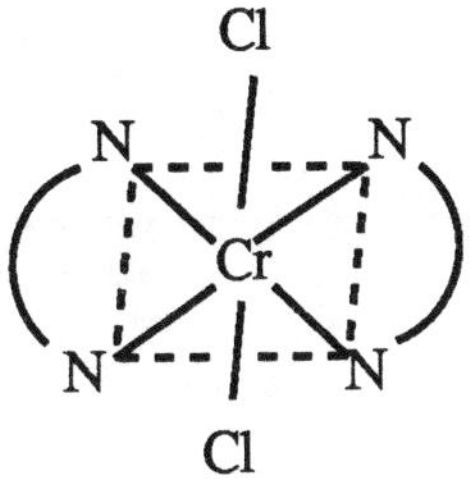

trans

21.83

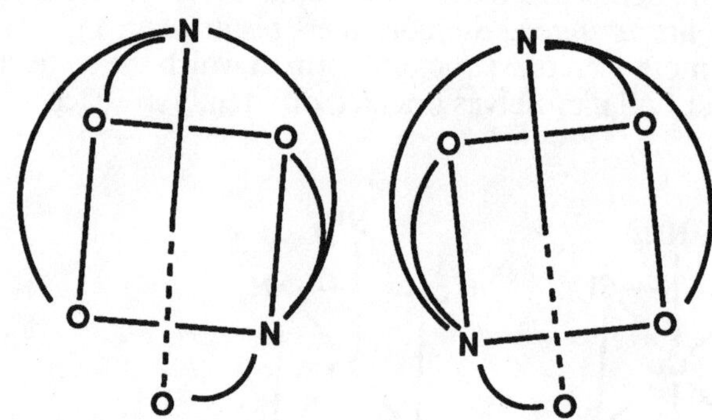

21.84 Enantiomers of a complex are two nonidentical mirror image isomers; e.g. $[Co(en)_3]^{3+}$. A racemic mixture contains an equal mixture of two enantiomers and, therefore, shows no optical activity.

21.85 (a) four (high spin)
(b) two (low spin)

21.86 See Figure 21.25

21.87

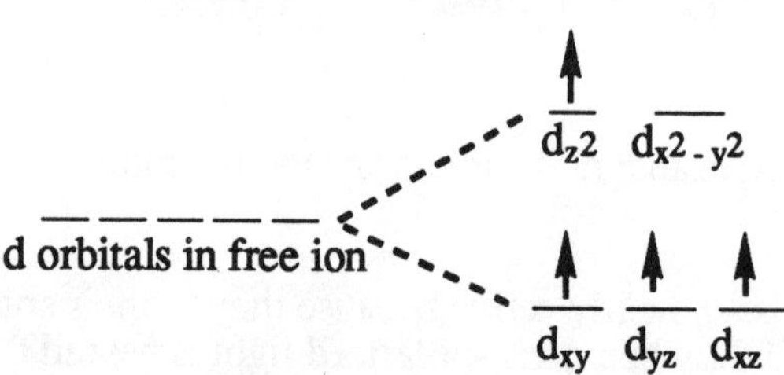

21.88 A portion of visible light is absorbed in promoting electrons from the lower t_{2g} levels to the higher e_g levels in complexes. The observed color of the complex depends on the magnitude of Δ and is due to the wavelengths that are not absorbed.

21.89 If $\Delta < P$, the electrons occupy all orbitals before pairing and a paramagnetic complex will be formed unless the orbital set is filled; when $\Delta > P$, the electrons pair up in the lower energy levels before occupying the higher energy levels; this increases the possibility of a diamagnetic complex (See Question 21.92 c).

21.90 A high-spin complex is one that possesses the maximum number of possible unpaired electrons. A low-spin complex possesses electrons paired in low energy orbitals before occupying high energy orbitals. Inner orbital complexes would be low-spin and outer orbital complexes would be high-spin.

21.91 (a) See Figure 21.33 (b) See Figure 21.35

21.92 (a)

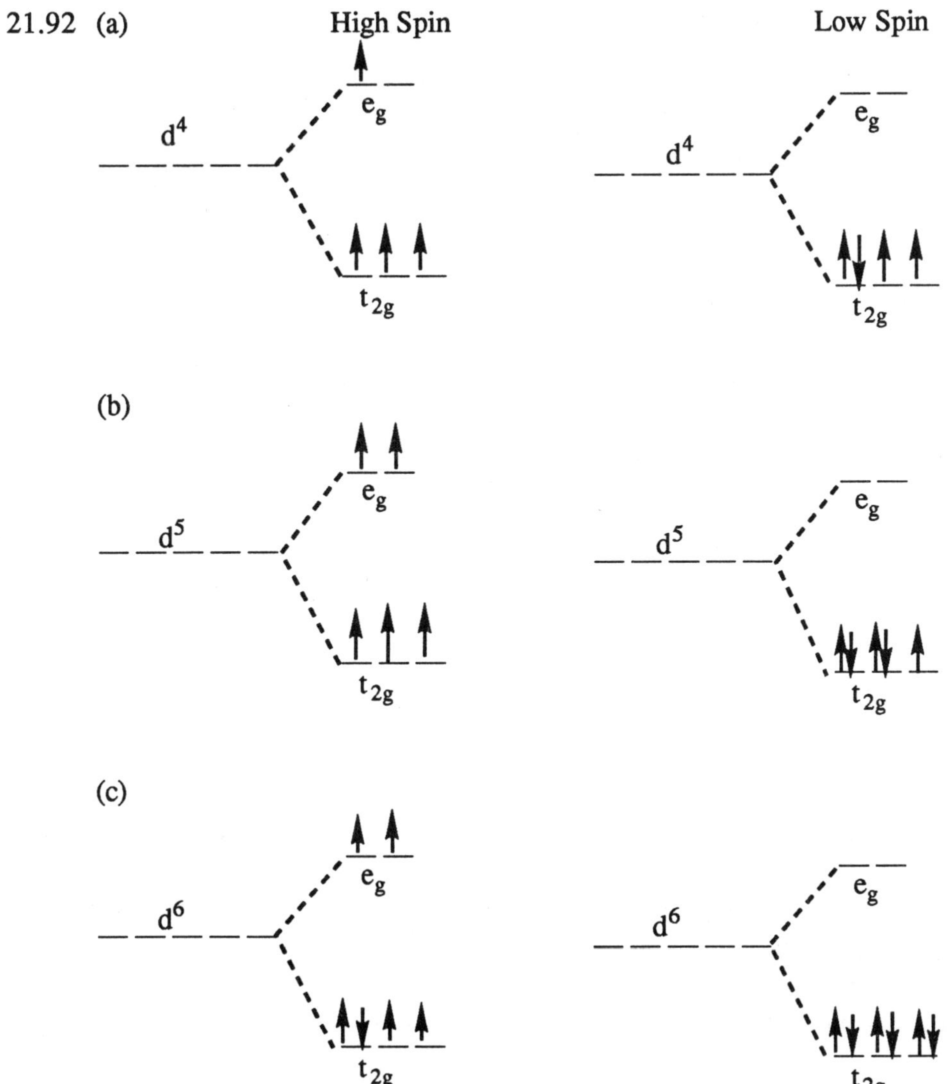

(continued)

21.92 (continued)

(d)

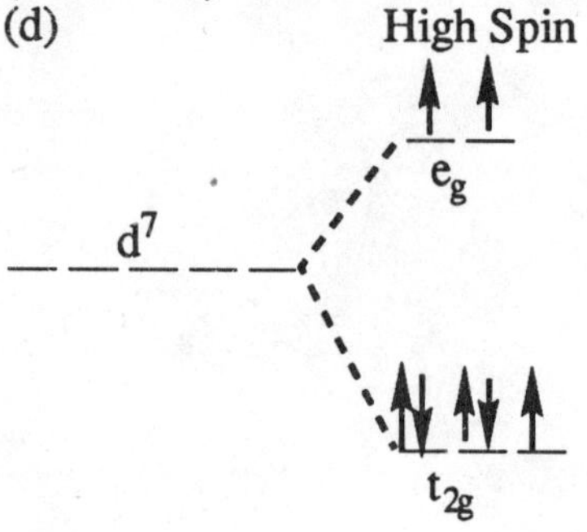

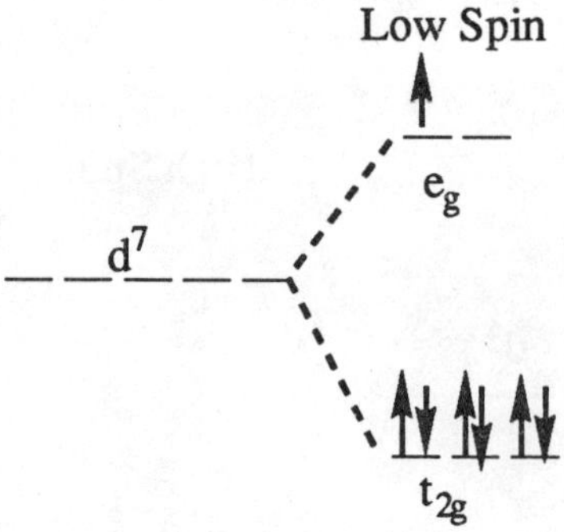

21.93 (a) 2 d-electrons -- weak ligand -- 2 unpaired electrons -- inner complex
(b) 8 d-electrons -- moderate ligand -- 2 unpaired electrons -- outer complex
(c) 5 d-electrons -- strong ligand -- 1 unpaired electron -- inner complex
(d) 6 d-electrons -- strong ligand -- no unpaired electrons -- inner complex
(e) 3 d-electrons -- weak ligand -- 3 unpaired electrons -- inner complex

21.94 $[Co(NO_2)_6]^{4-}$

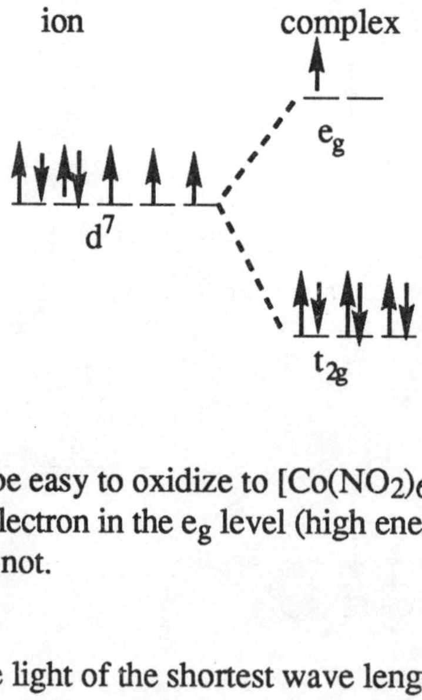

21.95 $[Co(NO_2)_6]^{4-}$ should be easy to oxidize to $[Co(NO_2)_6]^{3-}$ because the cobalt in $[Co(NO_2)_6]^{4-}$ has an electron in the e_g level (high energy electron) while the cobalt in $[Co(NO_2)_6]^{3-}$ does not.

21.96 $[Co(NO_2)_6]^{3-}$ Visible light of the shortest wave length is the visible light of the highest energy. The complex with the larger Δ is the one with the metal ion in the higher oxidation state.

22 HYDROGEN, OXYGEN, NITROGEN, AND CARBON, AND AN INTRODUCTION TO ORGANIC CHEMISTRY

22.1 The official name for H_2 is dihydrogen.

22.2 Hydrogen is the most abundant element in the universe.

22.3 Hydrogen is less abundant on Earth than the rest of the universe because the Earth's gravity wasn't strong enough to hold on to the hydrogen.

22.4 There is very little free hydrogen in the Earth's atmosphere because it reacts so readily with oxygen.

22.5 Advantage: less dense than He, so it has greater lifting power.
Disadvantage: hydrogen is extremely flammable but helium doesn't burn

22.6 The properties of hydrogen differ from the rest of the Group IA elements because there are no electrons below hydrogen's valence shell. Hydrogen is in Group IA because of its valence shell configuration.

22.7 $^{1}_{1}H$, protium; $^{2}_{1}H$ or D, deuterium; $^{3}_{1}H$ or T, tritium. Tritium is radioactive and is dangerous because it can replace ordinary hydrogen in molecules that can be incorporated into the body.

22.8 Advantages: It burns cleanly during a highly exothermic reaction with oxygen and is available in almost infinite amounts from water.
Disadvantages: It is difficult to store and carry and must first be extracted from water.

22.9 (a) $CH_4 + H_2O \xrightarrow[\text{catalyst}]{\text{heat}} CO + 3H_2$ $\quad CO + H_2O \xrightarrow[\text{catalyst}]{\text{heat}} CO_2 + H_2$

(b) $C + H_2O \xrightarrow{1000°C} CO + H_2$

22.10 $2CO + O_2 \rightarrow 2CO_2$

$2H_2 + O_2 \rightarrow 2H_2O$

22.11 Hydrogen is a byproduct in the preparation of caustic soda.

22.12 Preparation of ammonia is the greatest single use of hydrogen. $3H_2 + N_2 \rightarrow 2NH_3$

22.13 $Zn + H_2SO_4 \rightarrow ZnSO_4 + H_2$ See Figure 22.2.

22.14 Hydrogen can be used to manufacture methanol by the reaction:

$CO + 2H_2 \xrightarrow[\text{pressure, heat}]{\text{catalyst}} CH_3OH$

This reaction provides a route from coal to a liquid fuel.

22.15 $H{-}C{\equiv}C{-}H + 2H{-}H \longrightarrow CH_3{-}CH_3$

Hydrogenation of vegetable oils gives solid (or semi-solid) fats.

22.16 Hydrides are binary compounds of hydrogen. NaH is sodium hydride.

$2Na(s) + H_2(g) \rightarrow 2NaH(s)$

$NaH(s) + H_2O(\ell) \rightarrow NaOH(aq) + H_2(g)$

22.17 Hydrogen normally forms only one covalent bond. It has only one valence electron and needs only one more to fill its valence shell, therefore, it usually forms only the single bond.

22.18 (a) nonlinear (b) trigonal pyramidal (c) tetrahedral

22.19 As an <u>acid-base</u> reaction:

H^-	+	H_2O	→	H_2	+	OH^-
base		acid		acid		base

As a <u>redox</u> reaction:

H^- is oxidized to give H_2; H^+ in H_2O is reduced.

22.20 Linking together atoms of the same element to form chains is known as catenation. The tendency to catenate generally decreases going down a group in the periodic table (sulfur in Group VIA is an exception).

22.21 Some nonmetal hydrides have positive $\Delta G°_f$ and cannot be made by direct combination of elements. Those that can be made in high yield by direct combination are: CH_4, NH_3, H_2O, H_2S, HF, HCl, and HBr.

22.22 X^{n-} become stronger bases from right to left in a period (e.g., C^{4-} is a stronger base than F^-), and from bottom to top in a group (e.g., O^{2-} is a stronger base than S^{2-}).

22.23 The atmosphere is the commercial source of oxygen and nitrogen.

22.24 Dioxygen is the correct IUPAC name for O_2.

20.25 In the laboratory oxygen is conveniently prepared by the catalytic decomposition of $KClO_3$.

20.26 The manufacture of steel and metal fabrication are the largest uses of O_2.

20.27

$$O{=}O{-}O \longleftrightarrow O{-}O{=}O$$

Ozone is made in the laboratory by electric discharge through O_2.
Ozone has the advantage over Cl_2 for purification of drinking water because O_3 doesn't form poisonous and carcinogenic compounds with impurities in the water, but it does kill bacteria.

20.28 Ozone is formed in the Earth's upper atmosphere by the reactions:
$O_2 \xrightarrow{hv} 2O$ then $O_2 + O \rightarrow O_3$

Ozone protects the earth by absorbing harmful UV radiation.
Removal of O_3 from the ozone layer can be caused by nitrogen oxide pollutants:
$(NO + O_3 \rightarrow NO_2 + O_2)$
and by Freons: $(CFCl_3 \xrightarrow{hv} CFCl_2 + Cl$ and $Cl + O_3 \rightarrow ClO + O_2)$

22.29 $O^{2-} + H_2O \xrightarrow{100\%} 2OH^-$

22.30 Removal of Fe_2O_3 (rust) from the surface of iron or steel by reaction with an acid is known as "pickling".

22.31 $Al_2O_3 + 6H^+ \rightarrow 2Al^{3+} + 3H_2O$
$Al_2O_3 + 2OH^- \rightarrow 2AlO_2^- + H_2O$

22.32 From elemental carbon: $C + O_2 \rightarrow CO_2$
From lower oxides: $2CO + O_2 \rightarrow 2CO_2$
From hydride: $CH_4 + 2O_2 \rightarrow CO_2 + 2H_2O$

22.33 The $\Delta G°_f$ for N_2O is positive; therefore, K_c <<1 for its formation and it cannot be prepared in high yields from its elements.

22.34 $4NH_3 + 3O_2 \rightarrow 2N_2 + 6H_2O$
If a platinum catalyst is used in the above reaction, NO will be formed.

22.35 (a) $Cu + 4HNO_3 \rightarrow Cu(NO_3)_2 + 2NO_2 + 2H_2O$ (concentrated)
(b) $3Cu + 8HNO_3 \rightarrow 3Cu(NO_3)_2 + 2NO + 4H_2O$ (dilute)

22.36 (a) $2Na + O_2 \rightarrow Na_2O_2$ (a metal peroxide)
(b) $K + O_2 \rightarrow KO_2$ (a metal superoxide)

22.37 $H-\ddot{\underset{\cdot\cdot}{O}}-\ddot{\underset{\cdot\cdot}{O}}-H$, see Figure 22.6.

22.38 $2H_2O_2(\ell) \rightarrow O_2(g) + 2H_2O(\ell)$

22.39 The lack of reactivity of N_2 is attributed to the strength of the nitrogen-nitrogen triple bond.

22.40 Commercial uses of N_2 include: making ammonia; as an unreactive gaseous blanket during the manufacture of chemicals; and as a refrigerant (liquid N_2).

22.41 Nitrogen-fixing bacteria remove N_2 from the air and make usable nitrogen compounds from it.

22.42 Small amounts of N_2 can be prepared in the laboratory by warming a solution containing NH_4^+ and NO_2^-.

$$NH_4^+(aq) + NO_2^-(aq) \xrightarrow{\text{warm}} N_2(g) + 2H_2O(\ell)$$

22.43 Lithium is the only element that reacts with N_2 at room temperature.

22.44 $Mg_3N_2(s) + 6H_2O(\ell) \rightarrow 3Mg(OH)_2(s) + 2NH_3(g)$

22.45 As a result of forming hydrogen bonds with water, ammonia is very soluble in water.

22.46 NH_3 is the principal nitrogen-containing species in a solution of ammonium hydroxide.

22.47 Ammonia can be prepared in the laboratory by the following reaction:

$2NH_4Cl + CaO \rightarrow 2NH_3 + H_2O + CaCl_2$

Ammonia can't be collected by displacement of water because it is too soluble in water.

22.48 To test whether or not a particular unknown is an ammonium salt, warm it with base. If it contains ammonium ion, NH_3 will be evolved. The NH_3 can be detected by its odor or its basic effect on moist pink litmus paper.

22.49 $4NH_3(g) + 5O_2(g) \xrightarrow{Pt} 4NO(g) + 6H_2O(g)$

$2NO(g) + O_2(g) \rightarrow 2NO_2(g)$

$3NO_2(g) + H_2O(\ell) \rightarrow 2HNO_3(\ell) + NO(g)$

22.50 Concentrated HNO_3 is usually a pale yellow color due to photodecomposition giving NO_2.

$4HNO_3 + h\nu \rightarrow 4NO_2 + O_2 + 2H_2O$

22.51 Nitric acid can be prepared in the laboratory by the following reaction:

$NaNO_3 + H_2SO_4 + \text{heat} \rightarrow NaHSO_4 + HNO_3$

22.52 Commercial uses of nitric acid and nitrates include the manufacture of fertilizers, explosives, and plastics; curing of meats; and as a powerful oxidizing agent.

22.53 When zinc reacts with nitric acid, NH_4^+ is formed.

22.54 H^+ is not a strong enough oxidizing agent to dissolve Ag or Cu. Nitric acid also contains NO_3^- as an oxidizing agent, which is able to oxidize Ag and Cu.

22.55 Aqua regia is one part HNO_3 and three parts HCl, by volume. Nitrate ion oxidizes the noble metal, and chloride ion acts as a complex ion-forming agent to shift the equilibrium toward the dissolved metal.

22.56 (a) NH_3 (b) NH_2OH (c) N_2O (d) N_2O_3 (e) N_2O_5

22.57 $N_2 + 3H_2 \xrightarrow[\text{catalyst}]{\text{iron}} 2NH_3$

Temp; 400-500°C. Pressure; several hundred atmospheres. These conditions were chosen to produce the maximum amount of NH_3 in the shortest time. Pressure shifts the equilibrium to more product and temperature increases the reaction rate.

22.58 The formula of potassium amide is: KNH_2. In water, the NH_2^- hydrolyzes immediately to give NH_3 because NH_2^- is a very strong base.

22.59 A disproportionation reaction is one in which the same chemical undergoes both oxidation and reduction.

22.60 Haber and Ostwald processes prolonged WW I by allowing Germany to make munitions without having to import nitrates from other countries.

22.61 H—N̈—N̈—H, see Figure 22.9

(each N also bonded to an H below)

22.62 Hydrazine is prepared by the following reaction:

$2NH_3 + NaOCl \rightarrow N_2H_4 + NaCl + H_2O$

Hydrazine has been used as a rocket fuel because the combustion of hydrazine is very exothermic.

22.63 Hydroxylamine is: H—N̈—Ö—H (N also bonded to an H below) Basicities increase: $NH_2OH < N_2H_4 < NH_3$.

22.64 Household ammonia and liquid laundry bleach can react to form N_2H_4, which is very poisonous.

22.65 The preparation of nitrous oxide involves:

$NH_4NO_3(\ell) \xrightarrow{heat} N_2O(g) + 2H_2O(g)$

The resonance structures for nitrous oxide are:

:N̈=N=Ö: ⟷ :N≡N—Ö:

22.66 The $\Delta G°_f$ of N_2O (+ 104 kJ/mol), NO (+ 86.8 kJ/mol), NO_2 (+ 51.9 kJ/mol) are all positive. Their decompositions, therefore, have negative $\Delta G°$, so their decomposition reactions should proceed far toward completion. Stabilities of the oxides result because their rate of decomposition is slow.

22.67 [Lewis structure of NO_2] and [Lewis structure of N_2O_4]

$2NO_2 \rightleftharpoons N_2O_4$

22.68 Lewis structure of N_2O_3: O=N–N(=O)–O

It is formed by: $NO + NO_2 \xrightarrow[< -30^\circ C]{cool} N_2O_3$ N_2O_3 is the anhydride of HNO_2.

22.69 $3HNO_2 \rightarrow HNO_3 + H_2O + 2NO$

22.70 Lewis structures of NO_2 and $[NO_2]^-$. NO_2 should have the larger O–N–O bond angle because the single electron on the nitrogen in NO_2 doesn't offer as much resistance to increasing the O–N–O bond angle as does the pair of electrons on N in NO_2^-.

22.71 (a) As a reducing agent: $H^+ + 5HNO_2 + 2MnO_4^- \rightarrow 5NO_3^- + 2Mn^{2+} + 3H_2O$

(b) As an oxidizing agent: $2H^+ + 2HNO_2 + 2I^- \rightarrow I_2 + 2NO + 2H_2O$

22.72 $NaNO_3 + Pb \xrightarrow{heat} NaNO_2 + PbO$
$NaNO_2$ retards the growth of harmful bacteria and preserves the red color of the meat.
$NaNO_2$ may also cause cancer if HNO_2 produced in the stomach forms nitrosoamines with amines, but the risk of food poisoning without the use of $NaNO_2$ may outweigh the risk of cancer if $NaNO_2$ is present in the meat.

22.73 Lewis structure O=N(–O)–O–N(=O)–O (vapor); $NO_2^+ NO_3^-$ in solid

22.74 Dehydration of HNO_3 can be used to produce N_2O_5.

22.75 $N_2O_5 + H_2O \rightarrow 2HNO_3$

22.76 (a) $\Delta H = [8 \text{ mol } CO_2(g) (-394 \text{ kJ mol}^{-1}) + 9 \text{ mol } H_2O(g) (-242 \text{ kJ mol}^{-1})] - [1 \text{ mol } C_8H_{18}(\ell) (-255.1 \text{ kJ mol}^{-1}) - 0] = \mathbf{-5075 \text{ kJ}}$

(b) $\Delta H = [8 \text{ mol } CO_2(g) (-394 \text{ kJ mol}^{-1}) + 9 \text{ mol } H_2O(g) (-242 \text{ kJ mol}^{-1}) + 0] - [1 \text{ mol } C_8H_{18}(\ell) (-255.1 \text{ kJ mol}^{-1}) + 25 \text{ mol } N_2O(g) (81.5 \text{ kJ mol}^{-1})] = \mathbf{-7112 \text{ kJ}}$

22.77 A series of reactions that causes photochemical smog is:

$N_2 + O_2 \rightleftharpoons 2NO$

$2NO + O_2 \rightarrow 2NO_2$

$NO_2 \xrightarrow{hv} NO + O$

$O + O_2 \rightarrow O_3$

O_3 + hydrocarbons → PAN

22.78 NO_2, PAN is: (peroxyacylnitrates)

$$R-\overset{\overset{\Large :O:}{\|}}{C}-\ddot{\underset{..}{O}}-\ddot{\underset{..}{O}}-NO_2$$

22.79 Heating coal in the absence of air produces coke.

22.80 In diamond, C is sp^3 hybridized and gives a 3-dimensional tetrahedral interlocking network of covalent bonds. In graphite, C is sp^2 hybridized and arranged in planar sheets with a delocalized π-electron cloud covering upper and lower surfaces. Sheets are stacked one on another and slide over each other easily.

22.81 Graphite is used as a lubricant and in electrodes.

22.82 Its hardness, great thermal conductivity and unusual optical and acoustical properties make diamond very useful industrially.

22.83 Carbon formed by heating wood in the absence of air and then being finely pulverized is known as activated charcoal. Its large surface area per unit mass allows it to adsorb large numbers of molecules.

22.84 Carbon black is made by burning hydrocarbons in a very limited supply of O_2.

$CH_4 + O_2 \rightarrow C + 2H_2O$

It is used as a black pigment in inks and in making automobile tires.

22.85 $:C\equiv O:$, $\ddot{\underset{..}{O}}=C=\ddot{\underset{..}{O}}$

22.86 In the laboratory carbon monoxide can be prepared by:

$HCHO_2(\ell) \xrightarrow{H_2SO_4} H_2O(\ell) + CO(g)$

22.87 $Fe_2O_3(s) + 3CO(g) \xrightarrow{heat} 2Fe(s) + 3CO_2(g)$

22.88 Covalent compounds formed between a transition metal in a low (or zero) oxidation state and CO are known as metal carbonyl compounds. $Ni(CO)_4$ is an example.

22.89 Commercially CO_2 is made from limestone.

$CaCO_3 \xrightarrow{heat} CaO + CO_2$

In the laboratory it is made by the reaction:

$CaCO_3(s) + 2H^+(aq) \rightarrow Ca^{2+}(aq) + CO_2(g) + H_2O(\ell)$

22.90 Major industrial uses of CO_2 are the making of Na_2CO_3, beverage carbonation, and in refrigeration.

22.91 Green plants use CO_2 during photosynthesis to make glucose from which they make cellulose, starch and other chemicals.

22.92 The clearing of the rain forests in the Amazon basin removes a large user of CO_2 and leads to higher levels of CO_2 in the atmosphere.

22.93 Increased CO_2 in the Earth's atmosphere will absorb greater amounts of infrared radiation that would otherwise be radiated into space. The absorbed radiation will be converted to heat and cause a gradual warming known as the greenhouse effect.

22.94 $H_2O + CO_2(g) \rightleftharpoons H_2CO_3(aq)$

$CaCO_3(s) + H_2CO_3(aq) \rightleftharpoons Ca(HCO_3)_2(aq)$

22.95 As H_2O evaporates, the following equilibrium is shifted to the right: $Ca(HCO_3)_2(aq) \rightleftharpoons CaCO_3(s) + H_2O(\ell) + CO_2(g)$ and solid $CaCO_3$ forms the stalagmites and stalactites.

22.96 Hard water is water containing Ca^{2+}, Mg^{2+}, or Fe^{3+}. Addition of washing soda, $Na_2CO_3 \cdot 10H_2O$ or heating it, if it contains HCO_3^-, can remove hardness ions from hard water.

22.97 Boiler scale can be removed by washing with dilute acid to dissolve the $CaCO_3$.

22.98 Carborundum is SiC, a covalent binary compound. Si replaces half of the C atoms in the diamond structure.

$SiO_2(s) + 3C(s) \rightarrow 2CO(g) + SiC(s)$

22.99 $Al_4C_3(s) + 12H_2O(\ell) \rightarrow 4Al(OH)_3(s) + 3CH_4(g)$

22.100 $CaC_2(s) + 2H_2O(\ell) \rightarrow Ca(OH)_2(s) + C_2H_2(g)$

22.101 An interstitial carbide contains carbon atoms which are located between atoms of the host lattice. An example is tungsten carbide, WC.

22.102 $NH_3 + CH_4 \xrightarrow[Pt]{1200^\circ C} HCN + 3H_2$ HCN deactivates critical enzymes and binds irreversibly to iron in hemoglobin in the blood.

22.103 $\ddot{\underset{\cdot\cdot}{S}}{=}C{=}\ddot{\underset{\cdot\cdot}{S}}$; CS_2 is very flammable.

22.104 Organic compounds are hydrocarbons and those compounds derived from hydrocarbons by replacing H with other atoms.

22.105 Unsaturated hydrocarbons contain one or more double and/or triple bonds between the carbons. Saturated compounds contain only single bonds between carbons.

22.106 As the chain length increases, the London forces become stronger because each molecule is attracted to others at more points along the chain.

22.107 (a) $C_{30}H_{62}$ (b) $C_{27}H_{54}$ (c) $C_{33}H_{64}$

22.108 (a) $C_{17}H_{36}$ (b) $C_{17}H_{34}$ (c) $C_{17}H_{32}$ (d) $C_{17}H_{30}$ (e) $C_{17}H_{28}$

22.109 This compound is one of the 13 isomers of hexene. These are shown in the answer to Question 22.113.

22.110 An asymmetric (chiral) carbon atom (one with 4 different groups bonded to it) must be present for optical isomerism.

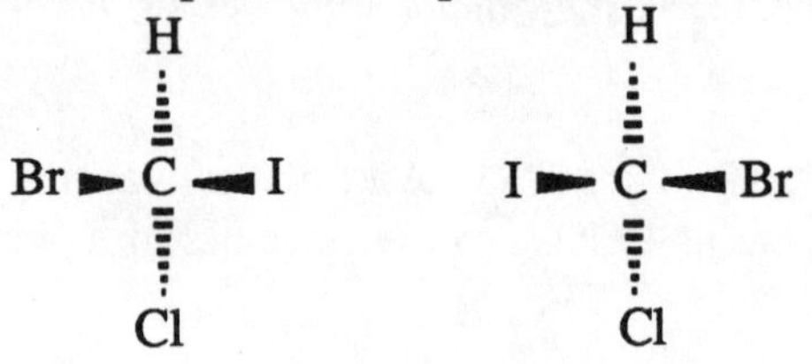

H H

Br ► C ◄ I I ► C ◄ Br

Cl Cl

22.111 CH_3—CH_2 CH_3 CH_3—CH_2 H

C=C C=C

H cis H H trans CH_3

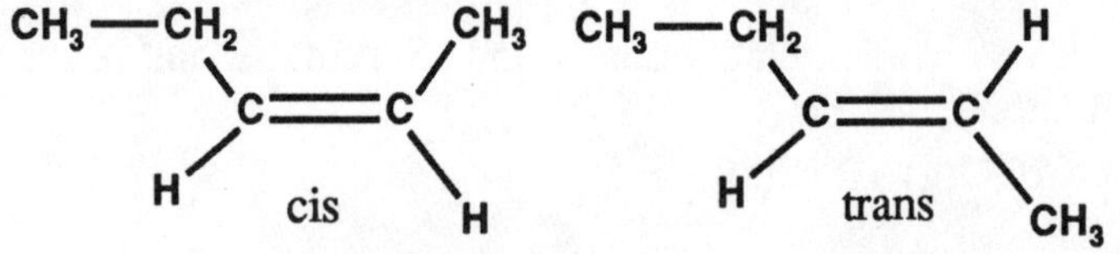

22.112 Asterisk indicates asymmetric carbon atom.

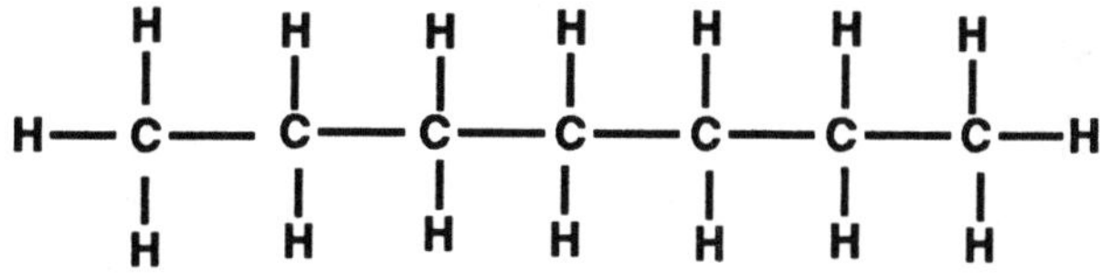

```
    H   H   H   H   H   H   H
    |   |   |   |   |   |   |
H — C — C — C — C — C — C — C — H
    |   |   |   |   |   |   |
    H   H   H   H   H   H   H
```

n-heptane

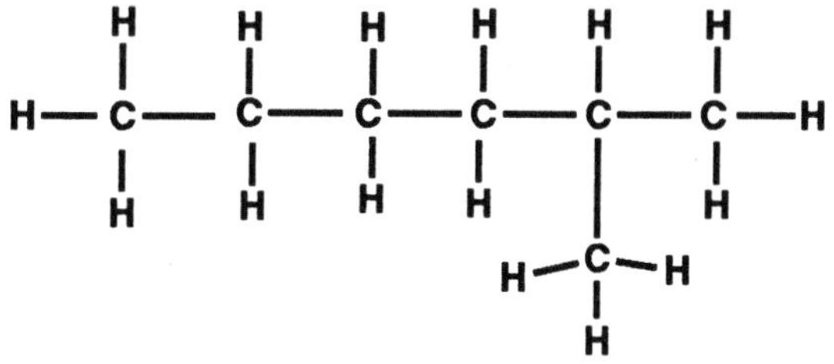

```
    H   H   H   H   H   H
    |   |   |   |   |   |
H — C — C — C — C — C — C — H
    |   |   |   |   |   |
    H   H   H   H   |   H
                    |
                H — C — H
                    |
                    H
```

2-methylhexane

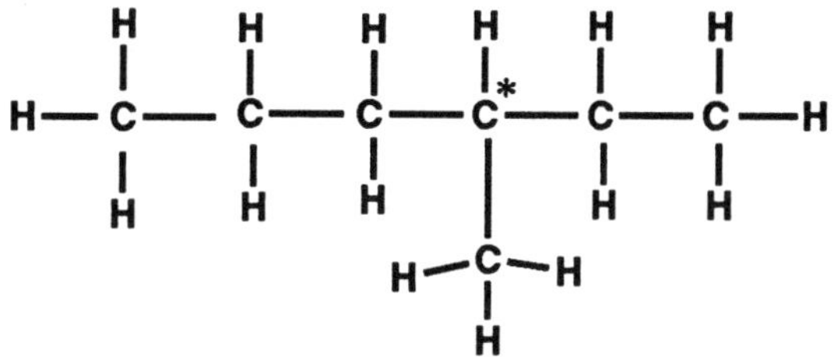

```
    H   H   H   H   H   H
    |   |   |   |*  |   |
H — C — C — C — C — C — C — H
    |   |   |   |   |   |
    H   H   H   |   H   H
                |
            H — C — H
                |
                H
```

3-methylhexane

```
    H   H   H   H   H
    |   |   |   |   |
H — C — C — C — C — C — H
    |   |   |   |   |
    H   |   H   |   H
        |       |
    H — C — H H — C — H
        |       |
        H       H
```

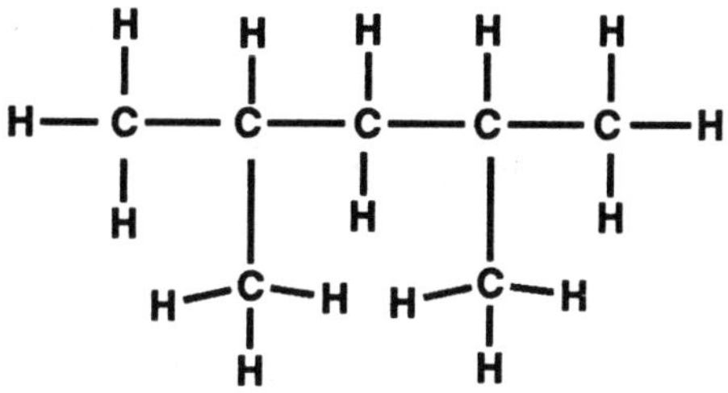

2,4-dimethylpentane

22.112 (continued)

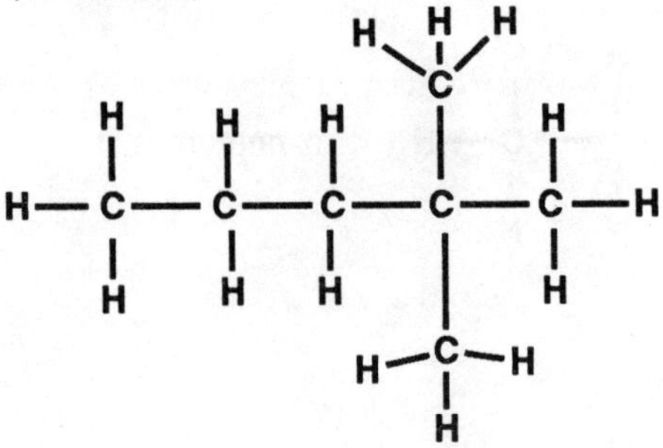

```
                      H  H  H
                       \ | /
                         C
    H      H     H       |      H
    |      |     |       |      |
H — C  —  C  —  C  —  C  —  C — H
    |      |     |       |      |
    H      H     H       |      H
                       H—C—H
                         |
                         H
```

2,2-dimethylpentane

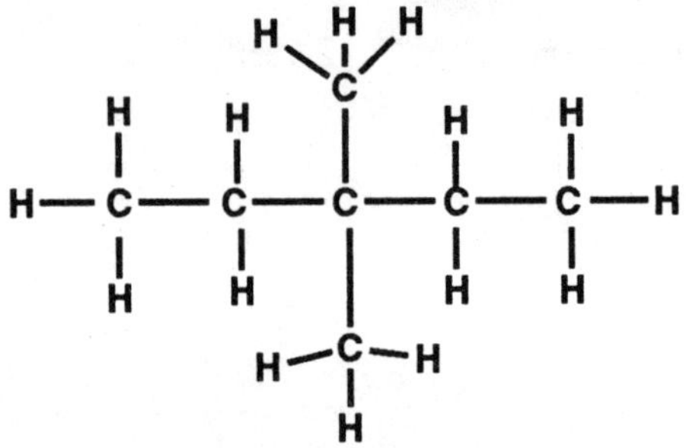

```
                 H  H  H
                  \ | /
                    C
    H      H        |      H      H
    |      |        |      |      |
H — C  —  C  —  C  —  C  —  C — H
    |      |        |      |      |
    H      H        |      H      H
                  H—C—H
                    |
                    H
```

3,3-dimethylpentane

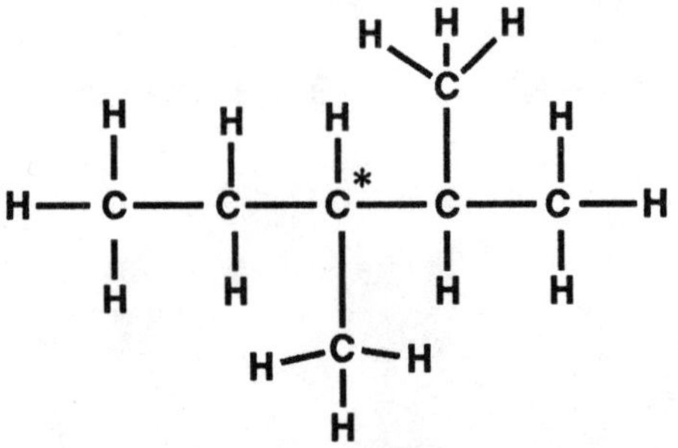

```
                         H  H  H
                          \ | /
                            C
    H      H     H          |      H
    |      |     |*         |      |
H — C  —  C  —  C  —  C  —  C — H
    |      |     |       |      |
    H      H     |       H      H
               H—C—H
                 |
                 H
```

2,3-dimethylpentane

```
    H      H     H      H      H
    |      |     |      |      |
H — C  —  C  —  C  —  C  —  C — H
    |      |     |      |      |
    H      H     |      H      H
               H—C—H
                 |
               H—C—H
                 |
                 H
```

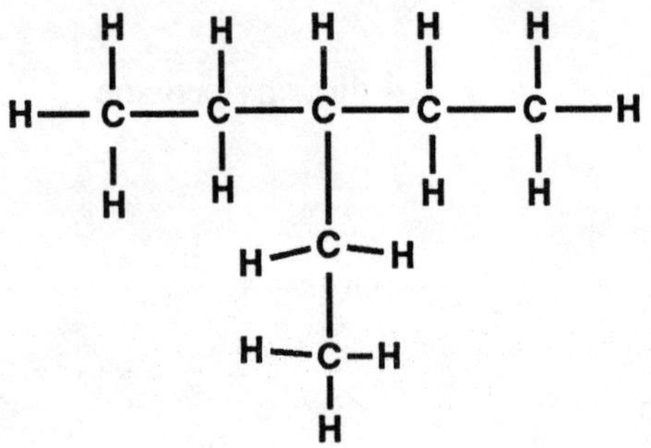

3-ethylpentane

(continued)

22.112 (continued)

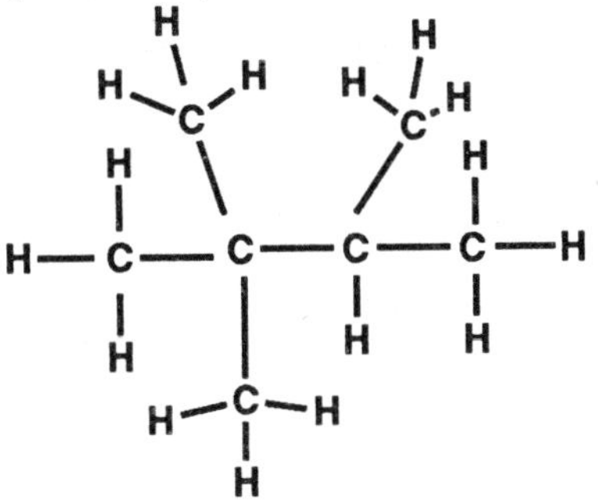

2,2,3-trimethylbutane

22.113 There are 13 isomers:

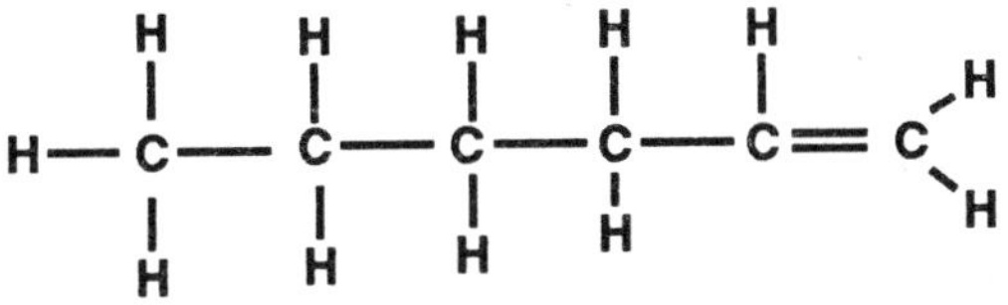

1-hexene

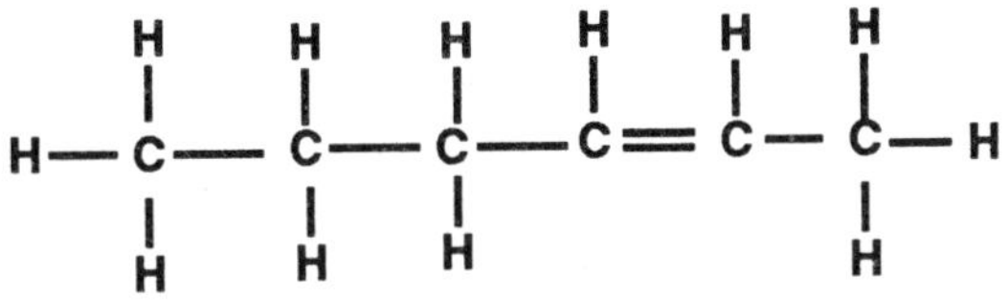

2-hexene; gives geometrical isomers

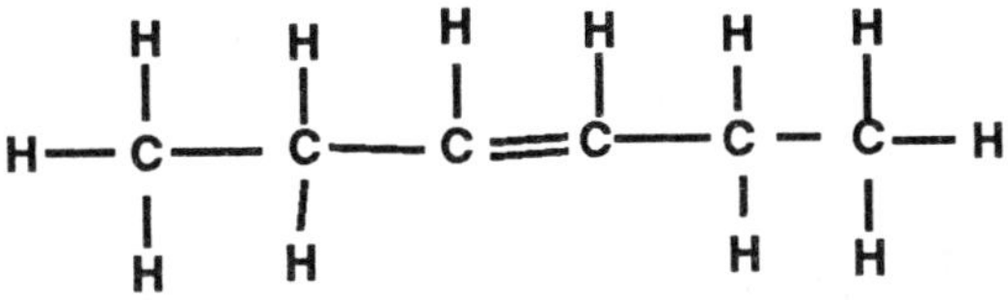

3-hexene; gives geometrical isomers

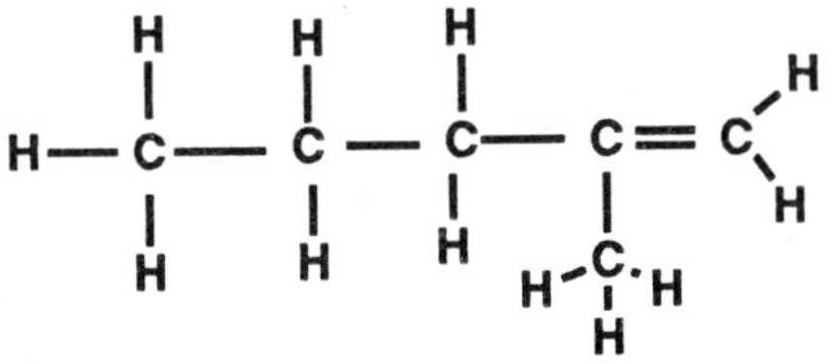

2-methyl-1-pentene

(continued)

22.113 (continued)

```
    H     H     H    H
    |     |     |    |     H
H — C — — C — — C — —C = C/
    |     |     |*        \
    H     H  H- C -H       H
                |
                H
```

3-methyl-1-pentene; gives optical isomers

```
    H     H     H    H
    |     |     |    |     H
H — C — — C — — C — —C = C/
    |     |     |         \
    H  H- C -H  H          H
          |
          H
```

4-methyl-1-pentene

```
    H     H     H          H
    |     |     |          |
H — C — — C — — C = C — — C — H
    |     |         |      |
    H     H      H- C -H   H
                    |
                    H
```

2-methyl-2-pentene

```
    H     H          H     H
    |     |          |     |
H — C — — C — — C = C — — C — H
    |     |     |          |
    H     H  H- C -H       H
                |
                H
```

3-methyl-2-pentene; gives geometrical isomers

```
    H     H          H     H
    |     |          |     |
H — C — — C — — C = C — — C — H
    |     |     |          |
    H  H- C -H  H          H
          |
          H
```

4-methyl-2-pentene; gives geometrical isomers

(continued)

22.113 (continued)

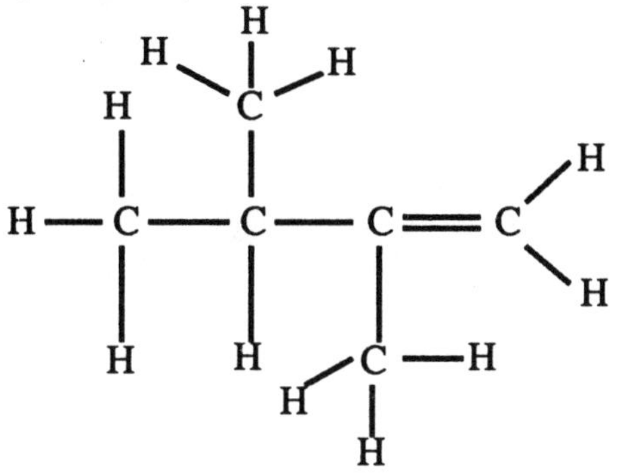

```
               H
          H    |    H
           \   |   /
     H         C
     |         |                 H
     |         |               /
H -- C ------- C ---- C == C
     |         |      |        \
     |         |      |          H
     H         H      C -- H
                    / |
                   H  H
```

2,3-dimethyl-1-butene

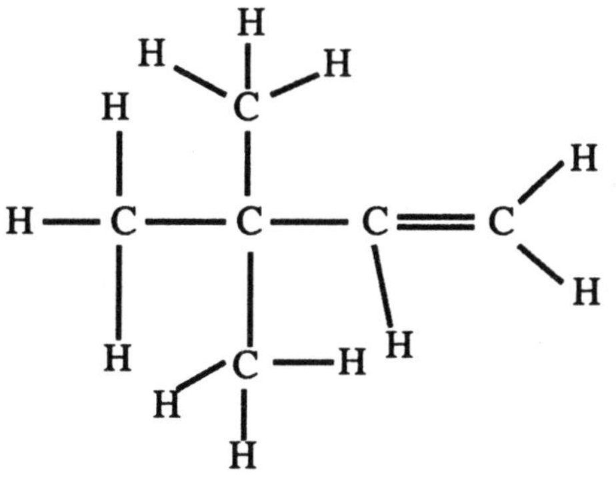

```
               H
          H    |    H
           \   |   /
     H         C
     |         |                 H
     |         |               /
H -- C ------- C ---- C == C
     |         |       \       \
     |         |        \        H
     H         C -- H    H
             / |
            H  H
```

3,3-dimethyl-1-butene

```
     H         H
     |         |                 H
     |         |               /
H -- C ------- C ---- C == C
     |         |      |        \
     |         |      |          H
     H         H  H - C - H
                      |
                  H - C - H
                      |
                      H
```

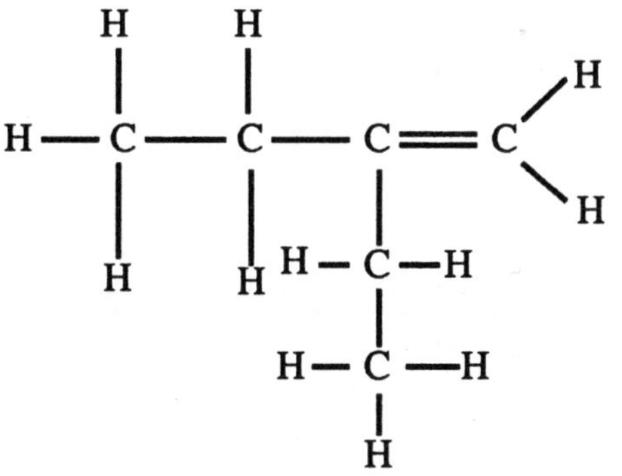

2-ethyl-1-butene

(continued)

22.113 (continued)

```
   H                 H
   |                 |
H—C——C═══C——C—H      2,3-dimethyl-2-butene
   |    |    |    \
   |    |    |     H
   H    |  H-C-H
     H-C-H   |
        \    H
        H
```

22.114

```
           F                        F
           ⁞                        ⁞
CH3 ▬ C ▬ H    and    H ▬ C ▬ CH3
           ⁞                        ⁞
          Cl                       Cl
```

These are nonsuperimposable mirror images (chiral).

22.115

```
           H                       H
           ⁞                       ⁞
Cl ▬ C ▬ Cl    and    Cl ▬ C ▬ Cl
           ⁞                       ⁞
          CH3                     CH3
```

These are mirror images that are superimposable; therefore, they are identical and not chiral

22.116 Tetrahedral

22.117 C_2Cl_4 is planar because of the double bond between the carbons. Each carbon uses sp^2 hybrid orbitals which gives a planar configuration and there is no rotation about the double bond.

22.118 The chain in butane arises from sp^3 hybridization on the carbon atoms. The hybridization is different in two of the carbons in 2-butyne. The middle two carbons are sp hybridized and, therefore, linear.

22.119 (a) 2,4-dimethylhexane (b) 3,5-dimethylheptane
(c) 5-ethyl-3-methyloctane (d) 5-methyl-3-heptene
(e) 2,4-dimethylhexane

22.120 (a) 4-ethyl-3,5-dimethyl-2,4-heptadiene (b) 5-methyl-3-heptyne
(c) 2,3,3,4,4-pentamethylhexane (d) 4-methyl-2-pentyne
(e) (*trans*, *cis*)-3,4-dimethyl-3,5-octadiene

22.121 (a) $CH_3—CH(CH_3)—CH_2—CH_2—CH_3$ (b) $CH_3—CH(CH_3)—CH(CH_3)—CH_3$

(c) $CH_3—C(CH_3)=C(CH_3)—CH_3$ (d) $CH_2=CH—CH=CH—CH=CH—CH_2—CH_3$

(e) $CH≡C—C(CH_3)_2—CH(CH_3)_2$

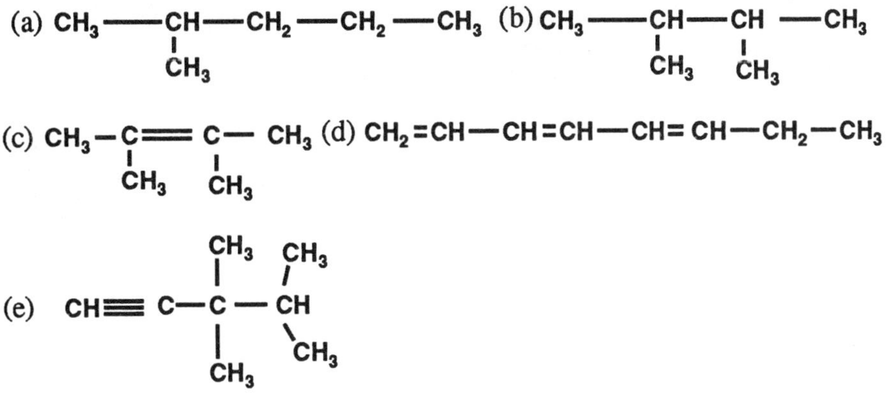

22.122 3,3,4,4,5-pentamethylheptane

22.123 A functional group is an atom or group of atoms that bestows some characteristic property to a molecule so that any molecule with the same grouping will react chemically in a similar fashion

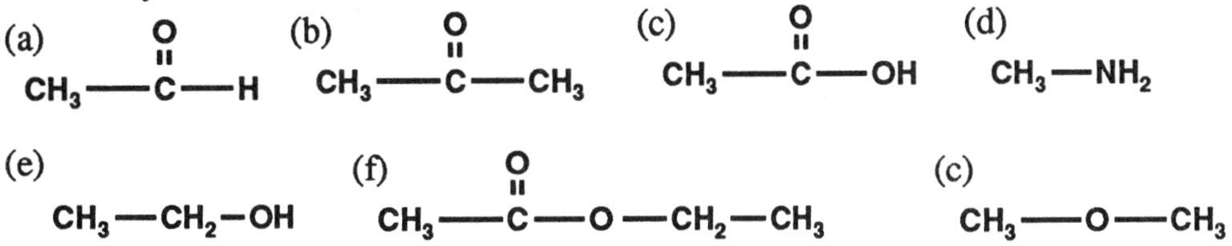

(a) $CH_3—C(=O)—H$ (b) $CH_3—C(=O)—CH_3$ (c) $CH_3—C(=O)—OH$ (d) $CH_3—NH_2$

(e) $CH_3—CH_2—OH$ (f) $CH_3—C(=O)—O—CH_2—CH_3$ (c) $CH_3—O—CH_3$

22.124 aldehyde (C=O), alcohol (OH) and ether (C-O-C)

22.125 (a) $CH_3—C(=O)—O—CH(CH_3)—CH_3$ (b) $CH_3—C(=O)—O—CH_2CH_2CH_2CH_2CH_3$

(c) $C_6H_5—C(=O)—O—CH_3$ (d) $H—C(=O)—O—CH_3$

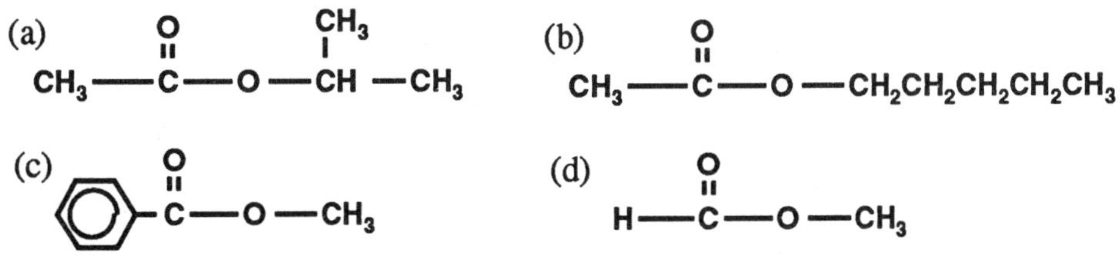

22.126 The lone pair of electrons on the nitrogen makes an amine a Lewis base.

$$(CH_3CH_2)_2NH + H_2O \rightleftharpoons (CH_3CH_2)_2NH_2^+ + OH^-$$

22.127 Esters

22.128 Amines

22.129 Alkenes and alkynes tend to undergo addition reactions whereas alkanes tend to undergo substitution reactions.

22.130 (a) $CH_3-\overset{O}{\overset{\|}{C}}-OH$ (b) $CH_3-\underset{CH_3}{\underset{|}{CH}}-\overset{O}{\overset{\|}{C}}-OH$ (c) $CH_3-\overset{O}{\overset{\|}{C}}-CH_3$

(d) $CH_3-CH_2-\overset{O}{\overset{\|}{C}}-OH$ (e) No Reaction (f) No Reaction (g) No Reaction

22.131 (a) $CH_3CH_2-\overset{O}{\overset{\|}{C}}-OCH_3 + NaOH \xrightarrow{H_2O} CH_3OH + NaO-\overset{O}{\overset{\|}{C}}-CH_2CH_3$

(b)
$CH_3CH_2O\text{-}\overset{O}{\overset{\|}{C}}CH_2CH_2\overset{O}{\overset{\|}{C}}\text{-}OCH_2CH_3 + 2NaOH \xrightarrow{H_2O}$
$2CH_3CH_2OH + NaO\overset{O}{\overset{\|}{C}}CH_2CH_2\overset{O}{\overset{\|}{C}}ONa$

22.132 An α-amino acid has the formula: $NH_2\text{-}\underset{R}{\underset{|}{CH}}\text{-}\overset{O}{\overset{\|}{C}}\text{-}OH$

Amino acids combine through amide linkages to form peptide bonds.

$NH_2\text{-}\underset{R}{\underset{|}{CH}}\text{-}\overset{O}{\overset{\|}{C}}\text{-}NH\text{-}\underset{R}{\underset{|}{CH}}\text{-}\overset{O}{\overset{\|}{C}}\text{-}OH$

peptide bond

23 PHOSPHORUS, SULFUR, THE HALOGENS, THE NOBLE GASES, AND SILICON, AND AN INTRODUCTION TO POLYMER CHEMISTRY

23.1 $Ca_3(PO_4)_2$

23.2 Phosphorus is in such biological compounds as DNA (genetic information storage), phospholipids (cell membranes), and phosphorus-oxygen compounds used for energy storage.

23.3 $[Ne]3s^23p^3$

23.4 Tetrahedral. The P–P–P bonds are strained (60°) and easily broken causing the white phosphorus to be very reactive.

23.5 $2Ca_3(PO_4)_2(s) + 6SiO_2(s) + 10C(s) \xrightarrow{1300^\circ C} 6CaSiO_3(\ell) + 10CO(g) + P_4(g)$

23.6 An electric furnace is one in which the contents are heated by passing an electric current through them.

23.7 **Red phosphorus** - believed to consist of P_4 tetrahedra joined to each other at their corners. **Black phosphorus** - layers of phosphorus atoms in which atoms in a given layer are covalently bonded to each other. Binding between layers is weak. Both of these forms are less reactive than white phosphorus.

23.8 (a) $P_4 + 5O_2 \rightarrow P_4O_{10}$ (b) $P_4 + 3O_2 \rightarrow P_4O_6$

23.9 (a) See Figure 23.2 (b) See Figure 23.3

23.10 (a) $P_4O_{10}(s) + 6H_2O(\ell) \rightarrow 4H_3PO_4(\ell)$ (b) $P_4O_6(s) + 6H_2O(\ell) \rightarrow 4H_3PO_3(aq)$

23.11 A desiccant is a substance that removes H_2O from a gas mixture. P_4O_{10} reacts with water to give H_3PO_4, thus giving a moisture-free gas.

23.12 $Ca_3(PO_4)_2(s) + 3H_2SO_4(aq) + 6H_2O(\ell) \rightarrow 3CaSO_4{\cdot}2H_2O(s) + 2H_3PO_4(aq)$

23.13 Concentrated H_3PO_4 that is about 85% H_3PO_4 by weight is called syrupy phosphoric acid.

23.14 The combustion of phosphorus with oxygen or air to form P_4O_{10} which is then dissolved in H_2O results in the purest preparation of phosphoric acid. This preparation is the one used to prepare the phosphoric acid used in food products.

23.15 The manufacture of fertilizers, food additives, and detergents have been three very large uses of phosphoric acid.

23.16 $Mg(H_2PO_4)_2$ magnesium dihydrogen phosphate
$MgHPO_4$ magnesium hydrogen phosphate $Mg_3(PO_4)_2$ magnesium phosphate

23.17 H_3PO_4

23.18 $H_2PO_4^-$ and HPO_4^{2-}. $H_2PO_4^-$ neutralizes base, HPO_4^{2-} neutralizes acid. These reactions help prevent large changes in the pH of the blood.

23.19 TSP is used as a water softener and as a cleansing agent. Solutions of Na_3PO_4 are basic because the hydrolysis of PO_4^{3-} makes the solution basic. $PO_4^{3-} + H_2O \rightarrow HPO_4^{2-} + OH^-$

23.20 See those parts of Section 23.1 on phosphoric acid and phosphorous acid.

```
      O                     O
      ||                    ||
HO — P — OH   and   HO — P — OH
      |                     |
      OH                    H
```

23.21 Superphosphate fertilizer is a mixture of calcium sulfate and calcium dihydrogen phosphate. It is made by the following reaction:

$$Ca_3(PO_4)_2 + 2H_2SO_4 + 4H_2O \rightarrow 2CaSO_4{\cdot}2H_2O + Ca(H_2PO_4)_2$$

23.22 $Ca_3(PO_4)_2$ is itself a poor fertilizer since it is insoluble and little phosphate enters solution to be available for absorption by plants.

23.23 (a) sodium hydrogen phosphate (b) sodium dihydrogen phosphate

23.24 $12Na(s) + P_4(s) \rightarrow 4Na_3P(s)$ $Na_3P(s) + 3H_2O(\ell) \rightarrow 3NaOH(aq) + PH_3(g)$

23.25 See the Lewis structures in the portion of Section 23.1 on "Polymeric Phosphoric Acids and Their Anions" that illustrates the formation of pyrophosphoric acid.

23.26 $\left(-O-\overset{\overset{\displaystyle O}{\|}}{\underset{\underset{\displaystyle OH}{|}}{P}}-\right)$ and $\left(-O-\overset{\overset{\displaystyle O}{\|}}{\underset{\underset{\displaystyle O^-}{|}}{P}}-\right)$

23.27 $(PO_3^-)_n$ is the metaphosphate ion. Its empirical formula is PO_3^-. It is formed by condensation of $H_2PO_4^-$. See Section 23.1 for an illustration of how it can be considered to be formed from phosphoric acid. For its Lewis structure the non-bonding electrons need to be added to an illustration like that referred to in Section 23.1.

23.28 See the structure near the end of the section on "Polymeric Phosphoric Acids and Their Anions." It is used in solid detergents.

23.29 3 mol NaH_2PO_4 to 2 mol Na_2HPO_4 (the HPO_4^{2-} units terminate the ends of the chains).

23.30 It promotes algae blooms which deplete the oxygen from the lake when the algae die and decompose. This kills fish and other aquatic life.

23.31 Phosphorous acid. $Mg(H_2PO_3)_2$ and $MgHPO_3$ (H_3PO_3 is a diprotic acid).

23.32 H_3PO_4 is a poor oxidizing agent. H_3PO_3 is a moderately good reducing agent.

23.33 See Figure 23.4. PCl_5 exists as $PCl_4^+PCl_6^-$ in the solid. In PCl_3, phosphorus uses sp^3 hybrid orbitals; in PCl_5 vapor or liquid it uses sp^3d hybrid orbitals. In solid PCl_4^+ it uses sp^3 and PCl_6^- it uses sp^3d^2.

23.34 $[Ne]3s^23p^4$

23.35 Brimstone is another name for elemental sulfur. It means "stone that burns."

23.36 Sulfur occurs in nature as deposits of elemental sulfur, as ionic sulfates in ores and the ocean, and as sulfides.

23.37 The two allotropic forms of sulfur are rhombic sulfur and monoclinic sulfur. They have different packing of S_8 rings in their crystals. When heated, solid sulfur melts to give an amber liquid containing S_8 rings. The rings break and join with S_x chains as the liquid darkens and thickens. The S_x chains break into smaller fragments at still higher temperatures and becomes less viscous again.

23.38 Plastic sulfur (another name for amorphous sulfur) is a form of sulfur produced when hot sulfur containing long S_x chains is suddenly cooled.

23.39 Superheated water is pumped into the sulfur deposit where it melts the sulfur. This is then foamed to the surface with compressed air. See Figure 23.5.

23.40 SO_2

23.41 $S(s) + O_2(g) \rightarrow SO_2(g)$ $\quad$ $2SO_2(g) + O_2(g) \xrightarrow{\text{catalyst}} 2SO_3(g)$

$SO_3(g) + H_2SO_4(\ell) \rightarrow H_2S_2O_7(\ell)$ $\quad$ $H_2S_2O_7(\ell) + H_2O(\ell) \rightarrow 2H_2SO_4(\ell)$

23.42 Without the presence of a catalyst the reaction of SO_2 with O_2 is slow; therefore, SO_2 is only slowly oxidized in air.

23.43 SO_2 can be conveniently prepared in the laboratory by reaction of a sulfite (e.g., Na_2SO_3) with an acid.

$Na_2SO_3(s) + H_2SO_4(aq) \rightarrow Na_2SO_4(aq) + H_2O(\ell) + SO_2(g)$

23.44 (a) $SO_2(g) + H_2O(\ell) \rightarrow H_2SO_3(aq)$ (b) $SO_3(s) + H_2O(\ell) \rightarrow H_2SO_4(aq)$

23.45 Rain falling through air that is polluted with SO_2 and SO_3 from the burning of sulfur-containing fuels becomes acidic because SO_2 and SO_3 react with water to form H_2SO_3 and H_2SO_4. It causes structural damage to buildings, causes corrosion of metals, and kills fish and plants.

23.46 H_2SO_4 is a stronger acid than H_2SO_3. The two lone oxygens on S in H_2SO_4 cause a greater polarization of the O–H bonds than does the single lone oxygen attached to S in H_2SO_3.

23.47 Sulfuric acid is used in the production of fertilizers, in refining petroleum, in lead storage batteries, in the manufacture of other chemicals, and by the steel industry.

23.48 (Lewis structures of SO_2 and SO_3 with resonance arrows)

bent 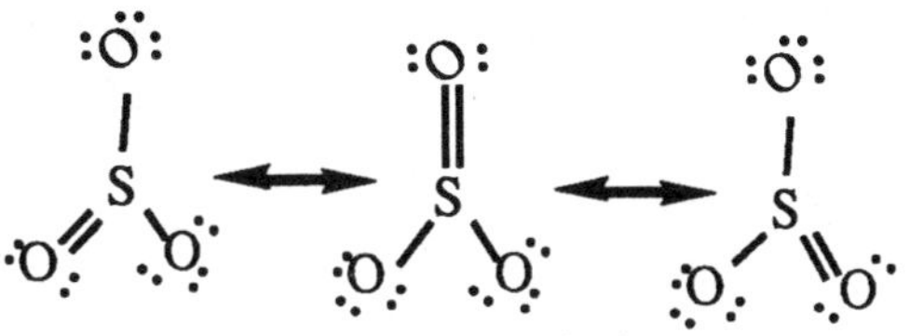 trigonal planar

23.49 Add concentrated H_2SO_4 to the water -- never the other way around.

23.50 The first step in the dissociation is complete; the second step proceeds about 10% toward completion. A solution having 1 mol H_2SO_4 contains more than 1 mol H_3O^+. H_2SO_4 is considered to be a strong acid since it is 100% ionized to H^+ and an anion.

23.51 $C_6H_{12}O_6 \xrightarrow[\text{conc.}]{H_2SO_4} 6C + 6H_2O$

23.52 (a) sodium sulfite (b) sodium hydrogen sulfate

23.53 Look at the structure of pyrophosphoric acid then give the Lewis structure of the formula, $H[OSO_2]_2OH$

23.54 $[:N{\equiv}C{-}\ddot{\underset{..}{S}}:]^-$

23.55 $CH_3CSNH_2 + 2H_2O \rightarrow CH_3CO_2^- + NH_4^+ + H_2S$

23.56 H_2S is poisonous.

23.57 The structure of the thiosulfate ion is given near the end of section 23.2.
$S(s) + SO_3^{2-}(aq) \rightarrow S_2O_3^{2-}(aq)$

23.58 In photography sodium thiosulfate is used to form a complex ion with Ag^+ and to help dissolve and remove unexposed silver halide from the film.

23.59 (a) $S_2O_3^{2-} + 4Cl_2 + 5H_2O \rightarrow 8Cl^- + 2SO_4^{2-} + 10H^+$
(b) $2S_2O_3^{2-} + I_2 \rightarrow S_4O_6^{2-} + 2I^-$

23.60 (a) I_2 and SO_2 (b) Cu^{2+} and SO_2 (c) S or H_2S and Zn^{2+}
(d) Zn^{2+} and H_2 (e) no reaction

23.61 F, $1s^22s^22p^5$; Cl, $1s^22s^22p^63s^23p^5$; Br, $1s^22s^22p^63s^23p^63d^{10}4s^24p^5$; I, $1s^22s^22p^63s^23p^63d^{10}4s^24p^64d^{10}5s^25p^5$

23.62 The name halogen comes from the Greek "halos" meaning salt. The name reflects the state in which the halogens are found in nature.

23.63 In the combined state in compounds, normally as halide ions.

23.64 Fluorine: CaF_2, Na_3AlF_6, $Ca_5(PO_4)_3F$
Chlorine: NaCl in sea water and mineral deposits.
Bromine: Sea water and brine wells.
Iodine: Sea water, brine wells, seaweed and as an impurity in saltpeter imported from Chile.

23.65 Electrolysis of HF dissolved in molten KF.

23.66 Lab: $MnO_2(s) + 4HCl(aq) \rightarrow MnCl_2(aq) + Cl_2(g) + 2H_2O$
Commercially: Electrolysis of molten NaCl or brine.
Uses of chlorine include: treating drinking water, making solvents and plastics, and manufacture of pesticides. The major use is to make chemical intermediates.

23.67 F_2: Pale yellow gas, b.p. of -188°C m.p. -233°C
Cl_2: Pale yellow-green gas, b.p. = -34.6°C m.p. -103°C
Br_2: Dark red liquid, b.p. = 58.8°C m.p. -7.2°C
I_2: Dark, metallic-looking solid, b.p. 184.4°C m.p. = 113.5°C

23.68 (a) $Cl_2 + 2KI \rightarrow I_2 + 2KCl$ (b) $F_2 + 2KBr \rightarrow Br_2 + 2KF$
(c) $I_2 + NaCl \rightarrow$ N.R. (d) $Br_2 + 2NaI \rightarrow I_2 + 2NaBr$

23.69 Elemental fluorine is very reactive as an oxidizing agent because the F–F bond splits easily to form 2 very reactive F atoms.

23.70 Bromine is recovered commercially from sea water and from brine from wells by the reaction: $2Br^-(aq) + Cl_2(aq) \rightarrow Br_2(aq) + 2Cl^-(aq)$. Blowing air through the water removes the volatile Br_2. Br_2 is used in making $C_2H_4Br_2$ (gasoline additive) and AgBr. In the lab, Br_2 can be made by the reaction: $MnO_2 + 2Br^- + 4H^+ \rightarrow Mn^{2+} + Br_2 + 2H_2O$

23.71 Iodine is recovered from seaweed and from $NaIO_3$, which is an impurity in Chilean saltpeter.

23.72 F_2 combines instantly and explosively with H_2 to form HF. Cl_2 combines with H_2 explosively if the mixture is heated or exposed to UV light. Reactions of H_2 with Br_2 and I_2 are less vigorous.

23.73 $CaF_2(s) + H_2SO_4(\ell) \rightarrow CaSO_4(s) + 2HF(g)$

23.74 $NaCl(s) + H_2SO_4(\ell) \rightarrow HCl(g) + NaHSO_4(s)$
HCl is used to remove rust from steel and in the manufacture of other chemicals.

23.75 $NaBr + H_3PO_4 \xrightarrow{heat} HBr + NaH_2PO_4$

$NaI + H_3PO_4 \xrightarrow{heat} HI + NaH_2PO_4$

23.76 When glass is exposed to HF, volatile SiF_4 is formed.
$SiO_2(s) + 4HF(aq) \rightarrow SiF_4(g) + 2H_2O(\ell)$

23.77 [Boiling Points] HCl < HBr < HI < HF
In the liquid state HF is hydrogen bonded into staggered chains.

23.78 HF is very dangerous to work with because it can cause very severe skin burns.

23.79 (a) hypobromous acid (b) sodium hypochlorite (c) potassium bromate
(d) magnesium perchlorate (e) periodic acid (f) bromic acid
(g) sodium iodate (h) potassium chlorite

23.80 A disproportionation reaction is one in which the same substance undergoes both oxidation and reduction; some of it is oxidized while the rest is reduced.

23.81 (a) $Cl_2 + H_2O \rightleftharpoons H^+ + Cl^- + HOCl$ (about 30% of chlorine as HOCl and Cl^-)

(b) $Cl_2 + 2OH^- \rightarrow OCl^- + Cl^- + H_2O$ (equilibrium lies further to the right)

23.82 $CaCl(OCl) + 2H^+ \rightarrow Ca^{2+} + H_2O + Cl_2$

23.83 OCl^- is stable, OBr^- reacts moderately fast, OI^- reacts very rapidly. Stability of OCl^- is related to slow rate of reaction rather than to thermodynamic stability.

23.84 $4KClO_3 \xrightarrow{heat} 3KClO_4 + KCl$ $2KClO_3 \xrightarrow[heat]{MnO_2} 2KCl + 3O_2$

23.85 $Ca(OCl)_2$

23.86 (a) nonlinear (b) trigonal bipyramidal (c) octahedral (d) trigonal pyramidal (e) T-shaped (f) unsymmetrical tetrahedral (g) square pyramidal (h) tetrahedral

23.87 Seven fluorine atoms cannot fit around the smaller chlorine atom.

23.88 Oxygen is too small to accommodate four fluorine atoms and in OF_4 there would be more than an octet of electrons in the valence shell of oxygen, which is not permitted because there are no d orbitals available in the valence shell.

23.89 $GeCl_4 + 2H_2O \rightarrow GeO_2 + 4HCl$ (like the reaction of $SiCl_4$)

23.90 Si has vacant d orbitals that can be used by attacking H_2O molecules, but C does not.

23.91 NH_3 is a good Lewis base; NF_3 is a poor Lewis base because the highly electronegative fluorine atoms make the nitrogen in NF_3 a poor electron pair donor.

23.92 A clathrate has atoms trapped in cagelike sites in a crystal lattice.

23.93 XeF_4, square planar with two nonbonding pairs of electrons; XeF_2, linear with three nonbonding pairs of electrons;.

23.94 They had believed that the completed octet was chemically inert.

23.95 $SiO_2(s) + 2C(s) \xrightarrow{\text{heat}} Si(s) + 2CO(g)$

23.96 A thin section of a bar of the substance to be refined is melted and the molten zone is gradually moved from one end of the bar to the other. The impurities collect in the molten zone. See Figure 23.14.

23.97 Si does not form stable π-bonds to other Si atoms.

23.98 Because Si does not form stable π-bonds to oxygen or any other atoms.

23.99 The structure of orthosilicate ion is given in section 23.5.

23.100 See the Lewis structure in the section on "Compounds With Silicon Oxygen Bonds."

23.101 Draw a Lewis structure like that in the section on "Compounds With Silicon Oxygen Bonds" with sufficient repeating SiO_3^{2-} units to give a total of six Si atoms (See Figure 23.16 b). It is found in beryl, $Be_3Al_2(Si_6O_{18})$ and emeralds.

23.102 See Figure 23.18 Repeating unit is $Si_4O_{11}^{6-}$.

23.103 See Figure 23.17

23.104 Planar sheet silicates. Some examples are soapstone and talc.

23.105 SiO_2 is the empirical formula for quartz.

23.106 Polymers are very large molecules formed by linking together a large number of smaller molecular fragments.

23.107 The chemical properties of polymers are similar to the chemical properties of the functional groups of the smaller, repeating units when they were not contained in such a large molecule. The physical properties of polymers are reflections of their enormous size.

23.108 Addition polymers are formed by adding monomer units together. Condensation polymers are formed when units combine with the elimination of small molecules such as water.

23.109 Addition polymerization reactions involve the opening of double bonds and the pairing of electrons between the monomer units to create the links. An initiator is used to get the process of polymerization started.

23.110

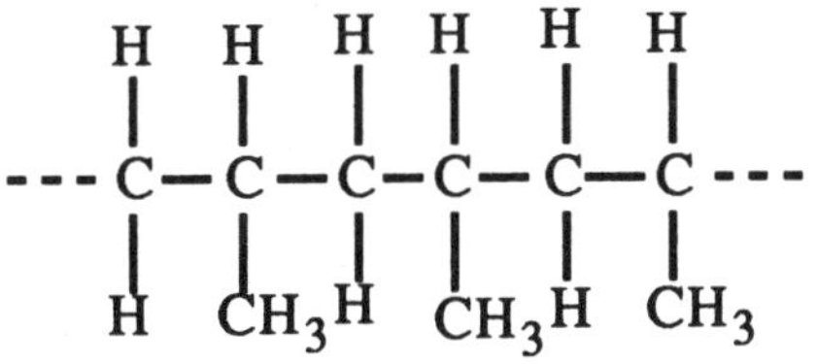

```
    H   H    H   H    H   H
    |   |    |   |    |   |
--- C — C  — C — C  — C — C ---
    |   |    |   |    |   |
    H   CH3  H   CH3  H   CH3
```

Catalysts can control the orientations of the methyl groups and, therefore, the properties of the polymers.

23.111 Teflon is an especially useful polymer because it is very resistant to chemical attack, very slippery and easy to cleanup (non-stick) because substances do not adhere to it.

23.112

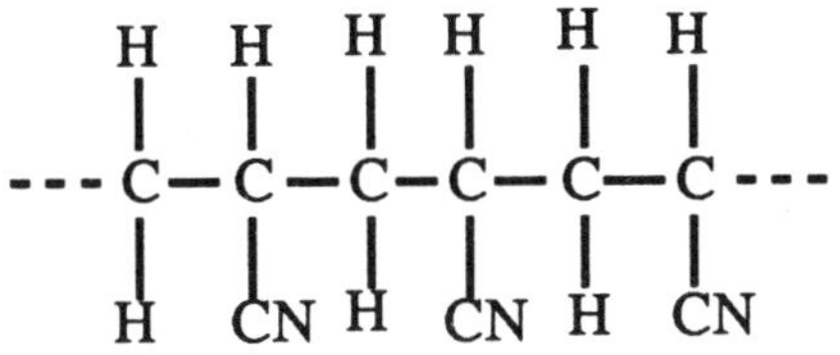

```
    H   H    H   H    H   H
    |   |    |   |    |   |
--- C — C  — C — C  — C — C ---
    |   |    |   |    |   |
    H   CN   H   CN   H   CN
```

Polyacrylonitrile is also called Orlon.

23.113 The bond between nitrogen and carbon in the NH-CO group (from an amine and a carboxylic acid) is the amide bond. The amide bond is found in peptides (proteins) as well as in polymers. The amide bond is present in nylon.

23.114 The structure of Dacron is given at the end of the section "Condensation Polymers"; it is formed by a condensation reaction in which methanol is eliminated from methyl terephthalate and ethylene gylcol.

23.115 Repeat a portion of the drawing found in the section on "Phosphazenes." If the groups attached to the backbone of a polyphosphazene polymer are polar, a water-soluble polymer is obtained; if they are nonpolar, the polymer is water-insoluble.

23.116 Poly(methyl methacrylate) is used to make transparent, hard sheets of plastic. It is also known as Lucite and Plexiglas.

23.117 A portion of the structure of a silicone polymer is shown in the section on "Silicone Polymers."

23.118

$$2\ CH_3-\underset{\displaystyle CH_3}{\overset{\displaystyle CH_3}{\underset{|}{\overset{|}{Si}}}}-OH \xrightarrow{-H_2O} CH_3-\underset{\displaystyle CH_3}{\overset{\displaystyle CH_3}{\underset{|}{\overset{|}{Si}}}}-O-\underset{\displaystyle CH_3}{\overset{\displaystyle CH_3}{\underset{|}{\overset{|}{Si}}}}-CH_3$$

23.119 Simethicone is a medicinally used polymer and is shown as the product of the last equation in the section on "Silicone Polymers."

23.120 The presence of either $SiCl_4$ or CH_3SiCl_3 would, after hydrolysis, provide more than two OH groups at which the polymer can grow. Draw a representation that shows the growth in a third direction or in third and fourth directions.

23.121 There is interest in polymers like poly(sulfur nitride) because of their ability to conduct electricity and to act as superconductors.

23.122 The monomers in polypeptides are the α-amino acids. The number of possible peptides is virtually infinite due to the use of about 20 different amino acids and the large number in the chains.

23.123 Polysaccharides are polymers of sugar molecules.

23.124 The difference of the orientation of some of the H and OH groups in starch and cellulose is shown in Figures 23.25 and 23.24. Both starch and cellulose yield glucose upon hydrolysis.

23.125 Nucleotides contain the three units, deoxyribose (a sugar), a nitrogenous base, and the phosphoric acid unit. They are linked as shown in Figure 23.29 with the phosphate unit as the bridge between the sugar units.

23.126 Hydrogen bonding interlocks the long strands that intertwine to form the necessary double helix.

24 NUCLEAR CHEMISTRY

24.1 Alpha particles, beta particles and gamma rays are the three main types of radiation emitted by radioactive nuclei. See Table 24.1

24.2 Beta particles and positrons are both particles with a mass equal to that of the electron, but they are opposite in charge.

24.3 (a) ${}^{81}_{36}Kr + {}^{0}_{-1}e \rightarrow \mathbf{{}^{81}_{35}Br}$

(b) ${}^{104}_{47}Ag \rightarrow {}^{0}_{1}e + \mathbf{{}^{104}_{46}Pd}$

(c) ${}^{73}_{31}Ga \rightarrow {}^{0}_{-1}e + \mathbf{{}^{73}_{32}Ge}$

(d) ${}^{104}_{48}Cd \rightarrow {}^{104}_{47}Ag + \mathbf{{}^{0}_{1}e}$

(e) $\mathbf{{}^{54}_{25}Mn} + {}^{0}_{-1}e \rightarrow {}^{54}_{24}Cr$

24.4 (a) ${}^{47}_{20}Ca \rightarrow {}^{47}_{21}Sc + \mathbf{{}^{0}_{-1}e}$

(b) ${}^{55}_{27}Co \rightarrow {}^{55}_{26}Fe + \mathbf{{}^{0}_{1}e}$

(c) ${}^{220}_{86}Rn \rightarrow {}^{216}_{84}Po + \mathbf{{}^{4}_{2}He}$

(d) ${}^{54}_{26}Fe + {}^{1}_{0}n \rightarrow {}^{1}_{1}H + \mathbf{{}^{54}_{25}Mn}$

(e) ${}^{46}_{20}Ca + {}^{1}_{0}n \rightarrow \mathbf{{}^{47}_{20}Ca}$

24.5 (a) ${}^{135}_{53}I \rightarrow {}^{135}_{54}Xe + \mathbf{{}^{0}_{-1}e}$

(b) ${}^{245}_{97}Bk \rightarrow {}^{4}_{2}He + \mathbf{{}^{241}_{95}Am}$

(c) ${}^{238}_{92}U + {}^{12}_{6}C \rightarrow {}^{246}_{98}Cf + \mathbf{4\,{}^{1}_{0}n}$

(d) ${}^{96}_{42}Mo + {}^{2}_{1}H \rightarrow {}^{1}_{0}n + \mathbf{{}^{97}_{43}Tc}$

(e) ${}^{20}_{8}O \rightarrow {}^{20}_{9}F + \mathbf{{}^{0}_{-1}e}$

24.6 (a) $^{35}_{17}Cl + ^{1}_{0}n \rightarrow ^{35}_{16}S + ^{1}_{1}H$ (b) $^{40}_{19}K \rightarrow ^{0}_{-1}e + ^{40}_{20}Ca$

(c) $^{98}_{42}Mo + ^{1}_{0}n \rightarrow ^{0}_{-1}e + ^{99}_{43}Tc$ (d) $^{229}_{90}Th \rightarrow ^{4}_{2}He + ^{225}_{88}Ra$

(e) $^{184}_{80}Hg \rightarrow ^{184}_{79}Au + ^{0}_{1}e$

24.7 (a) $^{11}_{5}B \rightarrow ^{4}_{2}He + ^{7}_{3}Li$

(b) $^{98}_{38}Sr \rightarrow ^{0}_{-1}e + ^{98}_{39}Y$

(c) $^{107}_{47}Ag + ^{1}_{0}n \rightarrow ^{108}_{47}Ag$

(d) $^{88}_{35}Br \rightarrow ^{1}_{0}n + ^{87}_{35}Br$

(e) $^{116}_{51}Sb + ^{0}_{-1}e \rightarrow ^{116}_{50}Sn$

(f) $^{70}_{33}As \rightarrow ^{0}_{1}e + ^{70}_{32}Ge$

(g) $^{41}_{19}K \rightarrow ^{1}_{1}H + ^{40}_{18}Ar$

24.8 A radioactive decay series is a series of nuclear changes in which one isotope decays to another and that to another, and so on until a stable (nonradioactive) isotope is formed. The ^{238}U series stops at ^{206}Pb because ^{206}Pb is the first nonradioactive isotope of an element to be formed in the decay series.

24.9 In a radioactive decay, the isotope that decays is called the parent isotope and the isotope that is formed is referred to as the daughter isotope.

24.10 41 (See Section 24.7 and the note in the margin)

24.11 $\ln \frac{[A]_0}{[A]_t} = kt$ (Equation 24.1) $[A]_t = 1/2[A]_0$ @ $t_{1/2}$

$$\ln \frac{[A]_0}{1/2[A]_0} = kt_{1/2}$$

$\ln (2) = kt_{1/2}$ or $t_{1/2} = \frac{\ln (2)}{k}$ $\quad t_{1/2} = \frac{0.693}{k}$ (Equation 24.2)

24.12 (a) $k = \frac{0.693}{t_{1/2}}$ $kt_{1/2} = 0.693$; therefore, kt = 0.693 x no. of half-life periods

$$\ln \frac{(^{60}Co)_o}{(^{60}Co)_a} = 1 \times 0.693 = \ln \frac{1.00 \text{ g } ^{60}Co}{(^{60}Co)_a} = 0.693$$

$$\frac{1.00 \text{ g } ^{60}Co}{(^{60}Co)_a} = 2.00$$

Therefore, $(^{60}Co)_a$ = **0.500 g**

(b) $$\ln \frac{(^{60}Co)_o}{(^{60}Co)_b} = \ln \frac{1.00}{(^{60}Co)_b} = 3 \times 0.693 = 2.08$$

$\ln(1.00 / ^{60}Co) = 2.08$

$1.00 / ^{60}Co) = 8.00$

$^{60}Co = 0.125$ $(^{60}Co)_b$ = **0.125 g**

(c) $$\ln \frac{(^{60}Co)_o}{(^{60}Co)_c} = 5 \times 0.693$$

$\ln(1.00 / ^{60}Co) = 3.46$

$1.00/ ^{60}Co = 32.0$ $^{60}Co = 0.0313$

$^{60}Co = 0.0313$ g $(^{60}Co)_c$ = **0.0312 g**

(An alternative method is shown below.)

When "n" is the number of half-life periods, the fraction of the original sample which remains = $1/2^n$.

(a) **1.00 g x $(1/2)^1$ = 0.500 g** (b) **1.00 g x $(1/2)^3$ = 0.125 g**

(c) **1.00 g x $(1/2)^5$ = 0.0312 g**

24.13 (See Problem 24.12)

(a) $\ln \frac{8.00\text{ g}}{X_a} = \frac{240}{120}$ (half-life periods) x 0.693 = 1.386

X_a **= 2.00 g**

(b) (4 half-life periods) X_b **= 0.500 g** (c) (8 half-lives) X_c **= 0.0312 g**

24.14 $t_{1/2} = \frac{0.693}{k} = \frac{0.693}{4.23 \times 10^{-3}\text{ days}^{-1}} = \mathbf{164\ days}$

24.15 $t_{1/2} = \frac{0.693}{k} = \frac{0.693}{2.30 \times 10^{-6}\text{ year}^{-1}} = \mathbf{3.01 \times 10^{5}\ years}$

24.16 $k = \frac{0.693}{27.72\text{ days}} \times \frac{1.00\text{ day}}{24 \times 60 \times 60\text{ s}} = \mathbf{2.89 \times 10^{-7}\ s^{-1}}$

24.17 $k = \frac{0.693}{t_{1/2}} = \frac{0.693}{470\text{ days}} = \mathbf{0.00147\ day^{-1}}$

24.18 (a) $k = \frac{1}{t} \times \ln \frac{[A]_o}{[A]_t}$ $k = \frac{1}{96.0\text{ hr}} \times \ln \frac{4720}{2560}$ $\mathbf{k = 6.37 \times 10^{-3}\ hr^{-1}}$

Or k can be obtained from the slope (from a plot of the data) = (8.46 - 7.85)/(96.0)

= **6.35 x 10^{-3} hr^{-1} = k** (b) $t_{1/2}$ = 0.693/k = **109 hr**

24.19 See Figure 24.4 and the first part of the section on "Measurement of Radioactivity".

24.20 (a) becquerel = 1 disintegration/s = Bq
(b) curie = 3.7 x 10^{10} disintegrations/s = 3.7 x 10^{10} Bq = Ci = no. of disintegrations per second per g of radium
(c) specific activity = disintegrations per second per gram = Bq/g
(d) gray = joules of energy absorbed per kilogram of absorbing material = J/kg = Gy
(e) rad (radiation absorbed dose) = 10^{-5} joules of energy absorbed per gram of absorbing material = 1/100 Gy

24.21 The rem (radiation equivalent for man) takes into consideration the amount of damage the various forms of radiation can cause to animal tissue. Rad does not differentiate between types of radiation.

24.22 (a) 1.24 Ci x (3.7 x 10^{10} Bq/Ci) = 4.6 x 10^{10} Bq

(b) 4.6 x 10^{10} Bq x (1/150 g) = 3.1 x 10^{8} Bq/g (specific activity)

24.23 (a) $$\frac{140\ \text{Bq}}{\text{g}} \times 1.0\ \text{mg} \times \frac{1.00\ \text{g}}{1{,}000\ \text{mg}} \times \frac{1.00\ \text{dis s}^{-1}}{\text{Bq}} \times 10\ \text{min}$$

$$\times \frac{60\ \text{s}}{\text{min}} \times \frac{1.80\ \text{MeV}}{\text{dis}} \times \frac{1.60 \times 10^{-13}\ \text{J}}{\text{MeV}} \times \frac{1}{\text{body weight}}$$

Answer = **2.4 x 10^{-11} J/body weight** If we assume the average body weight is 60 kg, then the answer is 4.0 x 10^{-13} J/kg = **4 x 10^{-13} Gy**

(b) $$\frac{(4.0 \times 10^{-13}\ \text{Gy})}{(\text{kg of body weight})} \times \frac{100\ \text{rad}}{\text{Gy}} = \mathbf{4.0 \times 10^{-11}\ \text{rad}}$$ (for a 60 kg weight person)

24.24 (a) β (b) γ

24.25 mol ^{40}Ar formed = mol ^{40}K decayed = 1.15 x 10^{-5} mol

$$k = \frac{0.693}{t_{1/2}} = \frac{0.693}{1.3 \times 10^{9}\ \text{yr}} = 5.3 \times 10^{-10}\ \text{yr}^{-1} \qquad t = \frac{2.303}{k} \times \log\frac{[A]_0}{[A]_t}$$

$[A]_0$ = (2.07 x 10^{-5} + 1.15 x 10^{-5}) mol ^{40}K $\qquad [A]_t$ = 2.07 x 10^{-5} mol ^{40}K

$$t = \frac{2.303}{5.3 \times 10^{-10}\ \text{yr}^{-1}} \times \log\frac{3.22 \times 10^{-5}}{2.07 \times 10^{-5}} = \mathbf{8.3 \times 10^{8}\ \text{yr}}$$

24.26 Since three half-life periods would reduce the ^{14}C to one eighth the initial value, the age of the wood equals 3 x 5770 yr (17,300 yr)

24.27 ON + O*NO ⇌ ONO*NO ⇌ ONO + *NO

bond breaking at this point to yield these products (ON + O*NO)

bond breaking at this point to yield these products (ONO + *NO)

24.28 Approximately half of the CH_3HgI should contain the labeled Hg. When the reaction proceeds in the reverse direction (if reversible), two molecules of CH_3HgI could form $(CH_3)_2Hg$ plus HgI_2. The forward reaction could result in producing labeled $CH_3{}^*HgI$ as half of the product and unlabeled CH_3HgI as the other half of the product if the $^*HgI_2$ reacted instead of a HgI_2.

24.29 One possible experiment would be to make the complex and allow the racemization to occur in a medium containing labeled $C_2O_4{}^{2-}$. If the racemization occurs by the loss and recombination with $C_2O_4{}^{2-}$, the complex should pick up labeled $C_2O_4{}^{2-}$ during the racemization.

24.30 ?mL (Assuming the density of the mixture is about the same as that of the coolant)

$$= 10.0 \text{ mL } CH_3OH \times \frac{0.792 \text{ g } CH_3OH}{1 \text{ mL } CH_3OH} \times \frac{580 \text{ cpm}}{1 \text{ g } CH_3OH}$$

$$\times \frac{1 \text{ mL (mixture)}}{\sim 0.884 \text{ g (mixture)}} \times \frac{1 \text{ g (mixture)}}{29 \text{ cpm}} = \mathbf{180\ mL\ (mixture)} \text{ (2 sign. fig.)}$$

The answer given above is for the volume of mixture after the 10.0 mL has been added. The original volume of coolant is: 180 -10 = **170 mL.** One can now calculate the density of the mixture (0.879) and show that upon recalculation, the answer is still 170 mL to 2 sign. figures.

24.31 $$?\text{mol Cr} = 165 \text{ cpm} \times \frac{1 \text{ g } K_2Cr_2O_7}{843 \text{ cpm}} \times \frac{2 \text{ mol Cr}}{294 \text{ g } K_2Cr_2O_7} = 1.33 \times 10^{-3} \text{ mol Cr}$$

$$?\text{mol } C_2O_4^{2-} = 83 \text{ cpm} \times \frac{1 \text{ g } H_2C_2O_4}{345 \text{ cpm}} \times \frac{1 \text{ mol } C_2O_4^{2-}}{90.0 \text{ g } H_2C_2O_4} = 2.67 \times 10^{-3} \text{ mol } C_2O_4^{2-}$$

Therefore, there are two oxalate ions bound to each Cr(III) in the complex ion.

24.32 All stable nuclei fall on the narrow band of stability. As the number of protons in the nucleus increases, there must be more and more neutrons present to help overcome the strong repulsive forces between the protons. Also, there seems to be an upper limit to the number of protons that can exist in a stable nucleus, that number being reached with bismuth. Nuclides above the band of stability must either lose neutrons or change neutrons into protons by β emission in order to achieve stability.

24.33 Elements higher than 83 must lose both neutrons and protons to achieve a stable n/p ratio. The only way this is possible is by α-emission or fission.

24.34 Nuclei that contain certain specific numbers of protons and neutrons possess a degree of extra stability. For protons these magic numbers are 2, 8, 20, 28, 50 and 82; for neutrons, 2, 8, 20, 28, 50, 82 and 126. The magic numbers for orbital electrons are 2, 8, 18, 36, and 54 (the number of electrons in filled electron shells).

24.35 e = even, o = odd, * = magic number

$${}^{4}_{2}He > {}^{58}_{28}Ni > {}^{39}_{20}Ca > {}^{71}_{32}Ge > {}^{10}_{5}B$$

(p,n) (e*,e*) (e*,e) (e*,o) (e,o) (o,o)

24.36 e = even, o = odd, * = magic number

$${}^{192}_{77}Ir < {}^{13}_{6}C < {}^{3}_{2}He < {}^{116}_{50}Sn < {}^{40}_{20}Ca$$

(p,n) (o,o) (e,o) (e*,o) (e*,e) (e*,e*)

24.37 Radiation emitted in quantum packets can be used to explain nuclear shells just as Bohr did in explaining his atomic theory. If protons move from shell to shell in the nucleus, each transition would result in an emission of a different energy.

24.38 Both Tc and Pm have an odd number of protons.

24.39 A nuclear transformation is a nuclear reaction in which a bombarding particle is absorbed and causes the absorbing nucleus to change into a nucleus of another element.

24.40 The cyclotron produces particles of very high velocities by the use of oscillating voltage to accelerate the particles and a magnetic field to help guide the particles into the target nuclei resulting in nuclear transformations.

24.41 Transuranium elements are elements 93 to 105 (elements with atomic numbers greater than that of uranium). Transuranium elements do not occur in nature, they have only been produced artificially.

24.42 (a) ${}^{27}_{13}Al + {}^{4}_{2}He \rightarrow {}^{1}_{0}n + {}^{30}_{15}P$ (b) ${}^{209}_{83}Bi + {}^{2}_{1}H \rightarrow {}^{1}_{0}n + {}^{210}_{84}Po$

(c) ${}^{15}_{7}N + {}^{1}_{1}H \rightarrow {}^{4}_{2}He + {}^{12}_{6}C$ (d) ${}^{12}_{6}C + {}^{1}_{1}H \rightarrow {}^{13}_{7}N + \gamma$

(e) ${}^{14}_{7}N + {}^{4}_{2}He \rightarrow {}^{1}_{1}H + {}^{17}_{8}O$

24.43 (a) $^{242}_{96}Cm + ^{4}_{2}He \rightarrow ^{245}_{98}Cf + ^{1}_{0}n$ (b) $^{108}_{48}Cd + ^{1}_{0}n \rightarrow ^{109}_{48}Cd + \gamma$
(c) $^{14}_{7}N + ^{1}_{0}n \rightarrow ^{14}_{6}C + ^{1}_{1}H$ (d) $^{27}_{13}Al + ^{2}_{1}H \rightarrow ^{25}_{12}Mg + ^{4}_{2}He$
(e) $^{249}_{98}Cf + ^{18}_{8}O \rightarrow ^{263}_{106}Unh + 4^{1}_{0}n$

24.44 Element 114 would fall under lead in Group IVA. It, therefore, would be a soft metal with a relatively low melting point. Its most stable oxidation state would be 2+ and would form such compounds as MO and MCl_2 and the 4+ as MO_2 and MCl_4. A likely spot to discover this element would be wherever lead ores are found and in meteorites.

24.45 (a) Na_2Uuh (b) H_2Uuh (c) $UuhO_2$ Since Po is a metalloid, element 116 (directly below it on the periodic table) would probably have metallic properties.

24.46 298 and 310 having 184 and 196 neutrons (magic numbers)

24.47 The bombarding nuclei have to contain a very large n/p ratio to place the products on the island of stability (Figure 24.9) no. p ≈ 16 and no. n ≈ 50 - 62. Light nuclei, however, contain n/p ratios of nearly 1.

24.48 (a) ununpentium, Uup (b) unbiseptium, Ubs

24.49 (a) 148 (b) 125 (c) 103 [This is lawrencium (Lr)]

24.50 Mass defect is the difference between the actual mass of a nucleus and the sum of the masses of its individual protons, neutrons and electrons.

24.51 ^{56}Fe = 26 p x 1.007277 u/p + 30 n x 1.008665 u/n + 26 e x 5.4859 x 10^{-4} u/e = 56.463415 u.
Since the actual atomic mass = 55.9349 u, the mass defect = 0.5285 u. What is the binding energy per nucleon?

$$?\text{MeV} = \frac{0.5285\ \text{u}}{56\ \text{nucleon}} \times \frac{931\ \text{MeV}}{1\ \text{u}} = \frac{\mathbf{8.79\ MeV}}{\mathbf{nucleon}}$$

Since this is the largest value of binding energy per nucleon (the highest point on the curve in Figure 24.10), neither fission nor fusion of iron 56 can yield energy, i.e., produce products with a greater binding energy per nucleon.

24.52 Δm = 2 x 2.014102 - 4.002603 = 0.025601 $E = \Delta m \times c^2$
E = (0.025601 u) (2.9979 x 10^8 m $s^{-1})^2$ x 6.022 x 10^{23} mol^{-1} x 1 kg/6.022 x 10^{26} u x 1 kJ/10^3 kg m^2 s^{-2} = **2.3009 x 10^9 kJ mol^{-1}**
(This answer was not limited to 4 sign. fig. by Avogadro's number; it cancelled out from the numerator and denominator.)

24.53 $E = mc^2$ $E = 2 \times 9.1096 \times 10^{-31}$ kg $(2.9979 \times 10^8 \text{ m s}^{-1})^2$ = **1.6374×10^{-13} J**
($1 \text{ kg m}^2 \text{ s}^{-2} = 1$ J)

24.54 $\Delta m\ ^7Li = (3 \times 1.007277 + 4 \times 1.008665 + 3 \times 0.0005486) - (7.01600)$
$= 0.042137$ u

$$E = \Delta mc^2 = 0.042137 \text{ g x mol}^{-1} \times (2.9979 \times 10^8 \text{ m s}^{-1})^2 \times \frac{1 \text{ kg}}{10^3 \text{ g}} = 3.7870 \times 10^{12} \text{ J mol}^{-1}$$

^{7}Li 3.787 x 10^9 kJ mol^{-1}, 39.2 MeV

for: **^{14}N 1.0098 x 10^{10} kJ mol^{-1}, 105 MeV**

^{19}F 1.4261 x 10^{10} kJ mol^{-1}, 148 MeV

24.55 Fission is the splitting of an atom into approximately equal parts. The fission of ^{235}U releases energy and neutrons that can cause other atoms of ^{235} U to undergo fission. The result could be a nuclear explosion.

24.56 In nuclear reactors the rate of fission is controlled by use of control rods which absorb neutrons and, thus, prohibit the chain reaction from getting out of control.

24.57 A breeder reactor is one that is designed to produce more nuclear fuel than it consumes. The nuclear equation for the reaction that generates the nuclear fuel that is produced is:

$$^{238}_{92}U + ^{1}_{0}n \rightarrow ^{239}_{94}Pu + 2\,^{0}_{-1}e$$

24.58 In a nuclear reactor, control rods are neutron absorbers that can be inserted into the reactor to slow the nuclear reaction. The rods are positioned to allow the nuclear reaction to take place at a desired rate. Moderators are substances that are used to slow fast neutrons. Water and graphite have been used as moderators. The slower neutrons are needed to initiate the fission processes.

24.59 Critical mass is the minimum amount of the fissile isotope required to sustain a chain reaction.

24.60 Thermal neutrons are neutrons that have been slowed down. Thermal neutrons can trigger fission processes. For example: $^{235}_{92}U + ^{1}_{0}n \rightarrow ^{139}_{56}Ba + ^{94}_{36}Kr + 3\,^{1}_{0}n$

24.61 The only naturally-occurring fissile isotope is U-235.

24.62 Fusion reactions are nuclear reactions in which two isotopes are brought together to form a heavier one. Fusion reactions release greater amounts of energy than do fission reactions.

24.63 It is difficult to build fusion reactors because fusion reactions have very high energies of activation and a very high temperature is required. A fusion reactor is difficult to construct since there is not an acceptable means of achieving the very high temperatures and containing this reaction.

24.64 A plasma is a reacting mass of charged particles. A plasma is difficult to contain because it is very reactive and at a very high temperature.

24.65 Fusion reactions might be better sources of energy in the future since they (1) they produce much more energy than do fission reactions, (2) they produce products that usually are not radioactive while most fission reactions produce radioactive products, and (3) the supply of fuel is virtually inexhaustible.

24.66 (a) **2.3009×10^9 kJ mol^{-1}** (See Problem 24.52)
(b) $2 \times 12.00000 - 23.98504 = 0.01496$; therefore, **E = 1.345×10^9 kJ mol^{-1}**
(Since carbon 12 is used as the basis of the atomic mass scale, the mass of C-12 used in the above calculation is exactly 12.)
Reaction (a) produces more energy per mole of product. On the basis of energy produced per gram of reactants, **reaction (a) wins by a large margin, slightly more than 10:1.**

24.67 $C_8H_{18}(\ell) + 25/2O_2(g) \rightarrow 8CO_2(g) + 9H_2O(\ell)$

$$8\ \Delta H_f(CO_2(g)) + 9\ \Delta H_f(H_2O(\ell)) - \Delta H_f(C_8H_{18}(\ell)) = \Delta H_{comb.}(C_8H_{18}(\ell))$$

$$8(-394\ \text{kJ/mol}) + 9(-286\ \text{kJ/mol}) - (-208.4\ \text{kJ/mol}) = -5520\ \text{kJ}$$

$$(C_8H_{18}(\ell)) = 1\ \text{mol}\ ^4\text{He} \times \frac{2.3009 \times 10^9\ \text{kJ}}{1\ \text{mol}\ ^4\text{He}} \times \frac{1\ \text{mol}\ C_8H_{18}}{5520\ \text{kJ}}$$

$$\times \frac{114\ \text{g}\ C_8H_{18}}{1\ \text{mol}\ C_8H_{18}} \times \frac{1\ \text{L}\ C_8H_{18}}{703\ \text{g}\ C_8H_{18}} \times \frac{1\ \text{gal}\ C_8H_{18}}{3.79\ \text{L}\ C_8H_{18}}$$

$$= \mathbf{1.78 \times 10^4\ gal\ C_8H_{18}(\ell)}$$

APPENDIX B: MOLECULAR ORBITAL THEORY

B.1 From the point of view of molecular orbital theory, the electronic structure of a molecule is similar to the electronic structure of an atom in that (1) no more than two electrons can populate a single orbital, (2) each electron will occupy the lowest energy orbital available, and (3) electrons are spread out as much as possible over orbitals of the same energy with unpaired spins if possible. The electronic structures of atoms and molecules differ in that (1) atomic orbitals and molecular orbitals are very different and (2) in molecular orbital theory, distinctly different bonding and antibonding orbitals exist unlike anything that exists in the electronic structure of an atom.

B.2 See Figure B. 1. The σ_{1s} molecular orbital will stabilize a molecule while the σ_{1s}^* will destabilize the molecule. The effect of the two will cancel each other. The σ_{1s}^* is said to be an antibonding orbital because it destabilizes a molecule when it contains electrons because the maximum electron density lies outside the region between the nuclei.

B.3 See Figure B. 2.

B.4 See Figure B. 4. The net result is one bonding sigma bond and two bonding pi bonds.

B.5 The molecular orbital diagram of N_2 can be seen in Figure B. 4. The species N_2^+ would have one less bonding electron and N_2^- would have an additional electron located in a π_{2p}^* molecular orbital. The net bond order of N_2 is 3 and of N_2^+ and N_2^- is each 2.5. Both N_2^+ and N_2^- are less stable than N_2 and both would have a longer bond length than N_2.

B.6 The molecular orbital diagram of O_2 can be seen in Figure B. 4. The species O_2^+ would have one less π_{2p}^* electron and the species O_2^- would have one additional π_{2p}^* electron. Net bond orders: $O_2 = 2$, $O_2^+ = 2.5$, and $O_2^- = 1.5$. The stability, therefore, would be: $O_2^+ > O_2 > O_2^-$. The bond lengths would be: $O_2^+ < O_2 < O_2^-$. Vibrational frequencies would be: $O_2^- < O_2 < O_2^+$.

B.7 See Figure B.4 . The O_2 molecule has the net bonding of a double bond while the O_2^{2-} ion has the net bonding of a single sigma bond. Therefore, the bond order of O_2 is 2 and that of O_2^{2-} is 1. The bond energy of O_2 would be greater, its bond length would be less, and its vibrational frequency would be greater than that of O_2^{2-}.

B.8

	Li_2	Be_2	B_2	C_2
$\sigma_{2p_z}^*$	—	—	—	—
$\pi_{2p_x}^* \pi_{2p_y}^*$	— —	— —	— —	— —
$\pi_{2p_x} \pi_{2p_y}$	— —	— —	— —	↑ ↑
σ_{2p_z}	—	—	↑↓	↑↓
σ_{2s}^*	—	↑↓	↑↓	↑↓
σ_{2s}	↑↓	↑↓	↑↓	↑↓

Be_2 should not exist, and C_2 is paramagnetic based upon this diagram.

[However, since $\pi_{2p_x} \pi_{2p_y}$ and σ_{2p_z} should switch positions in the energy diagram for these species, B_2 should be paramagnetic while C_2 should not.]

B.9 (a) Because a bonding electron is removed in each case, Li_2^+, B_2^+, and C_2^+ would be less stable than the neutral X_2 species. Be_2^+ would be more stable than Be_2 because an antibonding electron is removed.
(b) Li_2^- would be less stable, Be_2^-, B_2^-, and C_2^- would be more stable than the neutral X_2 species. The extra electron is antibonding in Li_2^- but is bonding in Be_2^-, B_2^-, and C_2^-.

B.10 It allows for the formation of molecular orbitals that extend over more than two nuclei. This delocalization is the equivalent of valence bond resonance.

B.11 H_2^+ Bond order = 1/2; $(\sigma_{1s})^1$
He_2^+ Bond order = 1/2; $(\sigma_{1s})^2 (\sigma_{1s}^*)^1$

B.12 Valence Bond Description: In all three species the central atom is sp^2 hybridized. In SO_2 the two resonance structures arise from alternating the double bond between the unhybridized p orbital on S and p orbitals on the oxygens. In NO_3^- three resonance structures arise because there are three O atoms which can π-bond to the unhybridized p orbital on N. In HCO_2^- the C-H bond is a single bond, and two resonance structures result by forming one of two possible double bonds as in the case of SO_2.
Molecular Orbital Description: Again, the central atom is sp^2 hybridized in all three cases. Delocalized π clouds are formed over three atoms in both SO_2 and HCO_2^- (i.e., over O-S-O and over O-C-O). In NO_3^- a delocalized π cloud extends over all 4 atoms as in the case of SO_3.